CBS PROBLEMS and SOLUTIONS SERIES

Problems and Solutions in Electrical Machines and Transformers

- ***Problems and Solutions in Electrical Machines and Transformers*** is suitable for examinations of all the Indian universities.
- Solutions of more than 340 problems are given in most simplified manner.
- All types of problems are included and the solutions given in detail.
- To explain the basics lucidly, computerised figures have been liberally added in the solutions.
- This volume contains more than 150 line diagrams which will help the reader in grasping the basics.
- Problems selected from the last ten years' Indian Engineering Service and Civil Service examinations are added till date in each chapter.

CBS Problems and Solutions Series

- Engineering Thermodynamics
- Strength of Materials
- Integrated Electronics
- Signals and Systems
- Communication Systems
- Network Analysis
- Control Systems
- Power Systems
- Power Electronics
- *plus many ... more new topics*

CBS PROBLEMS and SOLUTIONS SERIES

Problems and Solutions in Electrical Machines and Transformers

S. K. PRASAD

Engineer

CBS

CBS PUBLISHERS & DISTRIBUTORS PVT. LTD.

New Delhi • Bengaluru • Chennai • Kochi • Mumbai • Pune

ISBN: 81-239-1147-5

First Edition: 2004
Reprint: 2009, 2012, 2015

Published by:
Satish Kumar Jain for CBS Publishers & Distributors Pvt. Ltd.,
4819/XI Prahlad Street, 24 Ansari Road, Daryaganj, New Delhi - 110002
delhi@cbspd.com, cbspubs@airtelmail.in • www.cbspd.com
Ph.: 23289259, 23266861, 23266867 • Fax: 011-23243014

Corporate Office: 204 FIE, Industrial Area, Patparganj, Delhi - 110 092
Ph: 49344934 • Fax: 011-49344935
E-mail: publishing@cbspd.com • publicity@cbspd.com

Branches:
- ***Bengaluru:*** 2975, 17th Cross, K.R. Road, Bansankari 2nd Stage, Bengaluru - 70 Ph: +91-80-26771678/79 • Fax: +91-80-26771680 E-mail: cbsbng@gmail.com, bangalore@cbspd.com
- ***Chennai:*** No. 7, Subbaraya Street, Shenoy Nagar, Chennai - 600030 Ph: +91-44-26681266, 26680620 • Fax: +91-44-42032115 E-mail: chennai@cbspd.com
- ***Kochi:*** 36/14, Kalluvilakam, Lissie Hospital Road, Kochi - 682018 Ph: +91-484-4059061-65 • Fax: +91-484-4059065 E-mail: cochin@cbspd.com
- ***Mumbai:*** 83-C, Dr. E. Moses Road, Worli, Mumbai - 400018 Ph: +91-9833017933, 022-24902340/41 • E-mail: mumbai@cbspd.com
- ***Pune:*** Bhuruk Prestige, Sr. No. 52/12/2+1+3/2, Narhe, Haveli (Near Katraj-Dehu Road Bypass), Pune - 411041 Ph: +91-20-64704058/59, 32342277 • E-mail: pune@cbspd.com

Representatives:
- Hyderabad: 0-9885175004
- Nagpur: 0-9021734563
- Vijayawada: 0-9000660880
- Kolkata: 0-9831437309, 0-905115236
- Patna: 0-9334159340

Printed at:
J.S. Offset Printers, Delhi

Preface

Electrical Machines and Transformers

This is one of the fundamental subjects for the students of electrical engineering. The author feels great pleasure in presenting this intricate subject through typical problems and their solutions. This volume covers all major and important topics under this subject taught at the undergraduate level in an engineering college. All the problems are discussed at length, and detailed line diagrams and figures have been given extensively to help the reader in understanding the basics and grasping the logic. This book is suitable for preparing for all university examinations and also to face the competitions. Select questions of the past ten years from various competitive examinations like Indian Engineering Service and Civil Service examinations have been added till date. This feature is expected to help the reader in visualising the patterns of problems and the extent and level of difficulty in solving the problems likely to be encountered in the examination.

Although every care has been taken to ensure accuracy, yet some errors might have crept in. The author will be grateful if these are brought to his notice so that these could be rectified in the subsequent printings and editions of the book. Readers are requested to send their suggestions for further improvement and revision to me through the publishers at the following e-mail address: cbspubs@del3.vsnl.net.in.

S. K. PRASAD

Contents

Chapter 1

Transformers

TRANSFORMER

A transformer is a static device which transforms electrical energy from one circuit to another circuit through magnetic field without changing the supply frequency. It changes low-voltage level to high-voltage level or vice-versa.

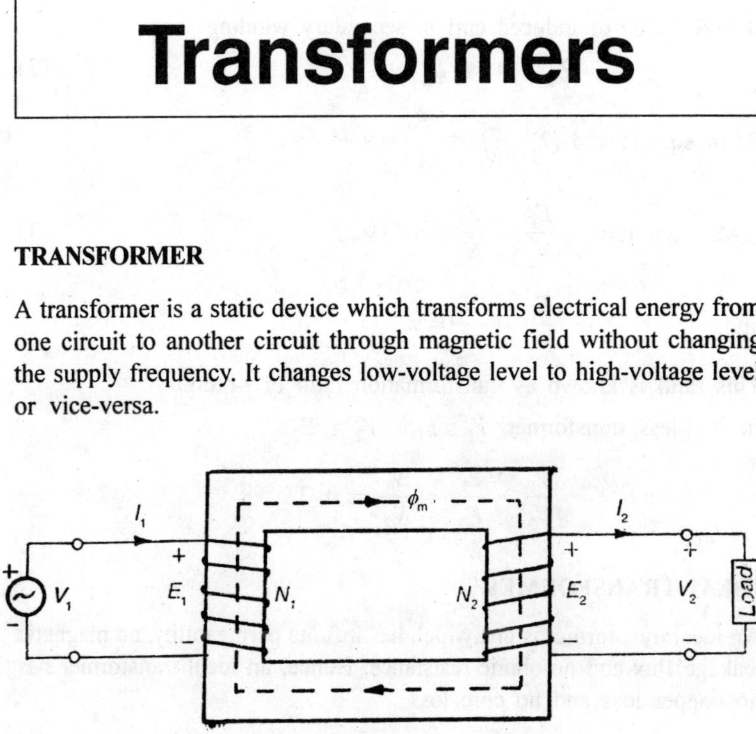

Fig. 1.1

In Fig. 1.1,

V_1 = Supply voltage

V_2 = Load voltage

f = Frequency of supply

N_1 = No. of turns in primary winding

N_2 = No. of turns in secondary winding

E_1 = Induced emf in primary winding

E_2 = Induced emf in secondary winding

ϕ_m = Maximum flux density in the core

EMF EQUATION OF TRANSFORMER

R.M.S. value of induced emf in primary winding.

$$E_1 = 4.44\ f\phi_m\ N_1 \quad (1)$$

R.M.S. value of induced emf in secondary winding

$$E_2 = 4.44 f\phi_m \mathrm{N}_2 \quad (2)$$

From eq. (1) and (2) $\dfrac{E_1}{N_1} = \dfrac{E_2}{N_2} = 4.44\ f\,\phi_m$

E.M.F. per turn $= \dfrac{E_1}{N_1} = \dfrac{E_2}{N_2} = 4.44\,\phi_m f \quad (3)$

and, $\dfrac{E_1}{E_2} = \dfrac{N_1}{N_2} = a$

This ratio is known as transformation ratio or *t*-ratio.

In loss-less transformer, $V_1 \cong E_1, \quad V_2 \cong E_2$

$$\frac{V_1}{V_2} = \frac{E_1}{E_2} = \frac{N_1}{N_2} = a$$

IDEAL TRANSFORMER

An ideal transformer is one which has infinite permeability, no magnetic leakage flux and no-ohmic resistance. Hence, an ideal transformer has no copper loss and no core loss.

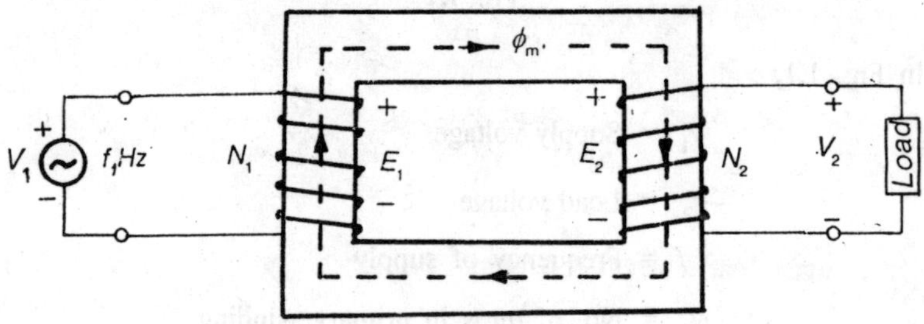

Fig. 1.2

In ideal transfomer,

$$\text{Power input} = \text{Power output}$$

$$V_1 I_1 = V_2 I_2 \Rightarrow \frac{V_1}{V_2} = \frac{I_2}{I_1}$$

Where I_1 = Primary current

I_2 = Secondary current

Transformation ratio $a = \frac{V_1}{V_2} = \frac{E_1}{E_2} = \frac{N_1}{N_2} = \frac{I_2}{I_1}$

The phasor diagram of ideal transformer is as below:

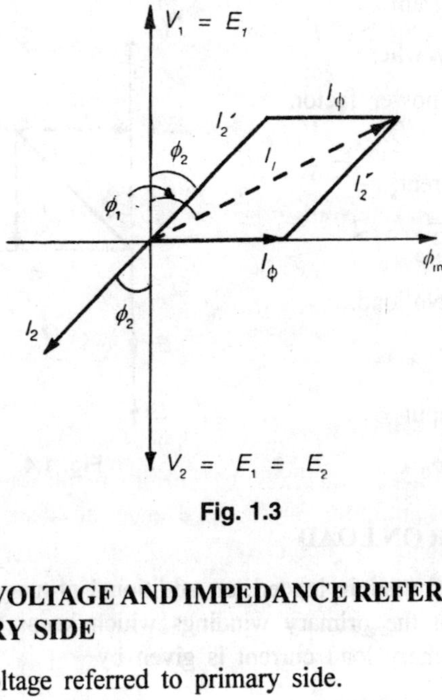

Fig. 1.3

CURRENT, VOLTAGE AND IMPEDANCE REFERRED TO PRIMARY SIDE

Secondary voltage referred to primary side.

$$V_2' = \left(\frac{N_1}{N_2}\right) V_1$$

Secondary load current referred to primary side.

$$I_2' = \left(\frac{N_2}{N_1}\right) I_1$$

Secondary impedance referred to primary side.

$$Z_2' = \left(\frac{N_1}{N_2}\right)^2 Z_2$$

TRANSFORMER ON NO-LOAD

When the primary winding is energized with secondary open, a very small current called exciting current (I_e), flows in the primary winding and establishes the flux in the core.

This exciting current has two components:

Magnetising component current (I_m) and coreloss component current (I_c).

The phasor diagram of transformer on no-load is as below:

From phasor diagram

$$I_c = I_e \cos\theta_0, \text{ where}$$

$$\cos\theta_0 = \text{no-load power factor,}$$

$$I_m = I_e \sin\theta_0$$

and, No-load current,

$$I_e = \sqrt{I_m^2 + I_c^2}$$

Power factor at No-load,

$$\cos\theta_0 = \frac{I_c}{I_e}$$

No-load power input

$$P_0 = V_1 I_e \cos\phi_0$$

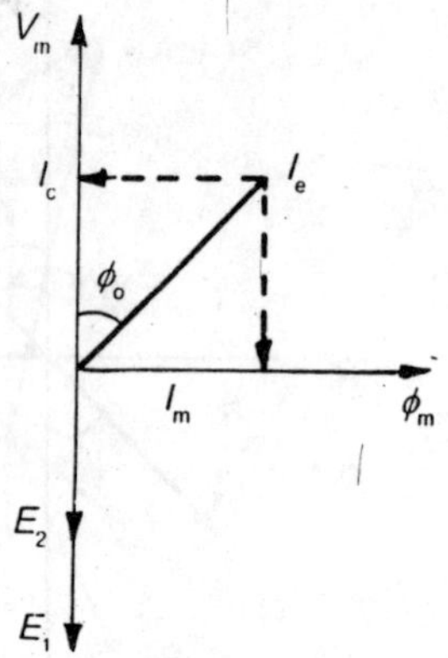

Fig. 1.4

TRANSFORMER ON LOAD

Whenever a transformer is loaded, an additional primary load current (I_1') flows through the primary windings which keeps the core flux constant. This primary load current is given by

$$N_1 I_1' = N_2 I_2$$

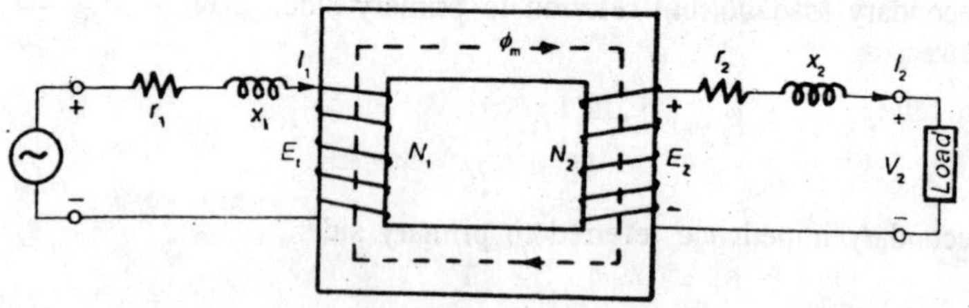

Fig. 1.5

In figure

r_1 and x_1 = Primary resistances and leakage reactances respectively

r_2 and x_2 = Secondary resistances and leakage reactances respectively

Under load conditions

$$\overline{E}_1 = \overline{V}_1 - \overline{I}_1 (r_1 + jx_1)$$

$$\overline{E}_2 = \overline{V}_2 + \overline{I}_2 (r_2 + jx_2)$$

and, $$\overline{I}_1' = \overline{I}_1' + \overline{I}_e$$

The phasor diagram for lagging p.f. can be drawn as below:

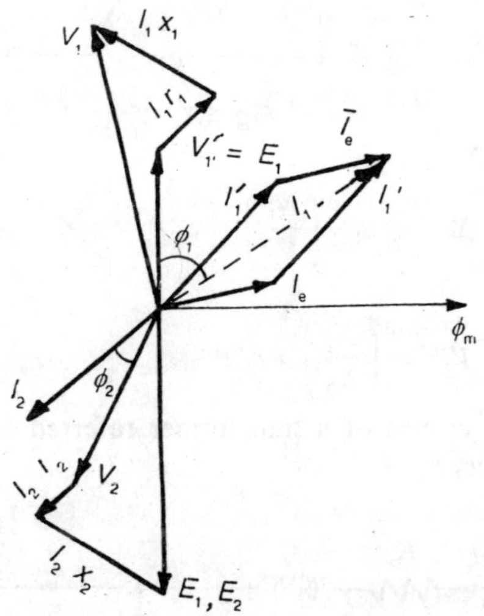

Fig. 1.6

Where θ_1 = Primary power factor angle

θ_2 = Load power factor angle

EQUIVALENT CIRCUIT OF A TRANSFORMER

The equivalent circuit of a transformer referred to primary side can be drawn as

In figure R_c = Core loss resistance

X_φ = Magnetising reactance

$$R_1 = r_1 + \left(\frac{N_1}{N_2}\right)^2 \times r_2$$

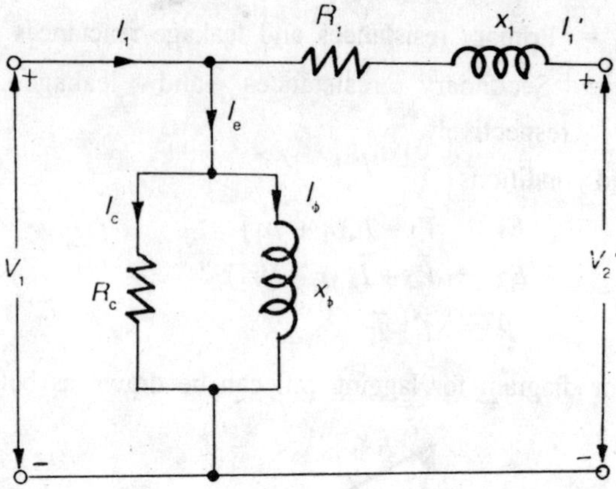

Fig. 1.7

$$X_1 = x_1 + \left(\frac{N_1}{N_2}\right)^2 \times x_2$$

$$V_2' = \left(\frac{N_1}{N_2}\right) \times r_2$$

The equivalent circuit of a transformer referred to secondary side can be drawn as

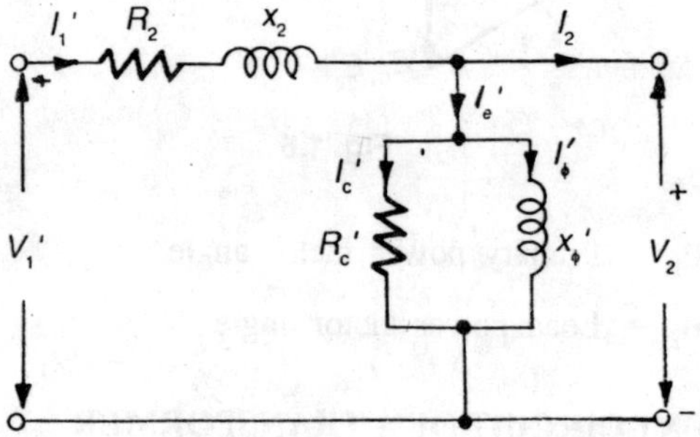

Fig. 1.8

In figure

$$R_2 = r_2 + \left(\frac{N_2}{N_1}\right)^2 \times r_1$$

$$X_2 = x_2 + \left(\frac{N_2}{N_1}\right)^2 \times x_1$$

$$I'_e = \left(\frac{N_1}{N_2}\right) I_e$$

$$R'_c = \left(\frac{N_2}{N_1}\right)^2 \times R_c$$

$$X'_\varphi = \left(\frac{N_2}{N_1}\right)^2 \times X_\varphi$$

$$I'_1 = \left(\frac{N_1}{N_2}\right) I_1$$

The most appropriate equivalent circuit of a transformer can be drawn as below:

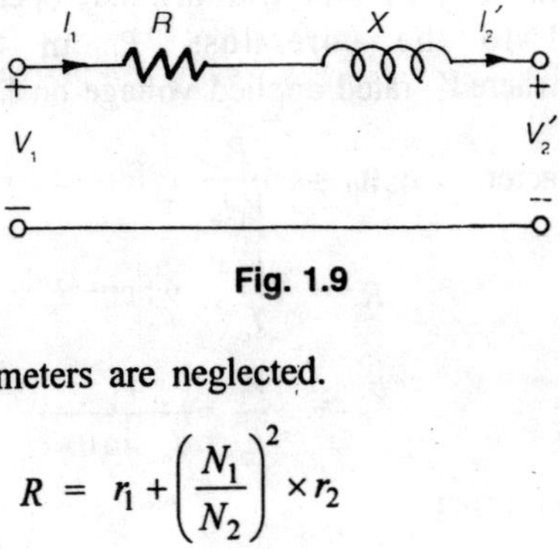

Fig. 1.9

Here shunt parameters are neglected.

$$R = r_1 + \left(\frac{N_1}{N_2}\right) \times r_2$$

$$X = x_1 + \left(\frac{N_1}{N_2}\right)^2 \times x_2$$

If resistance is neglected

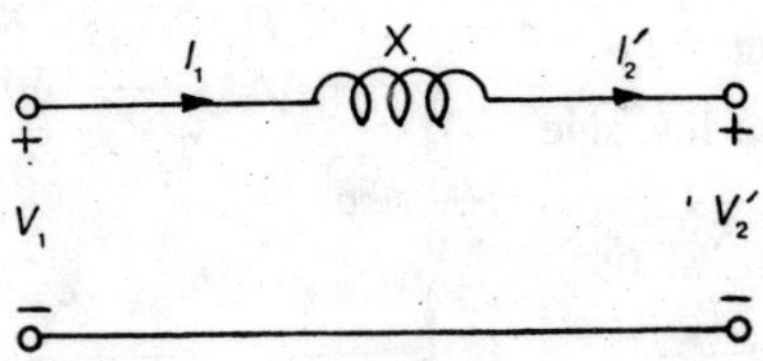

Fig. 1.10

OC TEST

The open circuit test of a transformer yields the shunt parameters. The equivalent circuit at no-load is given by

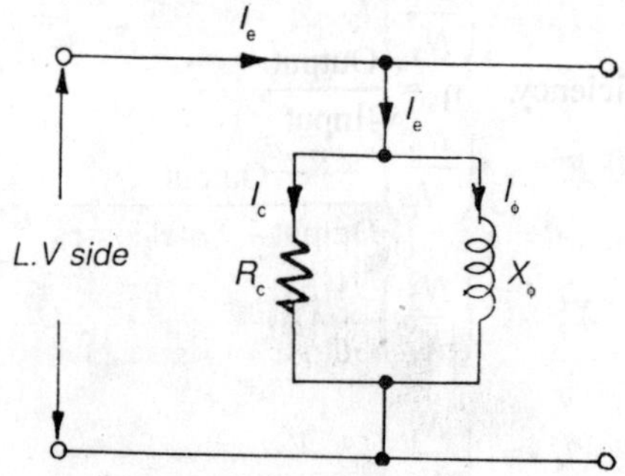

Fig. 1.11

Where $\quad I_e$ = No-load current or exciting current

This test is performed on l.v. side with h.v. side open. The total power input is equal to the core loss P_c in the transformer $P_c = V_1 I_e \cos\theta_0$, where V_1 rated applied voltage on l.v. side

$$\text{No-load power factor,} \quad \cos\theta_0 = \frac{P_c}{V_1 I_e}$$

$$R_c = \frac{V_1}{I_c}, \quad \text{where } I_c = \frac{P_c}{V_1} = \frac{V_1^2}{P_c}$$

and

$$X_\varphi = \frac{V_1}{I_\varphi} = \frac{V_1}{I_e \sin\theta_0}$$

SHORT-CIRCUIT TEST

This test yields the series parameters of the transformer. This test is performed on h.v. side with l.v. side short-circuited. This text yields total full load copper losses in both primary and secondary windings. The equivalent circuit can be drawn as

Total power input = P_{sc}

Current input = I_{sc}

Voltage applied to h.v. side

= V_{sc}

$$Z_{sc} = \frac{V_{sc}}{I_{sc}}, R_H = \frac{P_{sc}}{V_{sc}^{\ 2}}$$

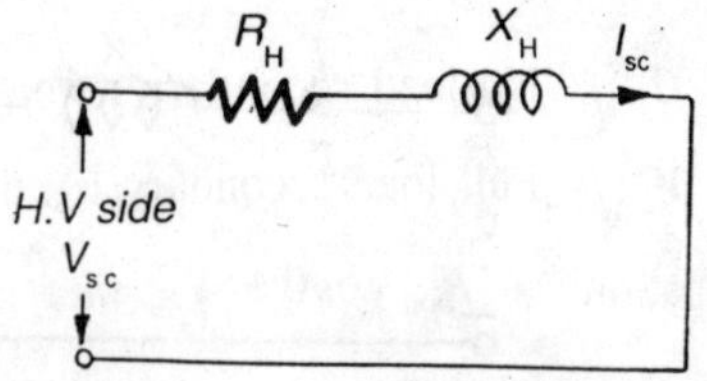

Fig. 1.12

$$\therefore \qquad X_H = \sqrt{Z_{sc}^2 - R_H^2}$$

EFFICIENCY OF TRANSFORMER

$$\text{Transformer efficiency,} \quad \eta = \frac{\text{Output}}{\text{Input}}$$

$$= \frac{\text{Output}}{\text{Output + Total losses}}$$

$$= \frac{\text{Output}}{\text{Output + Copper loss + Core loss}}$$

$$\eta = \frac{V_2 I_2 \cos\theta_2}{V_2 I_2 \cos\theta_2 + P_c + P_i}$$

Where
V_2 = Secondary terminal voltage
I_2 = Secondary load current
$\cos\theta_2$ = Load power factor
P_c = Full load copper loss
P_i = Core loss

For maximum efficiency,

Copper loss = Iron loss, $P_c = P_i$

$$I_2^2 R_2 = P_i$$

$$\text{Maximum efficiency,} \quad \eta_{max} = \frac{V_2 I_2 \cos\theta_2}{V_2 I_2 \cos\theta_2 + 2P_i}$$

Voltage-Regulation

The voltage-regulation of a transformer is given by

$$\text{Voltage regulation} = \frac{E_2 - V_2}{E_2}$$

Where
E_2 = No-load secondary terminal voltage
V_2 = Full load secondary terminal voltage.

$$\text{Voltage regulation} = R_{pu} \cos\theta \pm X_{pu} \sin\theta$$

where $R_{pu} = \dfrac{I_2 R}{V_2}$ = p.u. resistance drop

$$X_{pu} = \frac{I_2 X}{V_2} = pu \text{ reactance drop.}$$

$$\cos\theta = \text{Load power factor.}$$

All-day Efficiency

$$\text{All day efficiency} \quad (\eta_{All}) = \frac{\text{Output in kWh}}{\text{Input in kWh}}$$

$$\eta_{All} = 1 - \frac{\text{Losses in kWh}}{\text{Input in kWh}}$$

PARALLEL OPERATION OF TRANSFORMER

***(i)* Equal-Voltage Ratio**

The equivalent circuit can be drawn as.

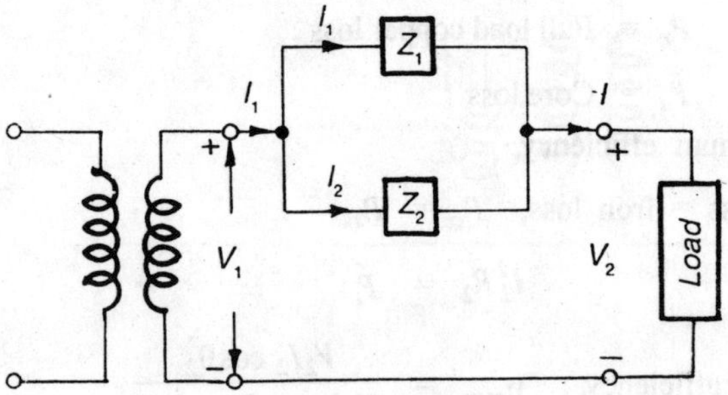

Fig. 1.13

The phasor diagram in given by

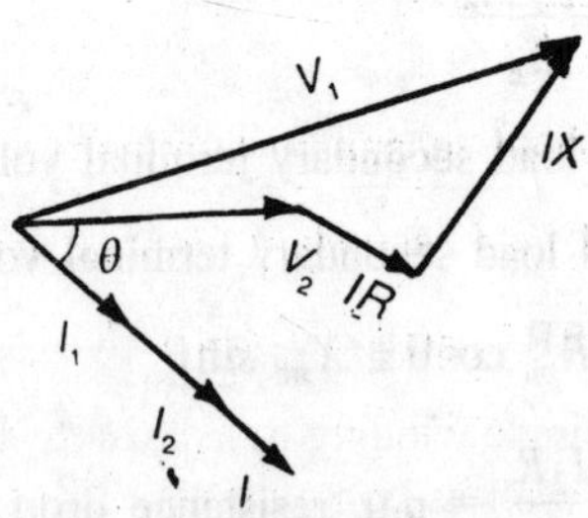

Fig. 1.14

$$I_1 = I \cdot \frac{Z_2}{Z_1 + Z_2}$$

$$\bar{S}_1 = \bar{S} \frac{\bar{Z}_1}{\bar{Z}_1 + \bar{Z}_2} = \text{load shared by } T_1$$

and
$$I_2 = I \cdot \frac{Z_1}{Z_1 + Z_2}$$

$$\bar{S}_2 = \bar{S} \frac{\bar{Z}_1}{\bar{Z}_1 + \bar{Z}_2} = \text{load shared by } T_2$$

where Z_1 = Equivalent impedance of Transformer 1.

Z_2 = Equivalent impedance of Transformer 2.

(ii) **Unequal-Voltage Ratio**

The circulating current is given by

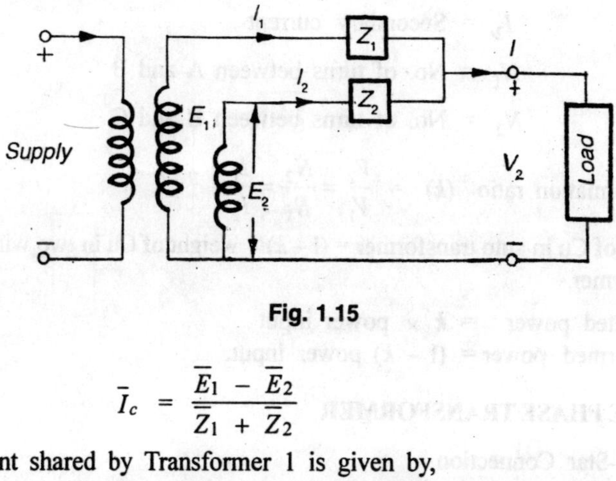

Fig. 1.15

$$\bar{I}_c = \frac{\bar{E}_1 - \bar{E}_2}{\bar{Z}_1 + \bar{Z}_2}$$

Current shared by Transformer 1 is given by,

$$\bar{I}_1 = \bar{I} \frac{\bar{Z}_2}{\bar{Z}_1 + \bar{Z}_2} + \frac{\bar{E}_1 - \bar{E}_2}{\bar{Z}_1 + \bar{Z}_2}, \text{similarly}$$

$$\bar{I}_2 = \bar{I} \frac{\bar{Z}_1}{\bar{Z}_1 + \bar{Z}_2} - \frac{\bar{E}_1 - \bar{E}_2}{\bar{Z}_1 + \bar{Z}_2}$$

AUTO-TRANSFORMER

Auto-transformer is a single winding transformer which trasforms power through inductively as well as conductively. The schematic diagram of a single phase auto transformer is given below

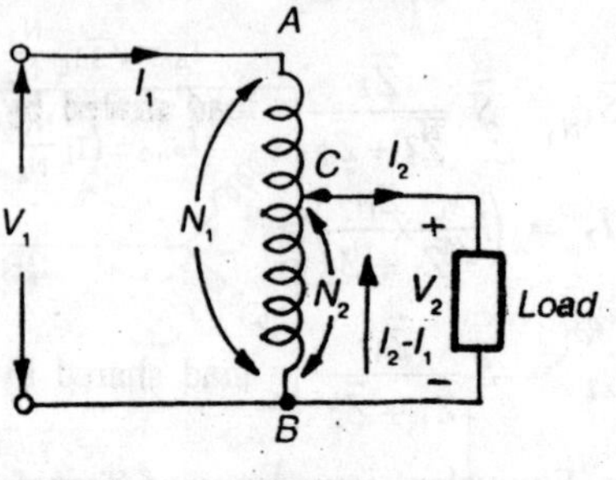

Fig. 1.16

In Fig. 1.16, V_1 = Primary applied voltage

V_2 = Secondary voltage across load

I_1 = Primary current

I_2 = Secondary current

N_1 = No. of turns between A and B

N_2 = No. of turns between B and C

$$\text{Transformation ratio } (k) = \frac{V_2}{V_1} = \frac{N_2}{N_1} = \frac{I_1}{I_2}$$

Weight of Cu in auto transformer = $(1 - k)$ × weight of Cu in two winding transformer.

Conducted power = k × power input

Transformed power = $(1 - k)$ power input.

THREE PHASE TRANSFORMER

(i) Star-Star Connection

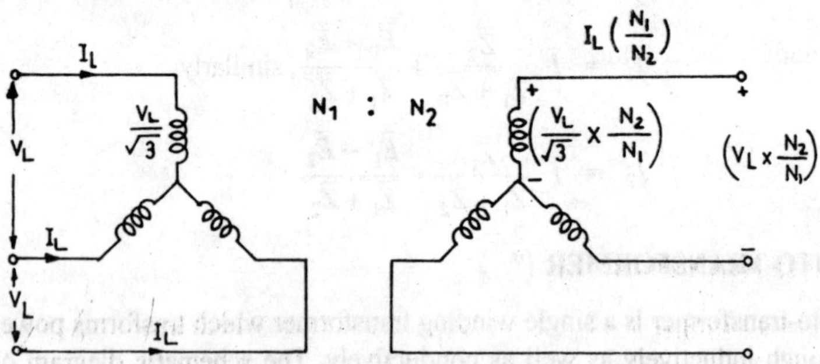

Fig. 1.17

(ii) Star-delta (Y – Δ) Connection

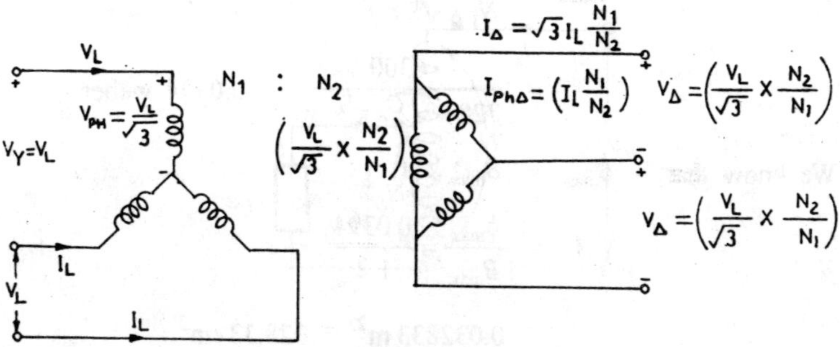

Fig. 1.18

$$\frac{V_\Delta}{V_Y} = \frac{N_2}{\sqrt{3}\ N_1} \text{ and } \frac{I_\Delta}{I_Y} = \sqrt{3}\frac{N_1}{N_2}$$

PROBLEMS

Problem 1.1 : A 6300/210 V. 50 Hz, single-phase transformer has per turn emf of about 9 volts maximum flux density of 1.2 T. Find the number of high-voltage and low-voltage turns and the net cross-sectional area of the core.

Solution :

Given, emf per turn,

$$E_t = 9\text{V/ turn},$$

Now, $B_{\max} = 1.2$ T, $V_1 = 6300$ V, $V_2 = 210$ V, $f = 50$ Hz

$$\text{Secondary no. of turns} = \frac{\text{secondary voltage}}{E_t}$$

$$N_2 = \frac{210}{9} = 23.33 \cong 24 \textbf{ Ans.}$$

$$\text{Primary no. of turns} \quad (N_1) = N_2 \times \frac{V_1}{V_2} 24 \times \frac{6300}{210} = 720 \textbf{ Ans.}$$

$$\text{Applying the relation,} \quad \frac{E_1}{N_1} = \frac{E_2}{N_2} = \text{emf per turn } \sqrt{2}\pi f \phi_{\max}$$

$$\therefore \quad \phi_{max} = \frac{E_1 \cong V_1}{N_1 \times \sqrt{2}\,\pi f}$$

$$= \frac{6300}{720 \times \sqrt{2} \times \pi \times 50} = 0.0394 \text{ weber}$$

We know that, $\phi_{max} = B_{max} \times A_c$

$$\therefore \quad A_c = \frac{\phi_{max}}{B_{max}} = \frac{0.0394}{1.2}$$

$$= 0.032833 \text{ m}^2 = 328.33 \text{ cm}^2$$

$\therefore$ Area of the core $= 328.33 \text{ cm}^2$ **Ans.**

Problem 1.2 : A 4600/230 V, 50 Hz, single-phase transformer with a core-section of 400cm² has maximum permissible flux density of 1.25 T. For iron space-factor of 0.9, calculate the number of primary and secondary turns, if the transformer is *(i)* shell type, and *(ii)* core type.

Solution :

Given $V_1 = 4600 \text{ V}, \; V_2 = 230 \text{ V}$

$B_{max} = 1.25 \text{ T}, \; f = 50 \text{ Hz}, \; A_c = 400$

$\text{cm}^2 = 400 \times 10^{-4} \text{ m}^2, \; A_i = 0.9$

(i) When the transformer is of shell type,

$$\text{Emf per turn} = \sqrt{2} \times \pi f \phi_{max} \times A_i$$

$$= \sqrt{2} \times \pi \times 50 \times 400 \times 10^{-4} \times 1.25 \times 0.9$$

$$\cong 10 \text{ V/turn}$$

Now, let us calculate no. of turns in low voltage side.
No. of turns on secondary side

$$(N_2) = \frac{V_2}{\text{Emf per turn}} = \frac{230}{10} = 23 \quad \textbf{Ans.}$$

No. of turns on primary side.

$$N_1 = \left(\frac{V_1}{V_2}\right) \times N_2, \quad N_1 = \frac{4600}{230} \times 23 = 460 \quad \textbf{Ans.}$$

(ii) When the transformer is of core type

$$\text{Emf per turn } (E_t) = \sqrt{2}\pi f \phi_{max} A_i$$

$$= \sqrt{2} \times \pi \times 50 \times 400 \times 10^{-4} \times 1.25 \times 0.9$$

$$\cong 10 \text{ V/turn}$$

Let us calculate no. of turns on l.v. side.

$$\text{No. of turns on l.v. side, } N_2 = \frac{V_2}{E_t} = \frac{230}{10} = 23$$

As we know that in core-type transformer, half of the primary turns, i.e. $\frac{23}{2} = 11.5$ are wound on each limb which is impossible. The no. of turns must be a whole number. Hence no. of turns on each limb must be 12 turns.

$\therefore$ Total No. of turns or secondary side,

$$N_2 = 12 + 12 = 24 \quad \textbf{Ans.}$$

No. of turns on primary side,

$$N_1 = \frac{V_1}{V_2} \times N_2 = \frac{4600}{230} \times 24 = 480 \quad \textbf{Ans.}$$

Problem 1.3 : In an ideal transformer, discuss how impedance is transferred from secondary to primary or vice-versa.

Solution :

Let A be the core area of transfomer A, and B be the core area of transformer B.

Given, $f_A = 50\text{ Hz}, \ f_B = 10\text{ kHz}$

According to Question,

Emf per turn of the transformer A = emf per turn of the trasformer B

$$\text{or} \quad \sqrt{2}\pi f_A \phi_{maxA} = \sqrt{2}\pi f_B \phi_{maxB}$$

$$\therefore \quad \frac{\phi_{max\,A}}{\phi_{max\,B}} = \frac{f_B}{f_A}$$

for the same core flux densities,

$$\frac{B_{max} \times A}{B_{max} \times B} = \frac{10 \times 1000}{50}$$

$$\therefore \quad A : B = 200 : 1 \quad \textbf{Ans.}$$

Problem 1.4 : A 1-phase trasformer has the following data:
Peak flux density in core = 1.40 T
Net core area = 0.012 m^2
Current density in conductors = 2.5 MA/m^2
Conductor dia. = 2.0 mm; Primary voltage = 230 V, 50 Hz
Calculate the kVA rating of the transformer and the number of turns on the primary winding.

Solution :

Given,
$$B_{max} = 1.40 \text{ T}$$
$$A_c = 0.012 \text{ m}^2$$
$$J = 2.5\times10^6 \ A/\text{m}^2$$
$$d = 2.0 \text{ mm} = 2\times10^{-3} \text{ m}$$
$$V_1 = 230 \text{ V}, \ f = 50 \text{ Hz}$$

Current through the conductor (I_1)
$$= \text{Current density } \times \text{Area}$$
$$= J\times\frac{\pi d^2}{4} = 2.5\times10^6\times\pi\left(\frac{d}{2}\right)^2$$
$$= 2.5\times10^6\times\pi\times1\times10^{-6} = 2.5\,\pi \text{ amp}$$

KVA rating of transformer $= V_1 I_1 = 230\times2.5\,\pi = 1.806$ kVA **Ans.**

Now, emf per turn $= \sqrt{2}\,\pi\, f\, B_{max}$

$$\therefore \text{No. of turns of primary} = \frac{V_1}{\text{Emf per turn}}$$
$$= \frac{230}{\sqrt{2}\times\pi\times50\times1.40\times0.012} = 61.63$$

Since number of the turns can't be fraction, we can take either 61 or 62. But more turns reduce the area of the core which is desirable.

Hence, No. of turns = 62 **Ans.**

Problem 1.5 : A 100 kVA, 2400/240-V, 50 Hz single phase transformer has an exiting current of 0.64 A and a core loss of 700 watts, when its high-voltage side is energised at rated voltage and frequency. Calculate the two components of the exciting current.

If the above transformer supplies a load current of 40 amperes at 0.8 power factor lag on its l.v. side, then calculate the primary current and its power factor. Ignore leakage impedance drops.

Solution :

Given,

$$f = 50 \text{ Hz}$$

$$P_0 = 700 \text{ W}$$

$$V_1 = 2400 \text{ V}$$

$$V_2 = 240 \text{ V}$$

$$I_e = 0.64 \text{ A}$$

The no-load phaser diagram of the transformer can be drawn as core loss component of I_e.

$$I_c = \frac{P_0}{V_1} = \frac{700}{2400} \text{ A}$$

$= 0.292$ A **Ans.**

From Fig. P1.5 (a)

Magnetising component of I_e,

$$I_m = \sqrt{I_e^2 - I_c^2}$$

$$= \sqrt{(0.64)^2 - (0.292)^2}$$

$= 0.5695$ **Ans.**

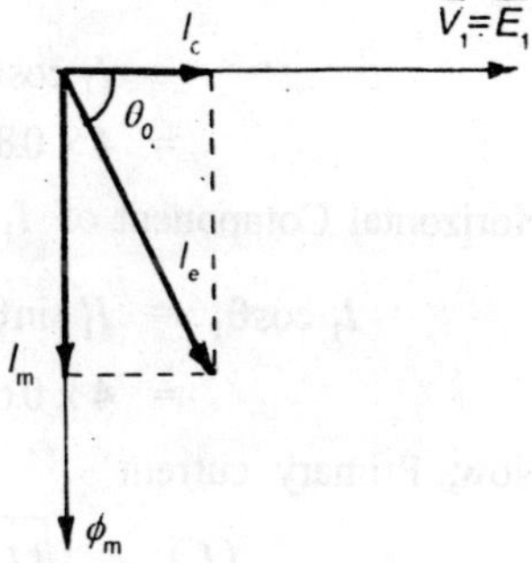

Fig. P. 1.5 (a)

When the transformer supplies a load current of 40 A at lagging p.f. 0.8 on low voltage side, a primary load component (I_1') flows in the primary side to neutralize the increase in core-flux. Hence.

$$I_1' N_1 = I_2 N_2$$

$$\therefore \quad I_1' = \frac{I_2 N_2}{N_1}$$

$$= \frac{40 \times 240}{2400} = 4\text{A}.$$

The load phasor diagram can be drawn as vertical component of I_1

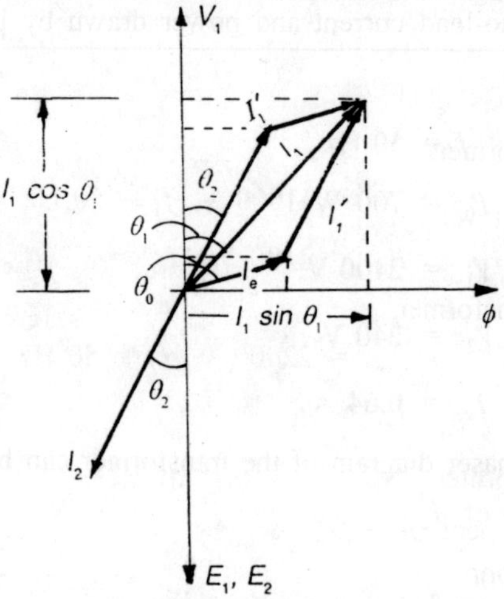

Fig. P. 1.5 (b)

$$= I_1 \cos\theta_1 = I_1' \cos\theta_2 = I_1' \cos\theta_2 + I_c$$
$$= 4 \times 0.8 + 0.292 = 3.492 \text{ A}$$

Horizontal Component of I_1

$$= I_1 \cos\theta_1 = I_1' \sin\theta_2 + I_e \sin\theta_0$$
$$= 4 \times 0.6 + 0.5695 = 2.9695 \text{ A}$$

Now, Primary current

$$(I_1) = \sqrt{(I_1 \cos\theta_1)^2 + (I_1 \sin\theta_1)^2}$$
$$= \sqrt{(3.492)^2 + (2.9695)^2} = 4.584 \text{ A } \textbf{Ans.}$$

Power factor on primary side

$$\cos\theta_1 = \frac{I_1 \cos\theta_1}{I_1} = \frac{3.492}{4.584} = 0.762 \text{ (lagging) } \textbf{Ans.}$$

Problem 1.6 : When the primary of a transformer is energised at rated voltage of 11,000V and at rated frequency of 50 Hz, it takes 3.2 A and 2400 watts at no-load. Another transformer has all its core dimensions $\sqrt{2}$ times the corresponding core dimensions of the first transformer.

Number of primary turns, type of core material and lamination thickness are the same in both the transformers. If the primary of the second transformer is energised from 22,000 V, 50 Hz supply, calculate the no-load current and power drawn by it.

Solution :

For first transformer,

primary voltage $= 11000$ V, $f_1 = 50$ Hz

$$P_{c1} = 2400 \text{ W}, \; I_{e1} = 3.2 \text{ A}$$

For second transformer,

Primary voltage $= 22000$ V, $f_2 = 50$ Hz

$$I_{e2} = ?, \quad P_{c2} = ?$$

For first transformer,

core-loss component of I_{e1}

$$I_{c1} = \frac{P_{c1}}{\text{primary voltage}}$$

$$= \frac{2400}{11000} = 0.2182 \text{ A}$$

$$\text{Flux } (\phi_{m1}) = \frac{\text{primary voltage}}{\sqrt{2}\,\pi f_1 N_1} = \frac{11000}{\sqrt{2}\,\pi f_1 N_1} \quad (1)$$

Magnetising component of I_{e1}

$$I_{m1} = \sqrt{I_{e1}^2 - I_{c1}^2}$$

$$= \sqrt{(3.2)^2 - (0.2182)^2} = 3.193 \text{ A}$$

As we know that core loss $\propto$ core volume

$$\frac{\text{core loss of 2nd transformer}}{\text{core loss of 1st transformer}} = \frac{\text{core volume of 2nd transformer}}{\text{core loss of 1st transformer}}$$

or $$\frac{P_{c2}}{P_{c1}} = \frac{\left(\sqrt{2}\right)^3}{1}$$

$\therefore$ $$P_{c2} = 2\sqrt{2} \times P_{c1}$$

or $$P_{c2} = 2\sqrt{2} \times 2400 = 6788.23 \text{ W} \;\; \textbf{Ans.}$$

Now, core-loss component, $I_{c2} = (I_2)^3 \times I_{c1} = 2\sqrt{2} \times 0.2182$

$$= 0.6172 \text{ A}$$

$$\text{Now,} \quad \phi_{m2} = \frac{V_2}{\sqrt{2}\ \pi f_2 N_2} = \frac{22000}{\sqrt{2}\ \pi f N_2} \qquad [\because f_1 = f_2 = f \text{ and } N_1 = N_2]$$

$$= \frac{2 \times 11000}{\sqrt{2}\ \pi\ f\ N_1}$$

$$\therefore \quad \phi_{m2} = 2\ \phi_{m1} \tag{2}$$

$$\text{Now,} \quad \Re_2 = \frac{\Re_1}{\sqrt{2}}$$

$$\therefore \quad \Re_1 = \sqrt{2} \times \Re_2 \tag{3}$$

We know that $\Re\phi = NI$

$$\therefore \quad \varphi_{m1} = \frac{N_1 I_1}{\Re_1}$$

Multiplying both sides by 2,

$$2\ \varphi_{m1} = \frac{2\ N_1\ I_{\phi 1}}{\Re_1}$$

$$\text{or} \quad \varphi_{m2} = \frac{2\ N_1\ I_{\phi 1}}{\Re_1}$$

$$\text{or} \quad \frac{I_{\phi 2} \times N_1}{\Re_2} = \frac{2\ N_1 \times I_{\phi 1}}{\Re_1}$$

$$\text{or} \quad I_{\phi 2} = \frac{2 \times \Re_2}{\Re_1} \times I_{\phi 2} = \frac{2 \times R_2}{\sqrt{2} \times R_2}\ I_{\phi 1}$$

$$= \sqrt{2} \times I_{\phi 1} = \sqrt{2} \times 3.193 = 4.515 \text{ A}$$

Now, No-load current of 2nd transformer,

$$I_{e2} = \sqrt{(I_{\phi 2})^2 + (I_{c2})^2}$$

$$= \sqrt{(4.515)^2 + (0.6172)^2} = 4.557 \text{ A} \ \textbf{Ans.}$$

Thus,

No-load current = 4.557 A, core loss = 6788.8 W **Ans.**

Problem 1.7 : A 50 Hz, 3-phase core-type star-delta transformer has a line voltage ratio of 11,000/440 volts. The cross-section of the core is square with a circumscribing circle of 0.4 m diameter. Determine the number of turns per phase on high-voltage and low-voltage windings for a maximum flux density of 1.25 T. Assume that the insulation occupies 10% of the total core area.

Solution :

Given, $V_1 = 11000$ V, $V_2 = 440$ V

$B_{max} = 1.25$ T

Insulation area = 10%

Core area = (100 – 10)% = 90%

$f = 50$ Hz,

Net core area, $A_c = (0.4 \cos 45^\circ)^2 \times 0.9 = 0.072$ Sq.m.

Now, applying the relation,

Emf per turn, $E_t = \sqrt{2}\,\pi f B_{\max} A_c$

$$= \sqrt{2}\,\pi \times 50 \times 1.25 \times 0.072 = 19.99 \cong 20 \text{ V/turn}$$

Low voltage turn per phase $= \dfrac{V_2}{E_t} = \dfrac{440}{20} = 22$ **Ans.**

High voltage turn per phase $= \left(\dfrac{V_1}{V_2}\right) \times N_2 = \left(\dfrac{11000}{\sqrt{3} \times 440}\right) \times 22$

$$= 317.54 \cong 318.$$

Since lower number of turns increases core area which is undesirable.

Hence high voltage turn per phase = 318 turns **Ans.**

Thus No. of primary turns/phase = 22

No. of secondary turns/phase = 318

Problem 1.8 : The constants of a single phase 50 Hz, 2200/220 V transformer are as follows.

h.v. side — $r_1 = 0.21\ \Omega$, $x_1 = 3.84\ \Omega$,

$R_{c1} = 4800\ \Omega$, $X_{\phi 1} = 3500\ \Omega$

l.v. side — $r_2 = 0.006\ \Omega$, $x_2 = 0.022\ \Omega$.

Find the equivalent circuit parameters referred to *(i)* h.v. side and *(ii)* l.v. side.

Solution :

Given $r_1 = 0.21\ \Omega,\quad x_1 = 3.84\ \Omega,$

$R_{c1} = 4800\ \Omega,\quad X_{\phi 1} = 3500\ \Omega,$

$r_2 = 0.006\ \Omega,\quad x_2 = 0.022\ \Omega.$

As we know that impedances are transfered as square of the turn ratio

(i) Equivalent circuit parameters reffered to h.v. side

$$r_{e1} = r_1 + \left(\frac{N_1}{N_2}\right)^2 \times r_2$$

$$= 0.21 + \left(\frac{2200}{220}\right)^2 \times 0.06 = 0.081 \quad \textbf{Ans.}$$

$$x_{e1} = x_1 + \left(\frac{N_1}{N_2}\right)^2 \times x_2$$

$$= 3.84 + \left(\frac{2200}{220}\right)^2 \times 0.022 = 6.04\ \Omega \quad \textbf{Ans.}$$

Shunt parameters on h.v. side remains same

$\therefore \quad R_{c1} = 4800\ \Omega,\quad X_{\phi 1} = 3500\ \Omega$

(ii) Equivalent circuit reffered to l.v. side

Equivalent resistance

$$r_{e_2} = r_2 + \left(\frac{N_2}{N_1}\right)^2 \times r_1$$

$$= 0.006 + \left(\frac{220}{2200}\right)^2 \times 0.21 = 0.0081\ \Omega = 48\ \Omega \quad \textbf{Ans.}$$

Equivalent leakage reactance

$$x_{e_2} = x_2 + \left(\frac{N_2}{N_1}\right)^2 \times x_1$$

$$= 0.022 + \left(\frac{220}{2200}\right)^2 \times 3.84 = 0.0604\ \Omega \quad \textbf{Ans.}$$

Core resistance reffered to l.v. side

$$R_{c_2} = \left(\frac{N_2}{N_1}\right)^2 \times R_{c_1}$$

$$= \left(\frac{220}{2200}\right)^2 \times 4800\,\Omega \quad \textbf{Ans.}$$

Magnetising reactance on l.v. side

$$X_{\phi 2} = \left(\frac{N_2}{N_1}\right)^2 \times X_{\phi 1}$$

$$= \left(\frac{220}{2200}\right)^2 \times 3500 = 35\,\Omega \quad \textbf{Ans.}$$

Problem 1.9 : A 100 k VA, 1000/400 volts, single-phase transformer, when excited at rated voltage on h.v. side, draws a no-load current of 3.0 A at 0.5 lagging power factor. If it is excited from the l.v. side at rated voltage, determine the no load current, power factor and power input.

Solution :

Given, $V_1 = 1000$ V, $V_2 = 400$ V

At rated voltage on h.v. side

$$\text{p.f.} = 0.5 \text{ lagging}$$

$$I_{e1} = 3.0 \text{ A}$$

When the transformer is excited from low voltage side at rated voltage.
No-load current on l.v. side

$$I_{e2} = I_{e1} \times \left(\frac{V_1}{V_2}\right) = 3.0 \times \left(\frac{1000}{400}\right) = 7.5 \text{ A} \quad \textbf{Ans.}$$

At No-load, p.f. remains same on both sides

$$\text{p.f.} = 0.5 \text{ (lag)} \quad \textbf{Ans.}$$

Power input $P_2 = V_2 I_{e2} \cos\theta = 400 \times 7.5 \times 0.5 = 1500 \text{ W}$ **Ans.**

Problem 1.10 : A 5 kVA, 1000/200 V, 50 Hz single-phase transformer gave the following test results:

Open-circuit test (1.V. side) — 200 V, 1.2 A, 90 W

Short-circuit test (h.v. side) — 50 V, 5A, 110 W.

Compute the parameters of the approximate equivalent circuit referred to 1.v. side. Also draw the exact equivalent circuit referred to 1.V. side.

Solution :

Given, $V_1 = 1000$ V, $V_2 = 200$ V,

$V_{oc} = 200$ V, $I_{oc} = 1.2$ A, $P_{oc} = 90$ W,

$V_{sc} = 50$ V, $I_{sc} = 5$ A, $P_{sc} = 110$ W.

Calculation for l.v. Side

As we known that OC test yields shunt parameters

∴ No-load power factor

$$\cos\theta_0 = \frac{P_{oc}}{V_{oc}\times I_{oc}} = \frac{90}{1.2\times 200} = 0.375$$

$$\therefore \quad \sin\theta_0 = \sqrt{1-\cos\theta_0^2} = \sqrt{1-(0.375)^2} = 0.927$$

Core loss resistance,

$$R_c = \frac{V_{oc}}{I_{oc}\cos\theta_0} = \frac{200}{1.2\times(0.375)} = 444.44\ \Omega$$

Magnetising reactance,

$$X_\phi = \frac{V_{oc}}{I_{oc}\sin\theta_0} = \frac{200}{1.2\times 0.927} = 179.80\ \Omega$$

Calculation for h.v. Side

As we know that SC test yields series parameters and is performed on h.v. side.

$$Z_{sc} = \frac{V_{sc}}{I_{sc}} = \frac{50}{5} = 10\ \Omega$$

$$r_{eq} = \frac{P_{sc}}{I_{sc}^2} = \frac{110}{25} = 4.4\ \Omega$$

Then, $$x_{eq} = \sqrt{Z_{sc}^2 - r_{eq}^2} = \sqrt{10^2-(4.4)^2} = 8.98\ \Omega$$

Equivalent resistance of Approx. equivalent circuit referred to l.v. side,

$$r_{e_2} = \left(\frac{V_2}{V_1}\right)^2 \times r_{eq} = \left(\frac{200}{1000}\right)^2 \times 4.4 = 0.176\ \Omega \quad \textbf{Ans.}$$

Equivalent leakage reactance of Approx equivalent circuit referred to l.v. side

$$x_{e2} = \left(\frac{V_2}{V_1}\right)^2 \times x_{eq} = \left(\frac{200}{1000}\right)^2 \times 8.98$$

$$= 0.359 \ \Omega \ \textbf{Ans.}$$

Equivalent core loss resistance referred to l.v. side

$$R_{c2} = 444.44 \ \Omega \ \textbf{Ans.}$$

Equivalent magnetising reactance referred to l.v. side

$$X_{\phi 2} = 179.80 \ \Omega \ \textbf{Ans.}$$

Calculation for Exact Equivalent Circuit Referred l.v.

$$r_2 = r_1' = \frac{r_{e2}}{2} = \frac{0.176}{2} = 0.088 \, \Omega \ \textbf{Ans.}$$

$$x_2 = x_1' = \frac{x_{e2}}{2} = \frac{0.359}{2} = 0.1795 \, \Omega \ \textbf{Ans.}$$

The exact equivalent circuit reffered to l.v. side can be drawn as

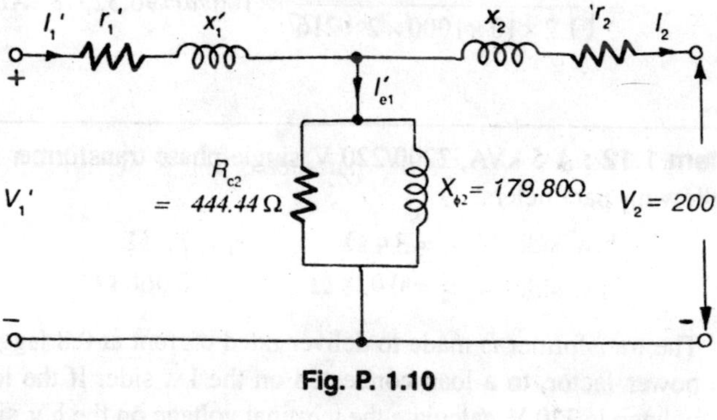

Fig. P. 1.10

> **Problem 1.11 :** A 10 k AV, 500/250 V, single-phase transformer gave the following test results:
>
> S.C. test (h.v. side) — 60 V, 20 A, 150 W
>
> The maximum efficiency occurs at unity pf and at 1.2 full-load current. Determine full-load efficiency at 0.8 pf. Also calculate the maximum efficiency.

Solution :

Given, Full load copper loss = 150 W

Full load primary current = 20 A

Since maximum efficiency occurs at unity p.f. and at 1.2 of full-load current

$\therefore$ Core loss = Copper loss at 1.2 full-load current

$= (1.2)^2 \times$ Full-load copper loss

$= 1.44 \times 150 = 216$ W

Now % Full load efficiency at 0.8 p.f.

$$= \left(\frac{\text{kVA} \times \text{p.f.}}{\text{kVA} \times \text{p.f.} + P_c + P_i}\right) \times 100\,\%$$

$$= \left(\frac{10 \times 1000 \times 0.8}{10 \times 1000 \times 0.8 + 216 + 150}\right) \times 100\% = 95.625\% \quad \textbf{Ans.}$$

It is given that maximum efficiency occurs at unity p.f. and 1.2 full-load current % Maximum efficiency

$$= \left(\frac{1.2 \times \text{kVA} \times \text{p.f.}}{1.2 \times \text{kVA} \times \text{p.f.} + 2 \times P_c}\right) \times 100\,\%$$

$$= \left(\frac{1.2 \times 10 \times 1000 \times 1}{1.2 \times 10 \times 1000 \times 2 \times 216}\right) \times 100\% = 96.525\% \quad \textbf{Ans.}$$

Problem 1.12 : A 5 kVA, 2200/220 V, single-phase transformer has the following parameters:

h.v. side – $r_1 = 3.4\,\Omega$ $\quad x_1 = 7.2\,\Omega$

l.v. side – $r_2 = 0.028\,\Omega$ $\quad x_2 = 0.060\,\Omega$

(a) The transformer is made to deliver rated current at 0.8 lagging power factor, to a load connected on the l.v. side. If the load voltage is 220 V, calculate the terminal voltage on the h.v. side.

(b) Repeat part *(b)* for a load pf of 0.8 leading.

(c) For a core loss of 30 watts at rated voltage and frequency, find the efficiency under the conditions of part *(a)* and *(b)*.

(d) The transformer exciting current is 3% of full-load current. Calculate voltmeter, ammeter and wattmeter readings for open-circuit test at rated voltage and short-circuit test at rated current. The instruments are connected on l.v. side for open-circuit test and on h.v. side for short circuit test.

Solution :

Given $V_1 = 2200$ V, $V_2 = 220$ V

Secondary rated current

$$(I_2) = \frac{\text{Rating}}{V_2} = \frac{5 \times 1000}{220} = 22.72 \text{ amp.}$$

Secondary rated current reffered to h.v. side

$$I_2' = \left(\frac{V_2}{V_1}\right) \times I_2 = \left(\frac{220}{2200}\right) \times \left(\frac{5 \times 100}{220}\right) = 2.27 \text{ A}$$

Equivalent resistance reffered to h.v. side

$$r_{el} = r_1 + \left(\frac{V_1}{V_2}\right)^2 \times r_2 = 3.4 + \left(\frac{220}{220}\right)^2 \times 0.028$$

$$= 3.4 + 2.8 = 6.2 \ \Omega$$

Equivalent reactance reffered to h.v. side

$$x_{el} = x_1 + \left(\frac{V_1}{V_2}\right)^2 \times x_2 = 7.2 + \left(\frac{2200}{220}\right)^2 \times 0.060 = 13.2 \ \Omega$$

Secondary terminal voltage reffered to h.v. side

$$V_2' = \left(\frac{V_1}{V_2}\right) \times V_2 = 2200 \text{ V}$$

The circuit diagram can be drawn as

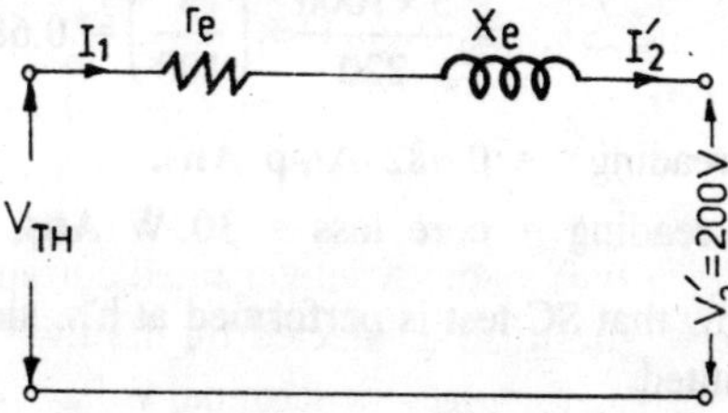

Fig. P. 1.12

(a) When the load is at 0.8 p.f. lagging

Terminal voltage on h.v. side

$$V_{1H} = V_2' + I_2' \,(r_{el} \cos\theta + x_{el} \sin\theta)$$

$$= 2200 + 2.27 \,(6.2 \times 0.8 + 13.6 \times 0.6)$$

$$= 2229.26 \text{ A} \quad \textbf{Ans.}$$

(b) When the load is at 0.8 p.f. leading

Teminal voltage on h.v. side

$$V_{1H} = V_2' + I_2' (r_{e1} \cos\theta - x_{e1} \sin\theta)$$
$$= 2200 + 2.27 (6.2 \times 0.8 - 13.2 \times 0.6)$$
$$= 2193.27 \text{ A } \textbf{Ans.}$$

(c) Total Copper loss $P_i = (I_2')^2 \times r_{e1} = (2.27)^2 \times 6.2 = 32 \text{ W}$

Total core loss, $P_c = 30 \text{ W}$

Power output $P_0 = V_2' I_2' \cos\theta = 2200 \times 2.27 \times 0.8$
$$= 3995.2 \text{ W}$$

$$\% \text{ Efficiency} = \left(\frac{\text{Output}}{\text{Output} + P_i + P_c}\right) \times 100\ \%$$

$$= \left(\frac{3995.2}{3995.2 + 30 + 32}\right) \times 100\% = 98.472\% \ \textbf{Ans.}$$

For leading power factor, the efficiency remains same as copper loss, core loss and terminal voltage are same.

(d) As we known that OC test is performed on low voltage side at rated voltage.

Voltmeter reading = 220 V **Ans.**

No. load current, $I_e = \left(\frac{\text{Power}}{V_2}\right) \times (3\%)$

$$= \frac{5 \times 1000}{220} \times \left(\frac{3}{100}\right) = 0.682 \text{ Amp } \textbf{Ans.}$$

Ammeter reading = 0.682 Amp **Ans.**

Wattmeter reading = core loss = 30 W **Ans.**

As we know that SC test is performed at h.v. side with l.v. sic short circuited

Rated current on h.v. side

$$I_{sc} = \frac{\text{Rating}}{V_1} = \frac{5 \times 1000}{2200} = 2.273 \text{ A } \textbf{Ans.}$$

Equivalent resistance reffered to h.v. side.

$$r_{sc} = r_1 + \left(\frac{V_1}{V_2}\right)^2 \times r_2 = 3.4 + \left(\frac{2200}{220}\right)^2 \times 0.028 = 6.2\ \Omega$$

Wattmeter reading $W_{sc} = (I_{sc})^2 \times r_{sc} = (2.273)^2 \times 6.2$

$= 32.08$ W **Ans.**

Equivalent reactance reffered to h.v. side.

$$x_{sc} = x_1 + \left(\frac{V_1}{V_2}\right)^2 \times x_2 = 13.2\,\Omega$$

$$\therefore \quad Z_{sc} = \sqrt{(x_{sc})^2 + (r_{sc})^2} = \sqrt{(13.2)^2 + (6.2)^2} = 14.58\,\Omega$$

Voltmeter reading $V_{sc} = I_{sc} \times Z_{sc} = 14.58 \times 2.273 = 33.14$ V **Ans.**

> **Problem 1.13 :** A 50 Hz, single-phase transformer draws a short circuit current of 30 A at 0.2 pf lag when connected to 16 V, 50 Hz source. What will be the short-circuit current, and it p.f. when the same transformer is energised from 16 V, 25 Hz source?

Solution :

Given, At 50 Hz $I = 30$ A, $V = 16$ V, $\cos\theta = 0.2$

Let r_2 and x_2 be the equivalent resistance and reactance of the transformer respectively.

We know that $$\cos\theta = \frac{r_e}{Z_e}$$

or $$0.2 = \frac{r_e}{\sqrt{r_e^2 + x_e^2}}$$

Squaring both sides,

or $$0.04 = \frac{r_e^2}{x_e^2 + r_e^2}$$

$$\therefore \quad x_e^2 = 24\, r_e^2 \qquad (1)$$

Also, $$I = \frac{V}{Z_e}$$

or $$Z_e = \frac{V}{I}$$

or $$\sqrt{r_e^2 + x_e^2} = \frac{16}{30}$$

or $$\sqrt{r_e^2 + 24\, r_e^2} = \frac{16}{30}$$

$\therefore$ $$r_e = 0.1067\ \Omega$$

Putting $$r_e = 0.1067\ \Omega \text{ in (1)}$$

$$x_e = 0.523\ \Omega$$

As we know that reactance changes as frequency changes,

$\therefore$ $$\frac{x_1}{f_1} = \frac{x_2}{f_2}$$

or $$\frac{0.523}{50} = \frac{x'_e}{25}$$

$\therefore$ $$x'_e = 0.2615\ \Omega$$

Since resistance is independent of frequency.

$\therefore$ $$r'_e = 0.1067\ \Omega$$

Now, Short circuit current

$$= \frac{V}{Z'_e} = \frac{V}{\sqrt{r'^2_e + x'^2_e}}$$

$$= \frac{16}{\sqrt{(0.1067)^2 + (0.2615)^2}} = 56.65 \text{ A } \textbf{Ans.}$$

$$\text{p.f.} = \cos\theta = \frac{r_e}{\sqrt{r_e^2 + x_e^2}}$$

$$= \frac{0.1067}{\sqrt{(0.1067)^2 + (0.2615)^2}} = 0.3378 \text{ (lag) } \textbf{Ans.}$$

Problem 1.14 : A 3300/230 V, single phase transformer has an equivalent resistance of 0.015 p.u. and an equivalent leakage reactance of 0.04 p.u. Find the secondary terminal voltage at full load and 0.8 p.f. lag, for a primary voltage of 3300 V.

Solution :

Given $r_{e(p.u.)} = 0.015$ p.u., $x_{e(p.u.)} = 0.04$ p.u.

$$\text{p.f. } (\cos\theta_2) = 0.8 \text{ (lagg)}$$

Let us take base voltage $V_b = 3300$ V

$$\therefore \quad V_H = \frac{V_1}{V_b} = \frac{3300}{3300} = 1 \text{ p.u.}$$

Full load current $(I_1 \cong I_2') = 1$ p.u.

The equivalent circuit diagram on h.v. side can be drawn as (neglecting shunt parameters)

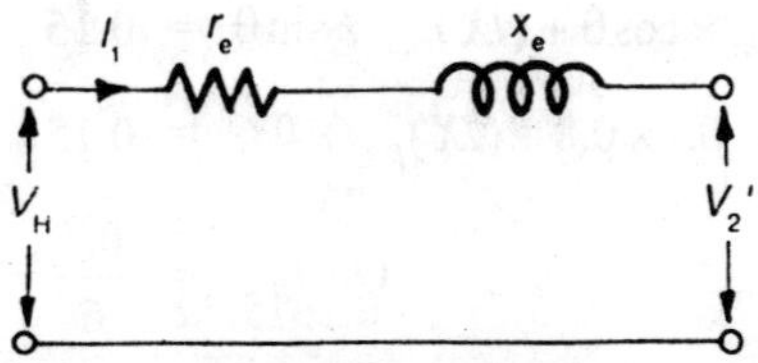

Fig. P. 1.14

From circuit diagram

$$V_H = V_2' + I_2'(r_e \cos\theta + x_e \sin\theta)$$

or

$$1 = V_2' + 1(0.015 \times 0.8 + 0.04 \times 0.6)$$

$$\therefore \quad V_2' = 0.964 \text{ p.u.}$$

Actual secondary voltage on primary side

$$= V_2' \times V_b = 0.964 \times 3300 \text{ V}$$

Actual secondary voltage on secondary side

$$= \left(\frac{V_2}{V_1}\right) \times V_2' \times V_b$$

$$= \left(\frac{230}{3300}\right) \times 0.964 \times 3300 = 221.72 \text{ V } \textbf{Ans.}$$

Problem 1.15 : A single-phase transformer has a regulation of 10% when delivering full load at unity pf and 15% when delivering the same load at 0.8 pf lagging. What would be the regulation if the transformer is delivering half-load at 0.8 p.f. leading?

Solution :

We know that % regulation

$$= IR_{pu} \times \cos\theta + IX_{pu} \times \sin\theta$$

Where θ = p.f. angle

At unit p.f. $\cos\theta = 1, \quad \sin\theta = 0$

$$\% \text{ regulation} = 0.10$$

or $$(IR)_{pu} \times 1 + (IX)_{pu} \times 0 = 0.10$$

$$\therefore \quad (IR)_{pu} = 0.10 \text{ p.u.} \qquad (1)$$

At 0.8 lagging, $\cos\theta = 0.8$, $\sin\theta = 0.6$, % regulation = 0.15

or $$(IR)_{p.u.} \times \cos\theta + (IX)_{p.u.} \times \sin\theta = 0.15$$

or $$0.1 \times 0.8 + (IX)_{p.u.} \times 0.6 = 0.15$$

$$\therefore \quad (IX)_{p.u.} = \frac{0.7}{6} \qquad (2)$$

Now % regulation at 0.8 leading p.f. at half load

$$= \left(\frac{(IR)_{p.u.}}{2} \times \cos\theta - \frac{(IX)_{p.u}}{2} \times \sin\theta\right) \times 100\,\%$$

$$= \left(\frac{0.10}{2} \times 0.8 - \frac{0.07}{6 \times 2} \times 0.6\right) \times 100\,\% = 0.5\% \quad \textbf{Ans.}$$

> **Problem 1.16 :** The maximum efficiency of a 500 kVA, 3300/500 V, 50 Hz single-phase transformer is 97% and occurs at 3/4 full load, unity power factor. If the impedance is 10%, calculate the regulation at full load, power factor 0.8 lagging.

Solution :

Given $\eta_{max} = 0.97$

Load current, $(I) = \frac{3}{4} \times I_{fl}$

Here $K = \frac{I}{I_{fl}} = \frac{3}{4}$

At maximum efficiency, Copper loss (P_i) = core loss (P_c)

Now, Maximum efficiency at 3/4 full load at unity p.f. = 0.97

$$\text{or} \qquad \frac{kV_2 I_2 \cos\theta_2}{kV_2 I_2 \cos\theta_2 + 2P_i} = 0.97$$

$$\text{or} \qquad \frac{\frac{3}{4} \times 500 \times 1000 \times 1}{\frac{3}{4} \times 500 \times 1000 \times 1 + 2P_i} = 0.97 \qquad [\cos\theta_2 = 1 \text{ for upf}]$$

$$\text{or} \qquad \frac{3}{4} \times 500 \times 1000 \times 1 + 2P_i = \frac{3 \times 500 \times 1000}{0.97 \times 4}$$

$$\text{or} \qquad 2P_i = \frac{3}{4} \times 500 \times 1000 \left(\frac{1}{0.97} - 1 \right)$$

$$\therefore \qquad P_i = \frac{3 \times 500 \times 1000 \times 0.03}{4 \times 2 \times 0.97} = 5798.969 \text{ W}$$

At maximum efficiency Copper loss = K^2 × Full load copper loss

$$\therefore \text{Full load copper loss } (P_{fl}) = \frac{16}{9} \times 5798.967 \text{ W}$$

$$\text{Equivalent resistance in p.u. } R_{pu} = \frac{\text{Full load copper loss}}{\text{Rating}}$$

$$= \frac{16 \times 5798.969}{9 \times 500 \times 1000}$$

$$= 0.0206 \text{ p.u.}$$

$$\text{Given,} \qquad Z_{pu} = 0.10 \text{ p.u.}$$

$$\therefore \qquad X_{pu} = \sqrt{Z_{pu}^2 - R_{pu}^2}$$

$$= \sqrt{(0.10)^2 - (0.0206)^2} = 0.0979 \text{ p.u.}$$

% Regulation at full load at 0.8 p.f. lagging

$$= \left(R_{pu} \times \cos\theta + X_{pu} \sin\theta \right) \times 100\%$$

$$= (0.0206 \times 0.8 + 0.0979 \times 0.6) \times 100\% = 7.522\% \textbf{ Ans.}$$

Problem 1.17 : The leakage impedance of a 10 kVA, 3300/220 V transformer as given on its name plate is 10%. Determine *(a)* the voltage that must be applied on h.v. side to circulate rated current with l.v. side shorted and *(b)* the maximum possible voltage regulation.

Solution :

Given, $V_1 = 3300$ V, $V_2 = 220$ V

Rating = 10 kVA, $Z_{p.u} = 10\% = 0.10$ p.u.

Rated current on h.v. side

$$I_1 = \frac{\text{Rating}}{V_1} = \frac{10 \times 1000}{3300}$$

Base impedance on h.v. side,

$$Z_{bH} = \frac{(kV)^2}{kVA} = \frac{(3300)^2}{10 \times 1000} = \frac{3300 \times 3300}{10 \times 1000} = 1089\ \Omega$$

$$\therefore \quad Z_{\text{actual}} = Z_{\text{base}}\ Z_{p.u.} = (1089) \times (0.1) = 108.9\ \Omega$$

Now, voltage applied to h.v. side $= I_1 \times Z_{\text{actual}}$

$$= \left(\frac{10 \times 1000}{3300}\right) 108.9 = 330 \text{ V} \quad \textbf{Ans.}$$

As we know that the magnitude of maximum voltage regulation is equal to the p.u. value of equivalent leakage impedance of the transformer.

Given, leakage impedance = 10% = 0.10 p.u.

∴ Maximum possible voltage regulation = 10%. **Ans.**

Problem 1.18 : *(a)* A non-sinusoidal voltage $U = 150 \sin 314 t - 75 \sin 1570\, t$ is applied to the 250 turn winding of a transformer. Find the core flux as a function of time.

(b) A voltage $v = 200 \sin 314\, t$ is applied to the transformer winding in a no-load test. The resulting current is found to be $i = 3 \sin (314t - 60°)$. Determine the core loss and r.m.s. value of the exciting current.

(c) A transformer has the following test data:

Test No. 1—100% voltage, 6% current, p.f. = 0.2

Test No. 2—8% voltage, 100% current, p.f. = 0.3

Identify the tests. Calculate efficiency and percentage regulation at full load upf and at 0.8 pf lagging.

[*Hint.* (*c*) core loss = 0.012

Full-load ohmic loss = 0.024, $Z = 0.08$, etc.]

Solution :

(a) Given, $v = 150 \sin 314\, t - 75 \sin 1570\, t$, $N = 250$ turns

flux linkage $(\Psi) = \phi \times N$

We know that the induced voltage is rate of change of flux linkage with time,

$$\therefore \qquad v = \left(-\frac{d\Psi}{dt}\right)$$

$$\text{or} \qquad v = \left(-N\frac{d\phi}{dt}\right), N\,d\phi = vdt$$

Integrating both sides,

$$N\phi = \int (150 \sin 314t - 75 \sin 1570t)\, dt$$

$$\text{or} \qquad \phi = \frac{1}{N}\left[-\frac{150}{314}\cos 314\,t + \frac{75}{1570}\cos 1570\,t\right] + c$$

At $t = 0$, $v = 0$ hence $\varphi = 0$

$$\therefore \qquad \phi = \frac{1}{250}\left[-\frac{150}{314}\cos 314\,t + \frac{75}{1570}\cos 1570\,t\right] \text{Wb}$$

$$= (-1.911 \cos 314\,t + 0.1911 \cos 1570\,t) \text{ mWb } \textbf{Ans.}$$

(b) Given, $V(t) = 200 \sin 314\,t$, $V_m = 200$ V

$$i(t) = 3 \sin(314\,t - 60),\ I_m = 3 \text{ A}$$

$$\text{core loss} = V_{rms} \times I_{rms} \times \cos\phi$$

$$= \frac{200}{\sqrt{2}} \times \frac{3}{\sqrt{2}} \times \cos 60° = \frac{200}{\sqrt{2}} \times \frac{3}{\sqrt{2}}$$

$$= 150 \text{ W } \textbf{Ans.}$$

Rms value of exciting current $= \dfrac{3}{\sqrt{2}} = 2.121$ A **Ans.**

(c) As we know that o.c. test is performed at rated voltage on l.v. side and s.c. test is performed at rated current on h.v. side. Hence Test1 is o.c. test and test 2 is s.c. test.

For o.c. test,

$$\text{core loss} = V_{oc} I_{oc} \cos\theta_o$$

$$= \frac{100}{100} \times \frac{6}{100} \times 0.2 = 0.012 \text{ p.u.}$$

$$\text{Full load ohmic loss} = V_{sc} I_{sc} \cos\theta$$

$$= \frac{8}{100} \times \frac{100}{100} \times 0.3 = 0.024 \text{ p.u.} = R_{pu}$$

$$Z_{pu} = \frac{V_{sc}}{I_{sc}} = \frac{8\,\%}{100\,\%} = 0.08 \text{ p.u.}$$

$$X_{pu} = \sqrt{2pu^2 - Rpu^2}$$

$$= \sqrt{(0.08)^2 - (0.024)^2} = 0.0764 \text{ p.u.}$$

Rated kVA = 1 p.u.

At full load unity power factor,

$$\% \text{ Efficiency} = \left(\frac{\text{kVA} \times \text{p.f.}}{\text{kVA} \times \text{p.f.} + p_c + p_i}\right) \times 100\,\%$$

$$= \left(\frac{1 \times 1}{1 \times 1 + 0.024 + 0.012}\right) \times 100\,\%$$

$$= 96.525\% \quad \textbf{Ans.}$$

At full load 0.8 p.f. lagging,

$$\% \text{ Efficiency} = \left(\frac{\text{kVA} \times \text{p.f.}}{\text{kVA} \times \text{p.f.} + p_c + p_i}\right) \times 100\,\%$$

$$= \left(\frac{1 \times 0.8}{1 \times 0.8 + 0.024 + 0.012}\right) \times 100\,\%$$

$$= 95.694\,\% \quad \textbf{Ans.}$$

Percentage voltage regulation at unity p.f.

$$= \left(R_{pu} \times \cos\theta + X_{pu} \sin\theta\right) \times 100\,\%$$

$$= (0.024 \times 1 + 0.0764 \times 0) \times 100\,\% = 2.4\ \% \quad \textbf{Ans.}$$

Percentage voltage regulation at full load 0.8 p.f. lagging

$$= \left(R_{pu} \times \cos\theta + X_{pu} \sin\theta\right) \times 100\,\%$$

$$= (0.024 \times 0.8 + 0.0764 \times 0.6) \times 100\,\% = 6.48\,\% \quad \textbf{Ans.}$$

Problem 1.19 : A 200 V, 60 Hz single-phase transformer has hysteresis and eddy current losses of 250 watts and 90 watts respectively. If the transformer is now energised from 230 V, 50 Hz supply, calculate its core losses. Assume Steinmetz's constant equal to 1.6.

Solution :

At $V_1 = 200$ V, $f_1 = 60$ Hz

$p_{h1} = 250$ W, $p_{el} = 90$ W

At $V_2 = 230$ V, $f_2 = 50$ Hz

$p_{h2} = ?$ and $p_{e2} = ?$

We know that $p_h = k_h v^x f^{1-x}$ (1)

and $p_e = k_e v^2$ (2)

From (1)

$$\therefore \quad \frac{p_{h2}}{p_{h1}} = \left(\frac{v_2}{v_1}\right)^x \left(\frac{f_2}{f_1}\right)^{1-x}$$

$$\therefore \quad p_{h2} = \left(\frac{230}{200}\right)^{1.6} \left(\frac{50}{60}\right)^{1-1.6} \times 250 = 348.8 \text{ W}$$

From (2) $$\frac{p_{e2}}{p_{e1}} = \left(\frac{V_2}{V_1}\right)^2$$

$$\therefore \quad p_{e2} = \left(\frac{V_2}{V_1}\right)^2 \times p_{e1} = \left(\frac{230}{200}\right)^2 \times 90 = 119 \text{ W}$$

$\therefore$ Total core loss $= p_{e2} + p_{h2} = 348.8 + 119$

$= 467.8$ W **Ans.**

Problem 1.20 : A 1-phase, 11/66 kv, 50 MVA transformer has primary and secondary leakage impedances of $0.01 + j\,0.06$ per unit each. The shunt branch impedance in p.u. is $80 + j\,0$ in parallel with $0 + j\,90$. All quantities are in per unit on the transformer base. Calculate, in ohms, the parameters of primary and secondary windings and the shunt branch referred to low-voltage side.

Solution :

Base impedance on l.v. side, Z

$$\text{base} = \frac{(kV_1)^2}{MVA} = \frac{(11 \times 1000)^2}{50 \times 10^6} = 2.42\,\Omega$$

Actual impedance on l.v. side.

$$Z_{Lv} = Z_{L\,base}\, Z_{pu}$$

$$= 2.42 \times (0.01 + j\,0.06)$$

$$= 0.242 + j\,0.1452\ \Omega = r_1 + jx_1$$

Primary winding resistance,

$$r_1 = 0.242\ \Omega\ \textbf{Ans.}$$

Primary winding leakage reactance

$$x_1 = 0.1452\ \Omega\ \textbf{Ans.}$$

Now, Base impedance on h.v. side,

$$Z_{H\,base} = \frac{(kV_2)^2}{MVA} = \frac{(66 \times 1000)^2}{50 \times 10^6} = 87.12\ \Omega$$

Actual impedance on h.v. side

$$Z_{HV} = Z_{H\,base} \times Z_{pu}$$

$$= 87.12 \times (0.01 + j\,0.06)$$

$$= 8712 + j\ 5.2272\ \Omega$$

Reffering secondary impedances on l.v. side,

$$Z'_{HV} = \left(\frac{v_1}{v_2}\right)^2 \times Z_{HV} = \left(\frac{11}{66}\right)^2 \times (0.8712 + j\,5.2272)$$

$$= 0.0242 + j0.1452$$

$$= r'_2 + ix'_2$$

$$r_1 = r'_2 = 0.0242\ \Omega\ \textbf{Ans.}$$

and $$x_1 = x'_2 = 0.1452\,\Omega\ \textbf{Ans.}$$

Thus, core loss resistance on l.v. side

$$R_c = R_c \text{ in p.u.} \times Z_{L\,\text{base}}$$

$$R_c = (80 + j0) \times 2.42$$

$$R_c = 193.60\ \Omega\ \textbf{Ans.}$$

Magnetising reactance on l.v. side

$$= X_\phi \text{ in p.u.} \times Z_{L\,base} = 90 \times 2.42$$

$$= 90 \times 2.42 = 217.80\ \Omega\ \textbf{Ans.}$$

Problem 1.21 : Fig. illustrates the equivalent circuit referred to the l.v. side of a 400/3200 V, 1-phase, 50 Hz, 16 kVA transformer. With primary voltage held fixed at 400 V, the h.v. is loaded to its rated current at p.f. 0.8 lag at its terminals. Compute

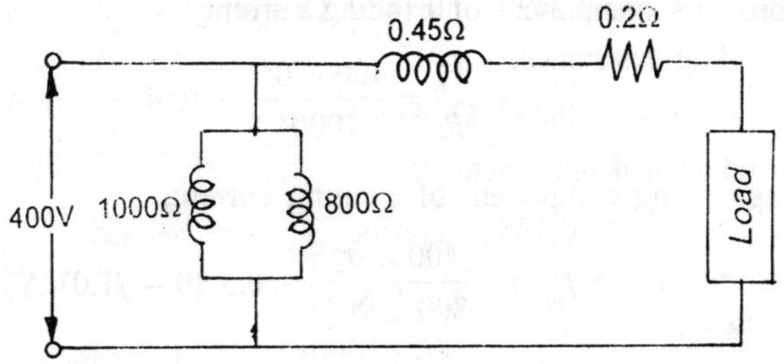

Fig. P. 1.21

(*a*) The terminal voltage at the h.v. terminals,
(*b*) The l.v. input current and its p.f., and
(*c*) The efficiency.

Solution :

(*a*) Equivalent series impedance reffered to h.v. side.

$$Z'_{L.v} = \left(\frac{V_2}{V_1}\right)^2 \times Z_{lv} = \left(\frac{3200}{400}\right)^2 \times (0.45 + j0.2)$$

$$= 28.8 + j\ 12.8\ \Omega = r_2 + j\, x_2$$

Secondary rated curent $I_2 = \dfrac{\text{kVA}}{v_2} = \dfrac{16 \times 1000}{3200} = 5\text{ A}$

Now, $E_2 - V_2 = I_2\,(R\cos\theta + X\sin\theta)$

At 0.8 p.f. lagging

$$V_2 = E_2 - I_2\,(X_2\cos\theta + X_2\sin\theta)$$

$$= 3200 - 5(28.8 \times 0.8 + 12.8 \times 0.6) = 3046.4\text{ V}\ \textbf{Ans.}$$

(*b*) The equiralent circuit can be drawn as

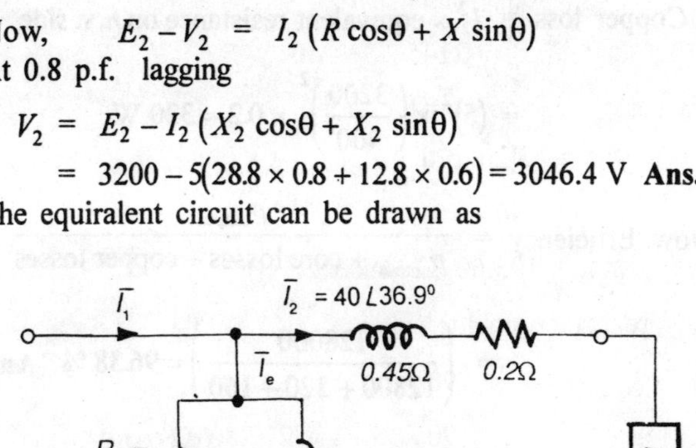

Fig. P. 1.21 (a)

Let us take V_t as reference phasor.

Core loss component of exciting current

$$\bar{I}_c = \frac{\bar{V}_t}{R_c} = \frac{400\angle 0°}{1000} = (0.4 + j\,0)\text{ A}$$

Magnetising component of exciting current,

$$\bar{I}_m = \frac{400\angle 0°}{800\angle 90°} = 0.5\,(0 - j1.0)\text{ A}$$

Now, Exciting current,

$$\bar{I}_e = \bar{I}_c + \bar{I}_m = 0.4 - j\,0.5$$

$$\text{Primary current} = \bar{I}_1 = \bar{I}_2 + \bar{I}_e$$

$$= 0.4 - j\,0.5\,(\cos 36.9° - j\sin 36.9°)$$

$$= 32.387 - j\,24.517 = 40.62\,\angle -37.13°$$

$$= 40.62\text{ A} \quad \textbf{Ans.}$$

$$\text{p.f.} = \cos 37.13° = 0.7973\text{ (lagging)} \quad \textbf{Ans.}$$

(c) Power output = kVA × p.f. = $16 \times 1000 \times 0.8 = 12800$ W

$$\text{Core loss} = \frac{V_t^2}{R_c} = \frac{400 \times 400}{1000} = 160\text{ W}$$

Copper loss = I_2^2 × equivalent resistance on *h.*v. side

$$= (5)^2 \times \left(\frac{3200}{400}\right)^2 \times 0.2 = 320\text{ W}$$

$$\text{Now Efficiency} = \frac{P_{output}}{P_{output} + \text{core losses} + \text{copper losses}}$$

$$= \left(\frac{128000}{12800 + 320 + 160}\right) = 96.38\,\% \quad \textbf{Ans.}$$

Problem 1.22 : A 200 kVA, 1-phase, 50 Hz, 2300/230 V transformer has a core loss of 300 W. At full load, I^2R loss is 800 W and the leakage reactance is 0.032 p.u. Calculate the total loss and the voltage regulation for *(a)* 1.2 p.u. load at pf 0.8 lagging and *(b)* 0.70 p.u. load at p.f. 0.6 leading.

Solution :

Given, core loss (p_c) = 300 W

Full load copper loss $(p_i) = I_{fl}^2 r$

$= 800$ W

Leakage reactance = 0.032 p.u.

Windings resistance in p.u. $= \dfrac{I^2 R \text{ loss}}{\text{kVA}} = \dfrac{800}{200 \times 1000} = 0.004$ p.u.

(*a*) Total loses at 1.2 p.u. load,

Here $k = 1.2$

$$= p_c + p_i = p_c + (1.2^2) \times \text{full load copper loss}$$

$$= 300 + (1.2)^2 \times 800$$

$$= 1452 \text{ W} \quad \textbf{Ans.}$$

Per unit voltage regulation

$$= k \ (R_{pu} \cos\theta + X_{pu} \sin\theta)$$

$$= 1.2\,(0.004 \times 0.8 + 0.032 \times 0.6)$$

$$= 0.02688 \text{ p.u.} \quad \textbf{Ans.}$$

(*b*) Here, $k = 0.7$, p.f. = 0.6 (leading)

Total losses $= p_c + k^2 \, p_i$

$$= 300 + (0.7)^2 \times 800 = 692 \text{ W} \quad \textbf{Ans.}$$

Per unit voltage regulation

$$= k[R_{pu} \times \cos\theta - X_{pu} \times \sin\theta]$$

$$= 0.7\,[\,0.004 \times 0.6 - 0.032 \times 0.8]$$

$$= -\,0.01624 \text{ p.u.} \quad \textbf{Ans.}$$

Problem 1.23 : A single-phase load is fed through a 66 kV feeder of impedance 100 + j320 ohms and a 66/6.6 kV transformer of equivalent leakage impedance 0.3 + j 1.2 ohms referred to l.v. side. For a load of 200 kW at pf 0.8 leading at 6.6 kV, calculate

(*a*) Voltage at the sending end of the feeder,

(*b*) Voltage at the primary terminals of the transformer, and

(*c*) Active and reactive powers and pf at the sending end of the feeder.

Solution :

The equivalent circuit diagram can be drawn as

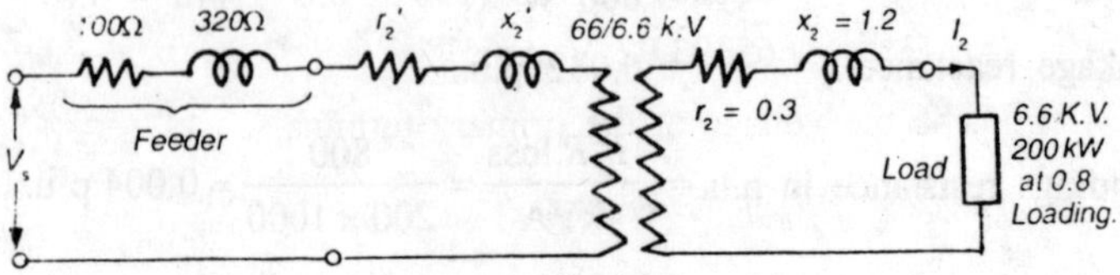

Fig. P. 1.23

Transformer equivalent circuit referred to h.v. side.

$$Z_{HV} = Z_{LV} \times \left(\frac{V_H}{V_L}\right)^2 = \left(\frac{66}{6.6}\right)^2 \times (0.3 + j\,1\cdot 2)$$

$$= 30 + j\,120\,\Omega = r_2' + j\,x_2'\,\Omega$$

Total impedances on h.v. side of the Transformer

$$\bar{Z}_{\text{Total}} = \bar{Z}_{HV} + \bar{Z}_f = 30 + j120 + 100 + j320$$

$$= 130 + j440\ \Omega = r_t + jX_t\ \Omega$$

secondary load curent, $I_2 = \dfrac{\text{power}}{\text{terminal voltage} \times \text{p.f.}}$

$$= \frac{200 \times 1000}{6.6 \times 1000 \times 0.8} = 37.88 \text{ A}$$

$$\text{p.f.} = 0.8 \text{ (leading)}$$

$$\therefore \quad \theta = \cos^{-1}(0.8) = 36.9°$$

In phasor form, $\bar{I}_2 = 37.88\ \angle 36.9°$

Secondary load current reffered to primary side of the transformer,

$$I_2' = I_2 \times \left(\frac{V_L}{V_H}\right) = 37.88 \times \frac{6.6}{66} = 3.788 \text{ A}$$

Secondary load voltage reffered to h.v. side,

$$V_2' = \left(\frac{V_H}{V_L}\right) \times V_2 = \frac{66}{6.6} \times 6.6 \times 1000$$

$$= 66 \text{ kV} = 66000 \text{ V}$$

(a) Let V_s be the sending end voltage.

Now, Applying the relation,

$$V_s = r_2' + I_2 (r_t \times \cos\theta - x_t \sin\theta)$$
$$= 66000 + 3.788 (130 \times 0.8 - 440 \times 0.6)$$
$$= 65393.92 \text{ V } \textbf{Ans.}$$

(b) Let V_1 be the voltage at the primary terminals.

$$\therefore \quad V_1 = V_2' + I_2' (r_2' \cos\theta - x_2' \sin\theta)$$
$$= 66000 + 3.788 (30 \times 0.8 - 120 \times 0.6)$$
$$= 65818.2 \text{ V } \textbf{Ans.}$$

(c) Active power loss $= I_2'^2 \times r_t = (3.788)^2 \times 130$

$= 1.865$ kW

Reactive power loss $= I_2'^2 \times x_t$

$= (3.788)^2 \times 440 = 6.314$ kW

Complex power loss $= (1.865 + j\ 6.314)$

Complex power at load end

$= 200 - j\ 200 \tan(\cos^{-1} 0.8)$

$= 200 - j\ 150$

Complex power at the sending end of the feeder = complex power at load end + complex power in feeder and transfomer.

$= (200 - j\ 150) + (1.865 + j\ 6.314)$

$= 201.865 - j143.68$

Active power $= 201.865$ kW **Ans.**

Reactive power $= 143.686$ kVAr **Ans.**

Since θ is +ve, p.f. will be leading.

$$\text{p.f.} = \cos\tan^{-1}\left(\frac{143.686}{201.865}\right)$$
$$= 0.8147 \text{ (leading) } \textbf{Ans.}$$

Problem 1.24 : A 100 kVA, 50 Hz, 440/11000 V, 1-phase transformer has an efficiency of 98.5% when supplying full-load current at 0.8 p.f. and an efficiency of 99% when supplying half-full load current at unity p.f. Find the iron losses and ohmic losses corresponding to full load current. At what value of load current will the maximum efficiency be attained? *(IAS 1987)*

Solution :

Let p_c be the core loss and p_{sc} be the full load copper loss

$$\text{We know that Efficiency} = \frac{\text{kVA rating} \times \text{p.f.}}{\text{kVA rating} \times \text{p.f.} + p_{sc} + p_c}$$

At full load 0.8 p.f., Efficiency = 0.985

$$\text{or} \quad \frac{100 \times 1000 \times 0.8}{100 \times 1000 \times 0.8 + p_c + p_{sc}} = 0.985$$

$$\text{or} \quad p_c + p_{sc} = 100 \times 1000 \times 0.8 \left(\frac{1}{0.985} - 1\right)$$

$$= 1218.27 \qquad (1)$$

At half of full load and unity p.f.,

$$\text{Efficiency} = 0.99$$

$$\text{or} \quad \frac{\frac{1}{2} \times 100 \times 1000 \times 1}{\frac{1}{2} \times 100 \times 1000 + p_c + \frac{1}{4} p_{sc}} = 0.99$$

$$\text{or} \quad p_c + \tfrac{1}{4} p_{sc} = \frac{1}{2} \times 100 \times 1000 \times \left(\frac{1}{0.99} - 1\right)$$

$$= 505.051 \text{ W} \qquad (2)$$

Subtracting (2) from (1)

$$p_{sc} - \frac{1}{4} p_{sc} = 1218.27 - 505.051$$

$$\text{or} \quad \frac{3}{4} p_{sc} = 713.219$$

$$\therefore \text{ Full load copper loss} = 713.219 \times \frac{4}{3} = 950.96 \text{ W} \textbf{ Ans.}$$

Putting $p_{sc} = 950.96$ W in eq (1),

$$p_c = 1218.27 - 950.96 = 267.31$$

Core loss = 267.31 W **Ans.**

We have, full load copper loss = 950.96 W

$$\text{or} \quad I_{fl}^2 \, r_{eq} = 950.96$$

$$\text{or} \quad \left(\frac{100 \times 1000}{11 \times 1000}\right)^2 \times r_{eq} = 950.96$$

$$\text{or} \quad r_{eq} = 11.51 \ \Omega$$

At maximum efficiency, copper loss = core loss

or $$I^2 \times r_{eq} = 267.31$$

$\therefore$ $$I = \sqrt{\frac{267.31}{11.51}} = 4.82 \text{ A}$$

Load current on l.v. side $$= I \times \left(\frac{v_H}{v_L}\right)$$

$$= 4.82 \times \left(\frac{11000}{440}\right) = 120.48 \text{ A } \textbf{Ans.}$$

> **Problem 1.25 :** The daily variation of load on a 100 kVA transformer is as follows:
>
> 8.00 AM to 1.00 PM — 65 kW, 45 kVAr
> 1.00 PM to 6.00 PM — 80 kW, 50 kVAr
> 6.00 PM to 1.00 PM — 30 kW, 30 kVAr
> 1.00 AM to 8.00 AM — no load
>
> This transformer has no load core loss of 370 watts and full load ohmic loss of 1200 watts. Determine the all day efficiency of the transformer.
>
> (*a*) What are the approximations involved in obtaining the equivalent circuit parameters from open-circuit and short-circuit tests? Discuss the validity of these tests.
>
> (*b*) Discuss the effect of frequency and voltage on the equivalent circuit parameters of a transformer.

Solution :

Given, No-load core loss (p_o) = 370 W

Full load ohmic loss $\left(I_{fl}^2 R\right)$ = 1200 W

We know that $$\text{kVA} = \sqrt{(kw)^2 + (kVAr)^2}$$

From 8.00 A.M. to 1.00 P.M. $$\text{kVA} = \sqrt{(65)^2 + (45)^2}$$

$$= 79.06 \text{ KVA}$$

During this 5 hours

Total ohmic loss $$= \left(\frac{79.06}{100}\right)^2 \times 1200 = 750.06 \text{ W}$$

W from 1 PM to 6 PM

$$\text{kVA} = \sqrt{(80)^2 + (50)^2} = 94.34 \text{ kVA}$$

kVA During this period of 5 hours

$$\text{Total ohmic loss} = \left(\frac{94.34}{100}\right)^2 \times 1200 = 1068 \text{ W}$$

From 6.00 PM to 1 PM

$$\text{kVA} = \sqrt{(30)^2 + (30)^2} = 42.42 \text{ kVA}$$

During this period of 6 hours

$$\text{Total ohmic loss} = \left(\frac{42.42}{100}\right)^2 \times 1200 = 215.93 \text{ W}$$

From 1.00 PM to 8 PM Total ohmic loss = 0 W

Daily energy lost as ohmic loss,

$$= \frac{1}{1000}(750.06 \times 5 + 1068 \times 5 + 215.93 \times 7) \text{ kWh}$$

$$= 10.6 \text{ kWh}$$

$$\text{Daily energy lost as core loss} = \frac{370 \times 24}{1000} = 8.9 \text{ kWh hr}$$

$$\text{Total energy loss} = (10.6 + 8.9) \text{ kWh}$$

$$= 19.5 \text{ kWh}$$

$$\text{Daily energy output} = 65 \times 5 + 80 \times 5 + 30 \times 7 + 7 \times 0$$

$$= 935 \text{ kWh}$$

∴ All day efficiency of the transformer

$$= \left(1 - \frac{\text{Total energy loss in kWh}}{(\text{Total output in kWh}) + \text{Total energy loss in kWh}}\right) \times 100\,\%$$

$$= \left(1 - \frac{19.5}{19.5 + 935}\right) \times 100\%$$

$$= 97.957\% \quad \textbf{Ans.}$$

Problem 1.26 : A 11,000 V, 50 Hz transformer has a flux density of 1.2 T and a core loss of 3000 watts at rated voltage and frequency. Now all the linear dimensions of the core are doubled; primary and secondary turns are halved and the new transformer is energised from 22,000 V, 50 Hz supply. Both the transformers have the same core material and the same lamination thickness.
Calculate the flux density and core losses for the new transformer.

Solution :

We know that

$$V \cong E = \sqrt{2}\,\pi f N \varphi_{max} = \sqrt{2}\,\pi\, f N B_{max}\, A_c$$

At $V_1 = 11000\text{ V}, f_1 = 250\text{ Hz}, B_{1max} = 1.2\text{ T}$

Let N_1 be the number of turns and A_1 be the core area.

$$\therefore \quad 11000 = \sqrt{2}\,\pi\, f_1\, N_1\, B_{max}\, A_1 \qquad (1)$$

At $V_2 = 22000\text{ V}, f_2 = 50\text{ Hz}$

Let N_2 be the number of turns and A_2 be the core area.

$$\therefore \quad 22000 = \sqrt{2}\,\pi\, f_2\, N_2\, B_{2max}\, A_2 \qquad (2)$$

According to Question $A_2 = (2)^2 \times A_1 = 4A_1, N_2 = \dfrac{N_1}{2}$

From (1) and (2) $$\frac{22000}{11000} = \frac{\sqrt{2}\,\pi \times 50 \times A_2 \times N_2 \times B_{2max}}{\sqrt{2}\,\pi \times 50 \times A_1 \times N_1 \times B_{1max}}$$

or $$2 = \frac{A_2\, N_2 \times B_{2max}}{A_1\, N_1\, B_{1max}}$$

$$\therefore \quad B_{2max} = \frac{2\, A_1\, N_1 \times B_{1max}}{A_2\, N_2} = \frac{2 \times A_1 \times N_1 \times 1.2}{A \times A_1 \times \dfrac{N_1}{2}}$$

$B_{2max} = 1.2$ T **Ans.**

As we know that Core loss $\propto (\text{Volume})^3$

$$\therefore \quad \frac{Pc_2}{Pc_1} = \frac{v_2^{\,3}}{v_1^3} \text{ or } \frac{Pc_2}{Pc_1} = \frac{2v_1^{\,3}}{v_1^3}$$

or $Pc_2 = 8 \times Pc_1 = 8 \times 3000 = 24{,}000$ W

$\therefore$ Core loss $= 24$ kW **Ans.**

Problem 1.27 : In a test for the determination of losses of a 440 V, 50 Hz transformer, the total iron losses were found to be 2500 watts at normal voltage and frequency. When the applied voltage and frequency were 220 V and 25 Hz, the iron losses were found to be 850 watts. Calculate the eddy current and hysteresis losses at normal voltage and frequency.

Solution :

We have Core loss (P_c) = hysteresis loss (P_h) + eddy current loss

$$(P_c) = k_h V^x f^{1-x} + k_c V^2$$

Where x = Steinmetz's constant = 1.6

When iron loss is 2500 W, V_1 = 440 V, f_1 = 50 Hz

$$\therefore \quad 2500 = k_h (440)^{1.6} \times (50)^{1-1.6} + k_e (440)^2$$

$$\text{or} \quad 2500 = 1622.342 \ k_h + 193600 \ k_e \qquad (1)$$

When iron loss is 850 W, V_2 = 220 V, f_2 = 25 Hz

$$\therefore \quad 850 = k_h (220)^{1.6} \times (25)^{1-1.6} + k_e (220)^2$$

$$\text{or} \quad 850 = 811.171 \ k_h + 48400 k_e \qquad (1)$$

Multiplying this equation by 4 both sides

$$\text{or} \quad 3400 = 3244.684 \ k_h + 193600 \ k_e \qquad (2)$$

From (1) and (2)

$$3400 - 2500 = (3244.684 - 1622.342) \ k_h + (193600 - 193600) \ k_e$$

$$\therefore \quad k_h = \frac{900}{1622.342} = 0.5548$$

Putting k_h = 0.5548 in (2)

We get k_e = 0.00826

Now From (1) Hysteresis loss at normal voltage and frequency

$$= 0.5548 \times (440)^{1.6} \times (50)^{1-1.6}$$

$$= 900 \text{ W} \quad \textbf{Ans.}$$

Eddy current loss at normal frequency

$$= k_e (440)^2 = 0.000826 \times (440)^2$$

$$= 1599.14 \text{ W} \quad \textbf{Ans.}$$

Problem 1.28 : Two similar 40 k VA, single-phase transformers gave the following results when tested by the back to back method:
W_1 in the supply line—800 watts.
W_2 in the secondaries series circuit at rated current—1000 watts.
Calculate the efficiency of each transformer at unity p.f.

Solution :

In back to back test of two similar transformers power input from supply line is the total core loss of both transformers.
Core loss of each transformer

$$= \frac{\text{Total core loss}}{2}$$

$$= \frac{800}{2} = 400 \text{ W}$$

The power consumed in the secondaries yields total copper loss of both the transformer
Ohmic loss of each transformer

$$= \frac{\text{Total ohmic loss}}{2}$$

$$= \frac{1000}{2} = 500 \text{ W}$$

Efficiency of each transformer at unity p.f.

$$= \left(\frac{\text{kVA} \times \text{p.f.}}{\text{kVA} \times \text{p.f.} + P_c + P_{sc}}\right) \times 100\,\%$$

$$= \left(\frac{40 \times 1000 \times 1}{40 \times 1000 + 500 + 400}\right) \times 100\%$$

$= 97.8\%$ **Ans.**

Problem 1.29 : A 200/100 V, single-phase transformer is rated 10 kVA. If this transformer is connected as an auto-transformer across a 50-V supply, what is the maximum output voltage that can be obtained from this transformer? If the load current is now 8 A, determine the kVA output and the currents in the various parts of the system.

Solution :

To get autotransformer with 50 V supply, the primary of the two windings transformer in connected in four parallel groups as shown below

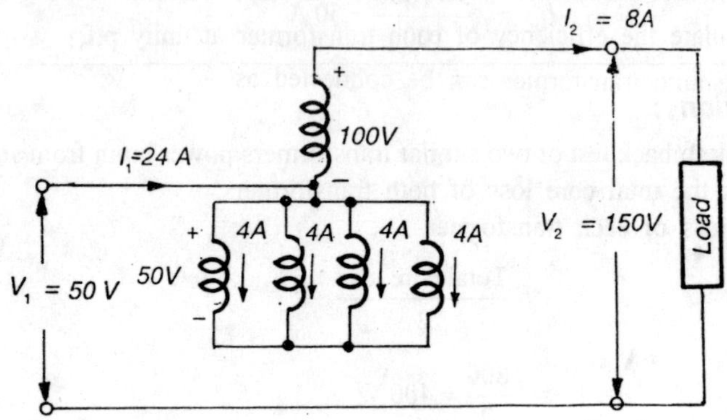

Fig. P. 1.29

From circuit diagram,

Maximum output voltage $= 100 + 50 = 150$ V **Ans.**

When load current is 8 A, 8 A flows through 100 V winding and 4 A flows through 50 V winding.

$$\therefore \quad \text{kVA output} = V_1 I_1 = V_2 I_2 = 24 \times 50 = 150 \times 8$$

$$= 1200 = 1.2 \text{ kVA } \textbf{Ans.}$$

Input current $(I_1) = 16 + 8 = 24$ A **Ans.**

Problem 1.30 : A 400/100 V, 5 kVA, single-phase two winding transformer is to be used as an auto-transformer to supply 400 volts from a 500 V voltage source. When tested as a two-winding transformer at rated load and 0.8 p.f. lagging, its efficiency was found to be 0.95.

(i) Determine its kVA rating as an auto-transformer. Also calculate the transformed kVA and conducted kVA.

(ii) Find its efficiency as an auto-transformer at rated load and at 0.8 p.f. lagging.

Solution :

Current rating of two winding transformer on H.V. side, i.e., 400 V side.

$$I_H = \frac{\text{Rating}}{V_H} = \frac{5 \times 1000}{100} = 12.5 \text{ A}$$

Current rating of two winding transformer on 100 V side,

$$I_L = \frac{5 \times 1000}{100} = 50 \text{ A}$$

The auto transformer can be connected as

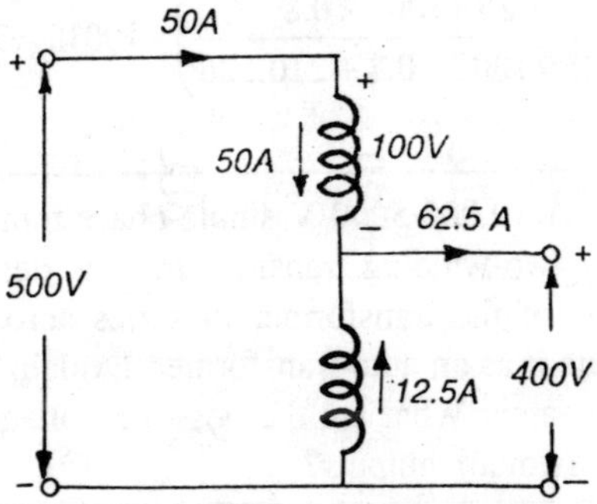

Fig. P. 1.30

The kVA rating of auto transformer

$$= V_1 I_1 = V_2 I_2$$

$$= 500 \times 50 = 62.5 \times 400 = 25 \text{ kVA } \textbf{Ans.}$$

Transformation ratio of auto transformer *(k)*

$$= \frac{V_2}{V_1} = \frac{400}{500} = 0.8$$

$$\text{Transformed kVA} = (1-k) \times VA_{TW}$$

$$= (1-0.8) \times 25 = 5 \text{ kVA } \textbf{Ans.}$$

$$\text{Conducted kVA} = k \times VA_{TW} = 0.8 \times 25$$

$$= 20 \text{ kVA } \textbf{Ans.}$$

(ii) For two winding transformer

$$\text{Efficiency} = 0.95, \quad \frac{\text{kVA} \times \text{p.f.}}{\text{kVA} \times \text{p.f.} + P_c + P_{sc}} = 0.95$$

or

$$= \frac{5 \times 1000 \times 0.8}{5 \times 1000 \times 0.8 + P_c + P_{sc}} = 0.95$$

or $$P_c + P_{sc} = 5 \times 1000 \times 0.8 \left(\frac{1}{0.95} - 1 \right) = 210.526 \text{ W}$$

Since auto transformer is operated at rated voltage and rated current, losses remain constant

$$\therefore \quad \eta_{auto} = \left(\frac{\text{kVA}_{\text{auto}} \times \text{p.f.}}{\text{kVA}_{\text{auto}} \times \text{p.f.} + P_c + P_{sc}} \right) \times 100\%$$

$$= \left(\frac{25 \times 1000 \times 0.8}{25 \times 1000 \times 0.8 + 210.526} \right) \times 100\% = 98.958\ \%\ \textbf{Ans.}$$

Problem 1.31 : A 10,000/5000 V single-phase transformer is rated at 100 kVA as a two-winding transformer. It is desired to connect the two windings of the transformer in series across 15,000 Volts supply and operate it as an auto-transformer. Explain the connections for achieving the same. What are the possible voltage ratios and the corresponding maximum outputs?

Solution :

For two winding transformer,

$$\text{H.V. side current rating, } I_H = \frac{\text{Rating}}{V_H} = \frac{100 \times 1000}{100000} = 10 \text{ A}$$

$$\text{L.V. side current rating, } I_L = \frac{100 \times 1000}{5000} = 20 \text{ A}$$

There are two possible connections of auto transformer for 15 kV supply as below

From Fig. P. 1.31 (a)

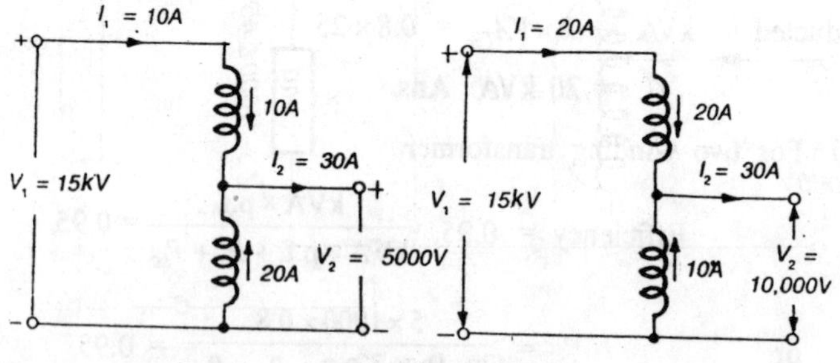

Fig. P. 1.31 (a), (b)

$$\text{Voltage ratio} = \frac{V_2}{V_1} = \frac{5000}{15000} = \frac{1}{3}$$

$$\text{kVA rating} = V_1 \times I_1 = V_2 \times I_2$$

$$= 15 \times 10 = 150 \text{ kVA} \quad \textbf{Ans.}$$

From Fig. P. 1.31 (b)

$$\text{Voltage ratio} = \frac{V_2}{V_1} = \frac{10{,}000}{15{,}000} = \frac{2}{3}$$

$$\text{kVA rating} = 10000 \times 30 = 300 \text{ kVA} \quad \textbf{Ans.}$$

Problem 1.32 : An ideal transformer has its secondary winding labelled as *a, b, c*. The primary has 100 turns. The number of turns on the secondary side are 400 between *a, b* and 200 between *b, c* so that total turns between *a, c* are 600. The transformer supplies a resistive load of 6 kW between *a* and *c*. In addition, a load of impedance 1000 $\angle 45°\ \Omega$ is connected between *a* and *b*. For a primary voltage of 1000 V, compute the primary current and the primary input.

Solution :

The diagram of auto transformer can be drawn as below

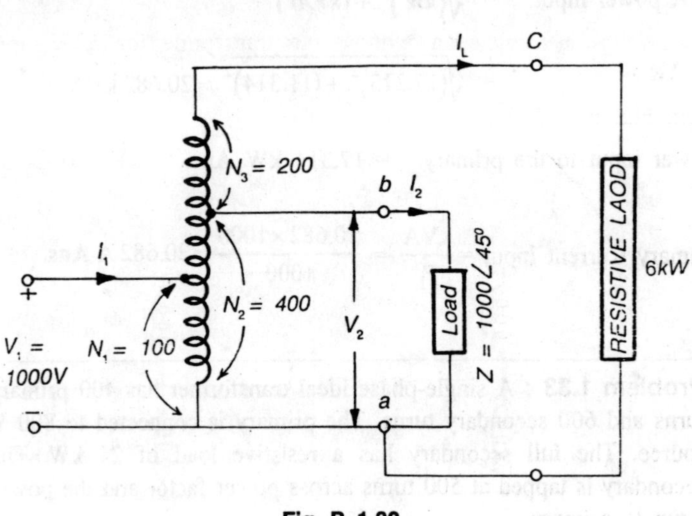

Fig. P. 1.32

Total number of turns between a and c

$$= N_2 + N_3$$

$$= 400 + 200 = 600$$

Voltage across terminals a and b

$$= \left(\frac{N_2}{N_1}\right)V_1 = \frac{400}{100} \times 1000 = 4000 \text{ V}$$

Let us take V_2 as reference phasor.
Complex power drawn by the load

$$= \frac{V^2}{Z\angle\theta} = \frac{4000 \times 4000}{1000}\angle -45°$$

$$= 16000\angle -45°$$

$$= 11313.709 - j\,11313.709$$

Total power input = Complex power drawn by the load + Power consumed in resistive load

$$= 11313.709 - j\,11313.709 + 6000\,(1 + j \cdot 0)$$

$$= 17313.709 - j\,11313.709$$

kVA power input $= \sqrt{(kW)^2 + (kVAr)^2}$

$$= \sqrt{(17.313)^2 + (11.314)^2} = 20.682 \text{ kVA}$$

Power input to the primary = 17.314 kW **Ans.**

$$\text{Primary Current input} = \frac{\text{kVA}}{V_1} = \frac{20.682 \times 1000}{1000} = 20.682 \text{ A } \textbf{Ans.}$$

Problem 1.33 : A single-phase ideal transformer has 400 primary turns and 600 secondary turns. The primary is connected to 800 V source. The full secondary has a resistive load of 24 kW. The secondary is tapped at 500 turns across power factor and the power input to primary.

Solution :

The transformer circuit can be drawn as

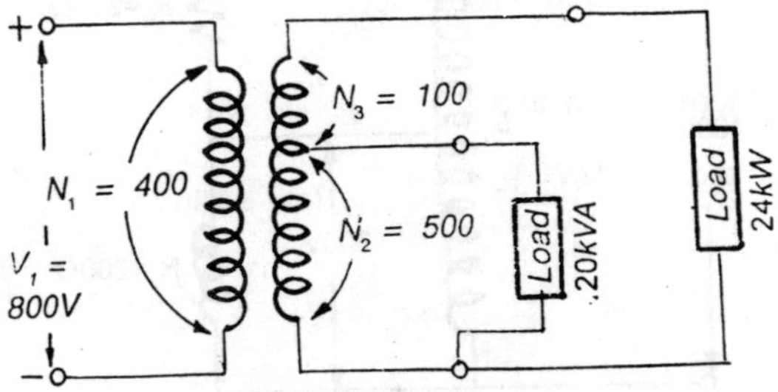

Fig. P. 1.33

$$\text{Total power output} = 24\angle 0° + 20\angle -90°$$
$$= 24 \text{ kW} - j\,20 \text{ kVAr}$$

Since the transformer is ideal.

$$\text{Total power input} = \text{Total power output}$$
$$= 24 \text{ kW} - j\,20 \text{ kVAr}$$

$$\text{Power input to primary} = 24 \text{ kW} \quad \textbf{Ans.}$$

$$\text{Total primary kVA rating} = (24)^2 + (20)^2 \text{ kVA}$$

$$\text{Primary current input} = \frac{\text{kVA}}{V_1}$$

$$= \frac{\sqrt{(24)^2 + (20)^2}}{800}$$

$$= 39.051\text{A} \quad \textbf{Ans.}$$

$$\text{Input power factor} = \cos\left[\tan^{-1}\left(\frac{20}{24}\right)\right]$$

$$= 0.7682\,(\text{lagging}) \quad \textbf{Ans.}$$

Problem 1.34 : Two resistive loads R_1 and R_2 each of 200 Ω are to be supplied with voltages of 100 V and 300 V respectively, from the available 400 V supply. To this end, one auto-transformer connected as shown in Fig. P. 1.34 is designed

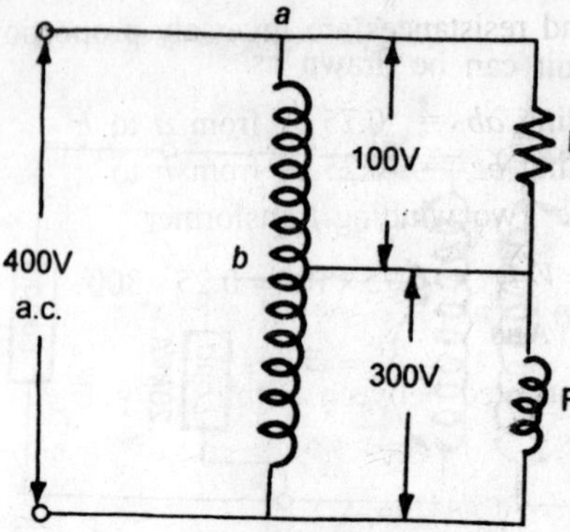

Fig. P. 1.34

Assuming the auto-transformer to be ideal, calculate the currents in its windings. What is the VA rating of a two winding transformer which would have approximately the same size as the auto-transformer. It is proposed to connect R_1 between terminals *d* and a suitable tap *c* on the winding, instead of across *a* and *b*, could it lead to a more economical design of the auto-transformer? Explain.

Solution :

(*a*) Current through $R_1 = \dfrac{100}{200} = 0.5$ A

Current through $R_2 = \dfrac{300}{200} = 1.5$ A

The direction of current flowing through the windings is shown below

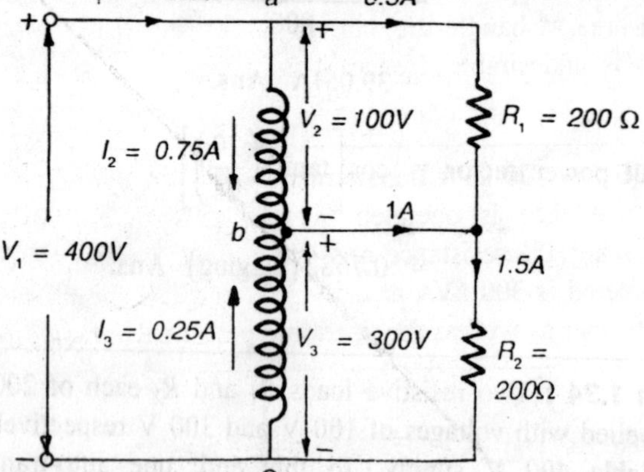

Fig. P. 1.34 (a)

Since current and resistances are inversely proportional to each other.

Current in winding *ab* = 0.75 A from *a* to *b*

Current in winding *bc* = 0.25 A from *b* to *c*

VA rating of the Two winding transformer

$$= V_2 I_2 = V_3 I_3 = 0.75 \times 100 = 0.25 \times 300$$

$$= 75 \text{ VA}$$ **Ans.**

(b) When R_1 is connected between terminal *d* and *c*

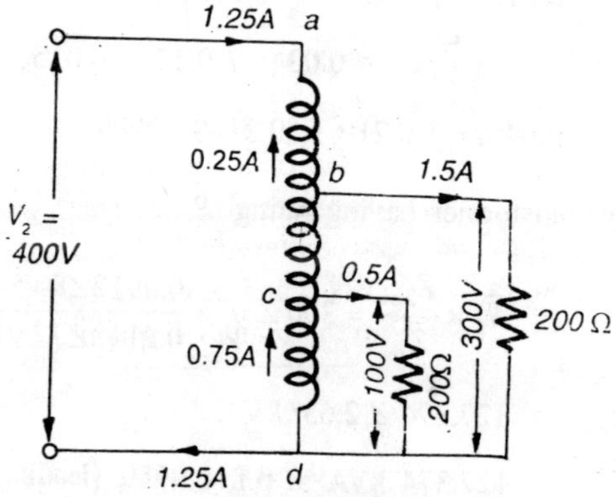

Fig. P. 1. 34 (b)

The current through winding *cd* = 0.75 A from *d* to *c*

The current through winding *cb* = 0.25 A from *c* to *b*

The current through winding *ab* = 1.25 A

The more current through winding *ab* requires windings having thicker cross-sectional area to handle this current thicker conductor increases the cost which is undesirable. Hence, it is uneconomical.

Problem 1.35 : A 100 kVA transformer having 1% resistance and 4% leakage reactance is operated in parallel with a 200 kVA transformer having 1% resistance and 6% leakage reactance. If the total load delivered is 300 kVA at unity p.f., calculate the kVA load on each transformer as well as the operating p.f. of each transformer.

Solution :

Given,

$$S_1 = 100 \text{ kVA}, \quad \bar{Z}_1 = (0.01 + j\,0.04)\,\Omega$$

$$S_2 = 200 \text{ kVA}, \quad \bar{Z}_2 = (0.01 + j\,0.06)\,\Omega$$

$$S_L = 300 \text{ kVA}$$

To solve this problem, let us first convert impedances into p.u. system. Let us take 300 kVA as reference kVA

$$\therefore \quad \bar{Z}_{1pu} = \bar{Z}_1 \times \frac{\bar{S}_b}{\bar{S}_1} = (0.01 + j\,0.04) \times \frac{300}{100}$$

$$= 0.03 + j\,0.12 = 0.1237 \angle 75.96° \text{ p.u.}$$

$$\bar{Z}_{2pu} = \bar{Z}_2 \times \frac{\bar{S}_b}{\bar{S}_2} = (0.01 + j\,0.06) \times \frac{300}{200}$$

$$= 0.015 + j\,0.090 = 0.0912 \angle 80.54°$$

$$\bar{Z} = \bar{Z}_{1pu} + \bar{Z}_{2pu} = 0.03 + j\,0.12 + 0.015 + j\,0.090$$

$$= 0.045 + j\,0.210 = 0.2148 \angle 77.91°$$

Load shared by transformer having rating S_1

$$= \bar{S}_L \times \frac{\bar{Z}_{2pu}}{\bar{Z}_{pu}} = 300 \angle 0° \times \frac{0.0912\angle 80.54°}{0.2148\angle 77.91°}$$

$$= 127.374 \angle 2.63° \text{ kVA}$$

$$= 127.374 \text{ kVA at p.f. } 0.9989 \text{ (leading)}$$ **Ans.**

Load shared by transformer having rating S_2,

$$= \bar{S}_L \times \frac{\bar{Z}_{1pu}}{\bar{Z}_{pu}} = 300 \angle 0° \times \frac{0.1237 \angle 75.96°}{0.2148 \angle 77.91°}$$

$$= 172.765 \text{ kVA} \angle -19.5°$$

$$= 172.765 \text{ kVA at p.f } 0.9994 \text{ (lagging)}$$ **Ans.**

Problem 1.36 : Two single-phase transformers in parallel supply a load of 500 amperes at 0.8 p.f. lagging and at 400 volts. Their equivalent impedances referred to secondary windings are $(2 + j\,3)$ ohms and $(2.5 + j\,5)$ ohms. Compute the current and kVA supplied by transformer and the corresponding power factors.

Solution :

Given,

$$\bar{Z}_1 = (2 + j\,3)\,\Omega = 3.606 \angle 56.31°$$

$$\bar{Z}_2 = (2.5 + j\,5)\,\Omega = 5.590 \angle 63.43°$$

$$\bar{Z} = \bar{Z}_1 + \bar{Z}_2 = 2 + j3 + 2.5 + j5 = (4.5 + j8)\ \Omega$$

$$= 9.1788 \angle 60.64°$$

$$\bar{I}_L = 500 \angle -36.9° \text{ A}, \ V_t = 400\text{V}$$

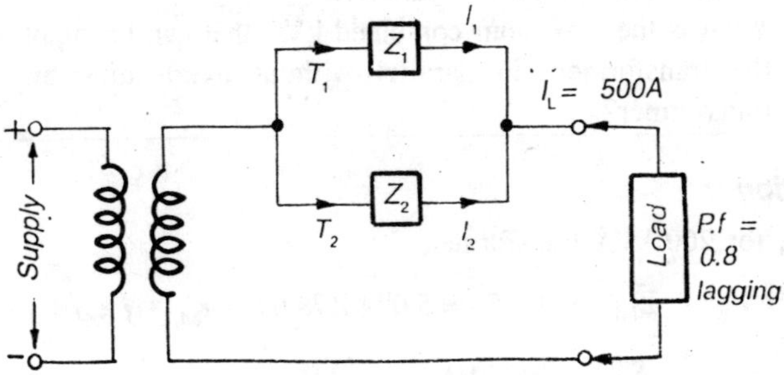

Fig. P. 1.36

Current shared by Transformer T_1

$$= \bar{I}_L \times \frac{\bar{Z}_2}{\bar{Z}_1} = 500\angle -36.9° \times \frac{5.590\angle 63.43°}{9.1788\angle 60.64°}$$

$$= 304.51 \angle -34.11°$$

$= 304.51$ A at p.f. 0.8280 lagging **Ans.**

kVA shared by transformer T_1

$= V_t \times I_1 = 400 \times 304.51 = 121.804$ kVA **Ans.**

Current shared by transformer T_2

$$= \bar{I}_L \times \frac{\bar{Z}_1}{\bar{Z}} = 500 \angle -36.9° \times \frac{3.606\angle 56.31°}{9.1788\angle 60.64°}$$

$$= 196.431 \angle -41.23°$$

$I_2 = 196.431$ A **Ans.**

p.f. $= \cos(41.23°) = 0.7521$ (lagging) **Ans.**

kVA shared by transformer

$T_2 = V_t \times I_2 = 400 \times 196.431 = 78.752$ kVA **Ans.**

Problem 1.37 : Two single-phase, 11000/ 440 V transformers have kVA ratings of 200 kVA and 100 kVA respectively. The equivalent resistance and reactance of the 200 kVA transformer are respectively

1.0 and 5.0 ohms, when referred to the 11,000 V side. The equivalent reactance of 100 kVA transformer referred to the 11,000 V side is 9.0 ohms.

(*i*) What should be the equivalent resistance of 100 kVA transformer if each transformer is to supply load in proportion to its kVA rating when operated in parallel?

(*ii*) What is the maximum combined kVA that can be supplied by the transformers in parallel without overloading any one transformer?

Solution :

Given, for 200 kVA transformer,

$$\overline{Z}_{1A} = 1 + 5j = 5.099\angle 78.69° = r_{eA} + j\,x_{eA}$$

$$S_A = 200 \text{ kVA}$$

For 100 kVA transformer, $\overline{Z}_{1B} = r_{eB} + j\,x_{eB} = \overline{r}_e + j\,9.0\,\Omega$

$$S_B = 100 \text{ kVA}$$

(*i*) For transformers having different ratings operating in parallel, the leakage impedance is inversly proportional to their kVA ratings

$$\therefore Z_{1A} \propto \frac{1}{S_A} \text{ and } Z_{1B} \propto \frac{1}{S_B}$$

$$\therefore \quad \frac{Z_{1A}}{Z_{1B}} = \frac{S_B}{S_A} \text{ or } \frac{\sqrt{(1)^2+(5)^2}}{\sqrt{r_{eB}^2+(9)^2}} = \frac{100}{200}$$

$$\text{or} \quad = \frac{26}{r_{eB}^2+81} = \frac{1}{4}$$

$$\therefore \quad r_{eB}^2 = 26\times 4 - 81$$

$$\therefore \quad r_{eB} = 4.796\,\Omega \textbf{ Ans.}$$

(*ii*) Here, $\dfrac{x_{eA}}{r_{eA}} \neq \dfrac{x_{eB}}{r_{eB}}$ and $Z_{1B} > Z_{1A}$

Now, For transformer T_A

$$\phi_A = \tan^{-1}\left(\frac{x_{eA}}{r_{eA}}\right) = \tan^{-1}\left(\frac{5}{1}\right) = 78.69°$$

For transformer T_B, $\phi_B = \tan^{-1}\left(\dfrac{x_{eB}}{r_{eB}}\right) = \tan^{-1}\left(\dfrac{9}{4.796}\right) = 61.95°$

Here, the angle between I_A and I_B in given by

$(\phi_A - \phi_B)$ *i.e.* $78.69° - 61.95° = 16.74°$

Maximum combined kVA

$$= \sqrt{\left[S_A + S_B \cos(\phi_A - \phi_B)\right]^2 + \left[S_B \, Sin(\phi_A - \phi_B)\right]^2}$$

$$= \sqrt{(200 + 100 \cos 16.74)^2 + (100 \sin 16.74)^2}$$

= 297.161 kVA **Ans.**

Problem 1.38 : A 11/33 kV, delta/star tap-changer transformer supplies a star-connected load of 1200 kW at 0.8 p.f. lag. Per-phase leakage impedances for l.v. and h.v. windings are respectively $0.5 + j5\,\Omega$ and $2.0 + j15\,\Omega$. Each line conductor has impedance of $2 + j5\,\Omega$. If l.v. winding is energised at 11 kV, find the tap-changer setting to give load terminal voltage of 33 kV.

Solution :

V_L = 11 kV, V_H = 33 kV

Per phase voltage at load terminal $= \dfrac{33 \times 1000}{\sqrt{3}} = 19.053$ kV

Load current at 0.8 p.f. $I_L = \dfrac{\text{Load Power in kW}}{\sqrt{3} \times V_H \times \text{p.f.}}$

$$= \frac{1200 \times 1000}{\sqrt{3} \times 33 \times 1000 \times 0.8} = 26.243 \text{ A}$$

L.V. impedance referred to h.v. side,

$$Z'_{LV} = Z_{LV} \times \left(\frac{V_H}{V_L}\right)^2$$

$$= (0.5 + j5) \times \left(\frac{33}{\sqrt{3} \times 11}\right)^2 = 1.5 + j15\,\Omega$$

$$\text{Total load side impedance } Z_L = Z'_{LV} + (2 + j\,15) + (2 + j\,5)$$
$$= (1.5 + j\,15) + (2 + j\,15) + (2 + j\,5)$$
$$= 5.5 + j\,35\,\Omega = r + j\,x\,\Omega$$

Total drop between load end and h.v. side of the transformer

$$= I_L\,(r \times \cos\theta + x \sin\theta)$$
$$= 26.243\,(5.5 \times 0.8 + 35 \times 0.6)\text{ V} = 666.572\text{ V}$$

Voltage at the h.v. side of the transformer

$$= 19053 + 666.572 = 19719.572\text{ V}$$

Change in h.v. side voltage

$$\Delta V_H = 19719.572 - 19053 = 666.572$$

Since the voltage increases on h.v. side.

$$\%\text{ tap up} = \frac{\Delta V_H}{V_H} \times 100\% = \left(\frac{666.572}{19053} \times 100\right)\%$$
$$= 3.5\%\ \textbf{Ans.}$$

> **Problem 1.39 :** A single-phase induction regulator has an effective turns ratio of 4 : 1. If the supply voltage is 240 V, determine the maximum and minimum values of output voltages.
> If the load current is 20 A and the secondary winding is set 60° away from the maximum boost position, calculate the primary winding current and the primary line current. Effect of no load current may be ignored.

Solution :

Given, Supply voltage = 240 V

Load current = 20 A

$\theta = 60°$

Magnitude of maximum secondary induced emf

$$= (\text{Turn ratio}) \times \text{Supply voltage}$$
$$= \frac{1}{4} \times 240 = 60\text{ V}$$

$$\text{Output voltage limit} = 240 \pm 60$$

$$\therefore \quad \text{Maximum Voltage} = 240 + 60 = 300\text{ V}$$
$$\text{Minimum Voltage} = 240 - 60 = 180\text{ V}\quad \textbf{Ans.}$$

Now, Primary winding current

$$= \frac{I_L \times \text{Maximum secondary emf} \times \cos\theta}{\text{Supply voltage}}$$

$$= \frac{20 \times 60 \cos 60°}{240} = 2.5 \text{ A } \textbf{Ans.}$$

Primary line current $= \text{Load current} + \text{Primary winding current}$

$$= 20 + 2.5 = 22.5 \text{ A } \textbf{Ans.}$$

Problem 1.40 : A 100/400 V, 2 kVA single-phase transformer has 5% impedance. If it is connected as a 500/400 V auto-transformer, deduce from fundamentals its kVA rating and percentage impedance as an auto-transformer

Solution :

The connection of two winding 100/400 V, 2 kVA single phase transformer can be connected as 500/400 V auto transformer as below.

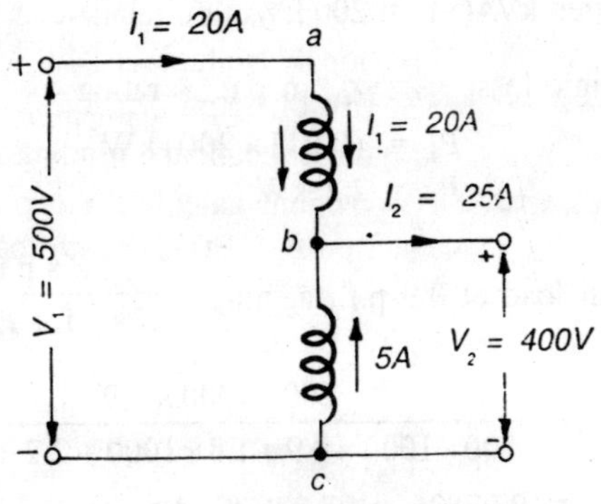

Fig. P. 1.40

Rated current on l.v. side of two winding transformer

$$= \frac{\text{Rating}}{H_{\text{LV}}} = \frac{2 \times 1000}{100} = 20 \text{ A flowing through winding } ab.$$

Rated current on H.V. side of two winding transformer

$$= \frac{\text{Rating}}{H_{\text{HV}}} = \frac{2 \times 1000}{400} = 5 \text{ A flowing through winding } bc$$

Rating of auto transformer $= V_1 I_1 = V_2 I_2$

$= 500 \times 20 = 400 \times 25 = 10\text{ kVA}$ **Ans.**

Turn ratio of auto transformer $(K) = \dfrac{V_2}{V_1} = \dfrac{400}{500} = \dfrac{4}{5} = 0.8$

% impedance of auto transformer $= (1-K)\,Z_{TW} = (1.08) \times 5\%.$

$= 0.2 \times 5\% = 1\%$ **Ans.**

> **Problem 1.41 :** A 2400/240 V, 200 kVA, 1-phase transformer has a core loss of 1.8 kW at rated voltage. Its equivalent resistance is 1.1 percent. Find the transformer efficiency at pf of 0.9 at full load. Find also the kVA load at which maximum efficiency occurs. Calculate, also, the maximum efficiency at pfs of 0.8, 0.9 and 1.

Solution :

Given, P_c (core loss) $= 1.8$ kW

$r_{eq} = 0.011$ p.u.

Rated kVA(S) $= 200$ kVA

$\therefore$ Full load ohmic loss $= r_{eq}$ in p.u. $\times$ rating

$P_i = (0.011 \times 200)\text{ kW}$

$P_i = 2.2\text{ kW}$

Efficiency at full load at 0.9 p.f. lagging $= \dfrac{S \times \text{p.f.}}{S \times \text{p.f.} + P_c + P_c}$

$$= \frac{200 \times 1000 \times 0.9}{200 \times 1000 \times 0.9 + 1.8 \times 1000 + 2.2 \times 1000}$$

$= 0.97826 = 97.826\ \%$ **Ans.**

At maximum efficiency, iron loss = variable I^2 R loss

or $$I^2 \times r_{eq} = 1.8 \times 1000$$

or $$I^2 \times \frac{\text{full load copper loss}}{(I_{fl})^2} = 1.8 \times 1000$$

or $$\left(\frac{I}{I_{fl}}\right) = \frac{1.8 \times 1000}{2.2 \times 1000}$$

$$\therefore \qquad \frac{I}{I_{fl}} = k = 0.905$$

$\therefore$ kVA load at maximum efficiency $S_{max} = k \times 5$

$$= 0.905 \times 200 \text{ kVA}$$

$$= 181 \text{ kVA } \textbf{Ans.}$$

Maximum efficiency at p.f 0.8 $= \dfrac{S_{max} \times \text{p.f.}}{S_{max} \times \text{p.f.} + 2P_c}$

$$= \frac{181 \times 0.8}{181 \times 0.8 + 2 \times 1.8}$$

$$= 0.97574 = 97.574\% \textbf{ Ans.}$$

Maximum efficiency at p.f 0.9 $= \dfrac{S_{max} \times \text{p.f.}}{S_{max} \times \text{p.f.} + 2P_c}$

$$= \frac{181 \times 0.9}{181 \times 0.9 + 2 \times 1.8}$$

$$= 0.97838 = 97.838\% \textbf{ Ans.}$$

Maximum efficiency at unity p.f. $= \dfrac{181 \times 1}{181 \times 1 + 2 \times 1.8}$

$$= 0.9805 = 98.05\% \textbf{ Ans.}$$

Problem 1.42 : In a single-phase transformer, zero voltage regulation occurs at pf of 0.9 leading, whereas its maximum possible voltage regulation is 4%. Find its voltage regulation at pf 0.8 lagging. If its ohmic losses are 136.5 W, find kVA rating of the transformer.

Solution :

We have, voltage regulation $= \dfrac{IR\cos\theta \pm IX\sin\theta}{V_2}$

$$= \left(\frac{IR}{V_2}\right)\cos\theta \pm \left(\frac{IX}{V_2}\right)\sin\theta$$

= (Resistance drop per unit) × cos θ ± (reactance drop per unit) × sin θ

$$= R_{pu} \times \cos\theta \pm X_{pu} \sin\theta$$

Given, At p.f. 0.9 leading voltage regulation = 0

$$R_{pu} \times 0.9 - X_{pu} \sin\theta = 0$$

$$R_{pu} \times 0.9 = X_{pu} \times 0.43589$$

$$\therefore \quad X_{pu} = \frac{0.9}{0.43589} R_{pu}$$

$$X_{pu} = 2.0647\, R_{pu} \qquad (1)$$

For maximum voltage regulation

$$\tan\theta = \frac{X}{R} = \frac{IX}{IR} = \frac{X_{pu}}{R_{pu}} = 2.047 \text{ p.u.}$$

Also Given, Maximum voltage regulation = 0.04

$$R_{pu} \cos\theta + X_{pu} \sin\theta = 0.04$$

or $$0.43589\, R_{pu} + 0.9 \times X_{pu} = 0.04$$

or $$0.43589\, R_{pu} + 0.9 \times 2.0647\, R_{pu} = 0.04$$

$$\therefore \quad R_{pu} = 0.017436 \text{ p.u.}$$

From (1) $$X_{pu} = 2.0647 \times 0.017436 = 0.0360 \text{ p.u.}$$

Now, Voltage regulation at p.f. 0.8 lagging

$$= (0.017436 \times 0.8 + 0.036 \times 0.6)$$

$$= 0.0355 = 3.55\% \textbf{ Ans.}$$

Now $$R_{pu} = \frac{\text{Full load ohmic loss}}{\text{kVA}}$$

$$\therefore \quad \text{kVA rating} = \frac{136.5}{0.017436 \times 0.8} = 9.8 \text{ kVA} \cong 10 \text{ kVA } \textbf{Ans.}$$

Problem 1.43 : For a 2000/200 V, 20 kVA, single-phase transformer, the load current at which maximum efficiency occurs is 90 A. Its equivalent resistance is 0.015 p.u. Find its half-rated load efficiency at p.f. 0.8.
Also calculate its voltage regulation at full load of unity p.f.

Solution :

Given, $S_{rated} = 20$ kVA, $R_{pu} = 0.015$

$$\text{Full load current} = \frac{S_{rated}}{V_2}$$

$$I_{fl} = \frac{200 \times 1000}{200} = 100 \text{ A}$$

$$\text{Full load ohmic loss } P_i = R_{pu} \times \text{Rating}$$

$$P_i = 0.015 \times 20 \times 1000$$

$$P_i = 300 \text{ W}$$

Given, Load current at maximum efficiency (I) = 90 A

At maximum efficiency, Core loss = Copper loss

$$P_e = I^2 r_{eq} = I^2 \times \frac{\text{Full load copper loss}}{I_{fl}^2}$$

$$= \left(\frac{I}{I_{fl}}\right)^2 \times \text{full load copper loss}$$

$$= \left(\frac{90}{100}\right)^2 \times 300 = 243 \text{ W.}$$

Half-rated load efficiency at p.f. 0.8

$$= \frac{\frac{1}{2} \times S_{rated} \times \text{p.f.}}{\frac{1}{2} \times S_{rated} \times \text{p.f.} + p_c + \left(\frac{1}{2}\right)^2 p_i}$$

$$= \frac{\frac{1}{2} \times 20 \times 1000 \times 0.8}{\frac{1}{2} \times 20 \times 1000 \times 0.8 + 243 + \frac{1}{4} \times 300}$$

$= 0.96177 = 96.177\%$ **Ans.**

Voltage regulation at full load at unity p.f.

$$= (0.015 \times 1 + x_{pu} \sin \theta)$$

$$= (0.015 \times 1 + x_{pu} \times 0)$$

$= 0.015 = 1.5\%$ **Ans.**

Problem 1.44 : For a 2300/230 V, 1-phase transformer, maximum possible voltage regulation is 5% and occurs at pf of 0.25. Find its load voltage at full-load power factors of 0.8 lead and 0.8 lag.

Solution :

We know that, Voltage regulation = $R_{pu}\cos\theta \pm X_{pu}\sin\theta$.

For maximum voltage regulation

$$\tan\theta = \frac{X}{R} = \frac{IX}{IR} = \frac{X_{pu}}{R_{pu}} = (\tan\cos^{-1} 0.25) = 3.873$$

$$X_{pu} = 3.873\, R_{pu} \qquad (1)$$

Given, Maximum possible voltage regulation = 0.05

$$R_{pu}\cos\theta + X_{pu}\sin\theta = 0.05$$

or $$R_{pu}\times 0.25 + X_{pu}\times 0.9682 = 0.05$$

or $$R_{pu}\times 0.25 + 3.873\times 0.9682\, R_{pu} = 0.05$$

$\therefore$ $$R_{pu} = 0.125$$

From (1), $$X_{pu} = 3.873 \times 0.0125 = 0.0484$$

Now, Voltage regulation at p.f. 0.8 leading

$$= R_{pu}\times 0.8 - X_{pu}\times 0.6$$
$$= 0.0125 \times 0.8 - 0.0484 \times 0.6$$
$$= -0.01904$$

or $$\frac{E_2 - V_2}{E_2} = -0.01904$$

$\therefore$ $$V_2 = E_2(1 + 0.01904)$$
$$= 230\ (1 + 0.01904) = 234.379 \text{ V} \textbf{ Ans.}$$

Voltage regulation at p.f. 0.8 lagging

$$= 0.0125 \times 0.8 + 0.6 \times 0.0484 = 0.03904$$

$$\frac{E_2 - V_2}{E_2} = 0.03904$$

$\therefore$ $$V_2 = E_2\,(1 - 0.03904) = 230\ (1 - 0.03904)$$
$$= 221.02 \text{ V} \textbf{ Ans.}$$

Problem 1.45 : A single-phase transformer has eddy current loss of 100 W and hysteresis loss of 120 W. For the same supply voltage, if both lamination thickness and operating frequencies are reduced by 10%, find the new value of core losses. Take Steinmetz's constant = 2.

Solution :

Given, $P_{e1} = 100$ W, $P_{h1} = 120$ W, Core area = A_1, No. of turns = N_1

Let f_1 be the frequency and B_{max} be the maximum flux density.

When frequency is reduced by 10%

$$f_2 = 0.9 f_1, \quad A_2 = A_1, \text{ No. of turns} = N_1,$$

Applying the relation $V = \sqrt{2}\,\pi\, f\, N\, B_{max}\, A_c$

$$\therefore \quad \frac{V_1}{V_2} = \frac{f_1}{f_2} \times \frac{B_{max1}}{B_{max2}}$$

For the same supply:

$$\frac{B_{max2}}{B_{max1}} = \frac{1}{0.9}$$

Also Applying $\quad P_h = k_h f B_{max}^x = \dfrac{P_{h2}}{P_{h1}} = \dfrac{f_2}{f_1} \times \left(\dfrac{B_{max2}}{B_{max1}}\right)^2$

$$= 0.9 \times \left(\frac{1}{0.9}\right)^2 = \frac{1}{0.9}$$

$$\therefore \quad P_{h2} = \frac{1}{0.9} \times 120 = 133.33 \text{ W}$$

Again $\quad P_e = k_e f^2 B_{max}^2$

$$\therefore \quad \frac{P_{e2}}{P_{e1}} = \frac{0.9\, k_{e1}}{k_{e1}} \left(\frac{f_2}{f_1}\right)^2 \times \left(\frac{B_{max2}}{B_{max1}}\right)^2$$

$$= 0.9 \times (0.9)^2 \times \left(\frac{1}{0.9}\right)^2 = 0.9$$

$$\therefore \quad P_{e2} = 0.9 \times 100 = 90 \text{ W}$$

Since lamination is reduced by 10%

$$\text{Reduced core loss} = 90 - \frac{90 \times 10}{100} = 81 \text{ W}$$

$\therefore$ Total core loss = $P_e + P_h$ = 133.33 + 81 = 214.33 W **Ans.**

> **Problem 1.46 :** For a 1-phase transformer, equivalent resistance is 1.5%. Its per unit core loss is 0.01. Find efficiency of this transformer at pf 0.8.

Solution :

Let us take rated kVA(S) as base kVA.

Given, p.f. $(\cos\theta) = 0.8$, $\dfrac{\text{core loss}}{\text{base kVA}} = 0.01$

Equivalent resistance per unit $= \dfrac{\text{Full load copper loss}}{\text{rated kVA}} = 0.015$

Efficiency at p.f. 0.8 is given by

$$= \frac{S\cos\theta}{S\cos\theta + P_i + P_c} = \frac{\cos\theta}{\cos\theta + \dfrac{P_i}{S} + \dfrac{P_c}{S}}$$

$$= \left(\frac{0.8}{0.8 + 0.01 + 0.015}\right) = 0.9697 = 96.97\% \ \textbf{Ans.}$$

> **Problem 1.47 :** A 50 kVA 10 KV/400 V, 3-phase transformer has its maximum possible efficiency of 96%, when it derives 40 kVA at unity power factor and rated voltage. The maximum possible voltage regulation of the transformer is 3%. Find the efficiency and voltage regulation of the transformer for rated kVA output at 0.8 p.f. lagging.

Solution :

Given, Load rating = 40 kVA at unity p.f.

$\eta_m = 0.96$, Maximum voltage regulation = 0.02

We know that Efficiency $= \dfrac{\text{Output}}{\text{Input}} = \dfrac{\text{Output}}{\text{Output + Losses}}$

$$\therefore \quad \text{Total losses} = \left(\frac{1}{\eta_m} - 1\right) \times \text{Output} = \left(\frac{1}{0.96} - 1\right) \times 40 \times 1000$$

$$= 1666.67 \ \text{W}$$

At maximum efficiency, Core loss = ohmic loss

$$\frac{\text{Total losses}}{2} = \frac{1666.67}{2} = 833.34 \ \text{W}$$

$$\text{Rated load ohmic loss} = \left(\frac{1}{0.8}\right)^2 \times 833.34 \ \text{W}$$

$$= 1302.1 \ \text{W}$$

Now, Transformer efficiency at rated kVA at 0.8 lagging

$$= \frac{50 \times 1000 \times 0.8}{50 \times 1000 \times 0.8 + 1302.1 + 833.34} = 0.9493$$

% Efficiency = 94.93%. **Ans.**

Given, Maximum possible voltage regulation = 0.02

$$R \text{ (in p.u.)} \times \cos\theta + X \text{ (in p.u.)} \times \sin\theta = 0.04 \quad (2)$$

Now $$R \text{ (in pu)} = \frac{\text{Ohmic loss}}{\text{rated kVA}} = \frac{1302.1}{50{,}000} = 0.026 \text{ pu}$$

$$Z \text{ (in pu)} = \text{maximum voltage regulation} = 0.03$$

$$\therefore \quad X \text{ (in pu)} = \sqrt{(0.03)^2 - (0.026)^2} = 0.015 \text{ pu.}$$

Voltage regulation at rated voltage and at 0.8 lagging =

$$R_{pu} \cos\theta + R_{pu} \times \sin\theta = 0.026 \times 0.8 + 0.015 \times 0.6 = 0.0298$$

$\therefore$ % Voltage regulation = 2.98% **Ans.**

Problem 1.48 : A 5 kVA, 60 Hz, single phase transformer has ratio 220/440 V. The data taken on the l.v. side at rated voltage show that the open circuit input wattage is 100 W. The mutual inductance between primary and secondary winding is 1.8 H. What will be the current taken by the transformer if the no-load test is connected on H.V. side at rated voltage? Neglect the effect of winding resistance and leakage inductance.

Solution :

As we know that the power input on the l.v. side at rated voltage with h.v. open circuit is equal to the core loss.

$$\therefore \quad P_{c1} = 100 \text{ W}$$

The core loss in a transformer remains same whether it is energized from l.v. side or h.v. side.

Core loss current on h.v. side $$I_{c2} = \frac{100}{440} = 0.227 \text{ A}$$

In a transformer $$E = \sqrt{2}\,\pi f N\phi_{max} = \sqrt{2}\,\pi f\, \Psi_{max}$$

$$[\because \text{flux } linkage\ (\Psi) = N\phi]$$

Maximum value of flux linkage with l.v. winding $$= \frac{V_1}{\sqrt{2}\,\pi f}$$

Now, Mutual inductance $= \dfrac{\text{Flux linkage with l.v. winding}}{\text{Current in h.v. winding}}$

$$M = \frac{V_1}{\sqrt{2}\,\pi f} \times \frac{1}{\sqrt{2}\,I_{\phi 2}}$$

Where $I_{\phi 2}$ = Magnetising current on h.v. side

$$\therefore \quad I_{\phi 2} = \frac{V_1}{\sqrt{2}\,\pi f} \times \frac{1}{\sqrt{2}\,M} = \frac{220}{2 \times 3.14 \times 60 \times 1.8} = 0.324 \text{ A}$$

Current taken by transformer when energized on h.v. side

$$I_{e2} = \sqrt{I_{c2}{}^2 + I_{\phi 2}{}^2} = \sqrt{(0.227)^2 + (0.326)^2} = 0.396 \text{ A} \quad \textbf{Ans.}$$

Problem 1.49 : A single phase 240/200 V auto transformer with secondary short circuited, takes a primary current of 50 A at 10 V and at p.f. of 0.22. It has $\dfrac{N_2}{N_1} = K = 0.83$. For an input voltage of 240 V, find the load voltage for a load current of 80 A at a p.f. of 0.8 lagging.

Solution :

Given, at short circuit condition $V_1 = 10$ V, $V_2 = 0$, $I_1 = 50$ A

Equivalent impedance reffered to primary side.

$$Z_{e1} = Z_1 + \left(\frac{1-K}{K}\right)^2 Z_2$$

Now, Applying the relation

$$V_1 = \frac{V_2}{K} + I_1\left[Z_1 + \left(\frac{1-K}{K}\right)^2 Z_2\right] \quad (1)$$

or
$$10 = 50\left[Z_1 + \left(\frac{1-K}{K}\right)^2 Z_2\right]$$

or
$$50\,Z_{e1} = 10$$

$$\therefore \quad Z_{e1} = 0.2\ \Omega$$

Equivalent resistance on primary side,

$$r_{e1} = 0.2 \times \text{p.f.} = 0.22 \times 0.2 = 0.044\ \Omega$$

Equivalent reactance on primary side

$$x_{e1} = \sqrt{z_{e1}^2 - r_{e1}^2} = \sqrt{(0.2)^2 - (0.044)^2} = 0.195\ \Omega$$

Since $\quad I_1V_1 = I_2V_2$

or $\quad I_1 = \frac{V_2}{V_1} \times I_2 = \frac{200}{240} \times 80 = 66.67$ A

To calculate terminal voltage at p.f. = 0.8 (lagging)

Applying the relation, $\quad V_1 = \frac{V_2}{\cos\theta} + I_1 (r_{e1} \cos\theta + x_{e1} \sin\theta)$

$$\Rightarrow \quad 240 = \frac{V_2}{0.8} + 66.67\,(0.044 \times 0.8 + 0.95 \times 0.6)$$

$$\therefore \quad V_2 = 183.88 \text{ V } \textbf{Ans.}$$

Problem 1.50 : A 150 kVA, 2500/250 V, single-phase two winding transformer is to be used as an auto transformer for stepping up the voltage from 2500 V to 2750 V. At rated load, the two winding transformer has 2.5% loss, 3% voltage regulation and 4% impedance. For the auto transformer, determine the followings:

(i) Voltage and current rating
(ii) kVA rating
(iii) Efficiency
(iv) Percentage impedance
(v) Regulation, and
(vi) Short circuit current on each side

Solution :

(i) To achieve auto transformer, the two winding transformer can be connected as shown in Fig. P. 1.50.

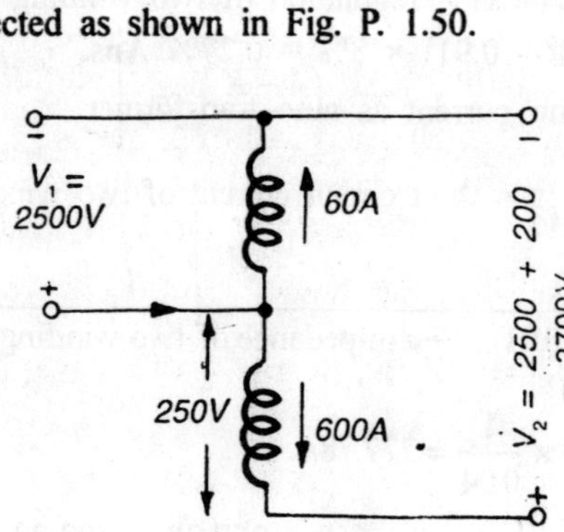

Fig. P. 1.50

For two winding transformer primary rated current,

$$I_1 = \frac{150,000}{2500} = 60 \text{ A}$$

Secondary rated current, $I_2 = \dfrac{150,000}{250} = 600 \text{ A}$

From diagram, for auto transformer

Primary voltage = 2500 V,

Primary current = 600 + 60 = 660 A

Secondary voltage = 2750 V,

Secondary current = 600 A **Ans.**

(ii) kVA rating $= V_1 I_1 = V_2 I_2$

$= 2500 \times 660 = 2750 \times 600 = 1650$ kVA **Ans.**

(iii) Here transformation ratio $k = \dfrac{2500}{2750} = 0.91$

Hence, Percentage full load losses in auto transformer

= $(1 - k)$ full load losses in two winding

= $(1 - 0.91) \times 2.5 = 0.225\%$

Efficiency of auto transformer = (100 − 0.225) %

= 99.775% **Ans.**

(iv) Percentage impedance of auto transformer

$= (1 - k) \times Z_{PTW}$

$= (1 - 0.91) \times 4\% = 0.36\%$ **Ans.**

(v) Regulation of auto transformer

= $(1 - k)$ × regulation of two winding transformer

= $(1 - 0.91) \times 3\% = 0.27\%$ **Ans.**

(vi) Short circuit current as auto transformer

$= \dfrac{1}{(1-k)} \times$ short circuit current of two windings transformer

$$= \frac{1}{(1-0.91)} \times \frac{1}{pu \text{ impedance of two winding transformer}}$$

$$= \frac{1}{0.09} \times \frac{1}{0.04} = 277.78\text{A}$$

Primary current = 660 × 277.78 = 183.33 kA **Ans.**

Secondary current= 600 × 277.78 = 166.67 kA **Ans.**

Problem 1.51 : The maximum efficiency of a single phase 1200/500 V, 600 kVA transformer is 99% and occurs at 90% of full load, unity power factor. The percentage impedance is 4%. Load power factor is no vaied while the load current and the supply voltage are held constant at their rated values. Determine the load power factor at which the secondary terminal voltage is minimum and find the value of latter.

Solution :

We know that Total losses $= \left(\dfrac{1}{\text{efficiency}} - 1\right) \times \text{Output}$

Given, $\eta = 0.99,\ P_{out} = 600 \times 1000$

Now, Total losses $= \left(\dfrac{1}{0.99} - 1\right) \times 600000 \times 0.9 \times 1 = 5445.54$ W

At maximum efficiency, core loss = copper loss $= \dfrac{1}{2} \times$ total losses
$= 2722.87$ W

Now, Full load ohmic loss $= 2722.87 \times \left(\dfrac{100}{90}\right)^2 = 3361.57$ W

Equivalent resistance on secondary side,

$$r_{e2} = \frac{3361.57}{600,000} \times 100 = 0.560 \text{ percent}$$

Given, $Z_{e2} = 4\ \%.$

Equivalent reactance on secondary sides

$$x_{eq} = \sqrt{4^2 - (0.560)^2} = 3.96\ \Omega\ \%$$

As we know that the load power factor at which secondary terminal voltage is minimum is given by

$$\cos\theta_2 = \frac{r_{e2}}{z_{e2}} = \frac{.560}{4} = 0.142 \text{ (lagging)}$$

$$\sin\theta = \sqrt{1 - (0.142)^2}$$

Now, Voltage regulation in pu $= r_{e\,pu} \cos\theta + x_{e\,pu} \times \sin\theta$

or $$\left(\frac{E_2 - V_2}{E_2}\right) \times 100 = 0.560 \times 0.142 + 3.96 \times 0.9893$$

or $\left(\frac{E_2 - V_2}{E_2}\right) = 0.04$

or $E_2 - V_2 = 0.04\, E_2$

or $V_2 = (1 - 0.04) \times E_2$

$\therefore$ $V_2 = (1 - 0.04) \times 500 = 480$ V **Ans.**

Problem 1.52 : The approximate equivalent circuit of a 2 kVA, 200/100 V single phase transformer referred to primary side is shown below. Evaluate its efficiency and regulation when it delivers full load at rated voltage and unity p.f.

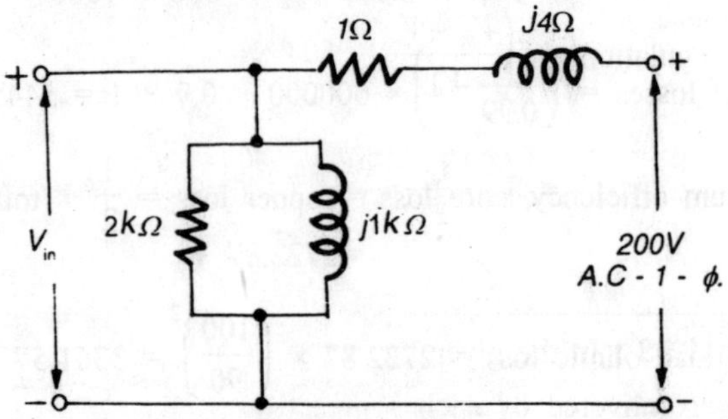

Fig. P. 1.52

Solution :

Give, $r_{e1} = 1\,\Omega, x_{el} = 4\,\Omega$

$r_c = 2 \times 10^3\,\Omega, x_\phi = 1 \times 10^3\,\Omega$

Primary rated load current is given by

$$I_1 = \frac{2 \times 1000}{200} = 10\text{ A}$$

Let us take 200 V as reference phaser. Applying the equation

$$V_{in} = \sqrt{(V_t \cos\theta + I_a r_a)^2 + (V_t \sin\theta + I_a x_a)^2}$$

$$\therefore \quad V_{in} = \sqrt{(200 \times 1 + 10 \times 1)^2 + (200 \times 0 + 10 \times 4)^2}$$

$$= \sqrt{(210)^2 + (40)^2}$$

$$= \sqrt{44100 + 1600} = \sqrt{45700} = 213.8\text{ V}$$

$$\text{Ohmic loss} = I_1^2 \times r_{eq} = (10)^2 \times 1 = 100 \text{ W}$$

$$\text{Core loss} = \frac{V_m^2}{r_c} = \frac{(213.8)^2}{2 \times 1000} = 11.43 \text{ W}$$

$$\text{We know that efficiency} = \frac{VI\cos\theta}{VI\cos\theta + P_c + P_i}$$

Here p.f. = 1, $I = I_{fl}$

$$\% \text{ Efficiency} = \frac{2000 \times 1}{2000 \times 1 + 11.43 + 100} = 94.72\% \textbf{ Ans.}$$

$$\% \text{ Voltage regulation} = \left(\frac{V_{in} - V_t}{V_{in}}\right) \times 100\%$$

$$= \left(\frac{213.8 - 200}{213.8}\right) \times 100 = 6.4\,\% \textbf{ Ans.}$$

Problem 1.53 : In the given circuit of ideal transformer calculate:

(a) power delivered by each source,

(b) the power dissipated in each resistor

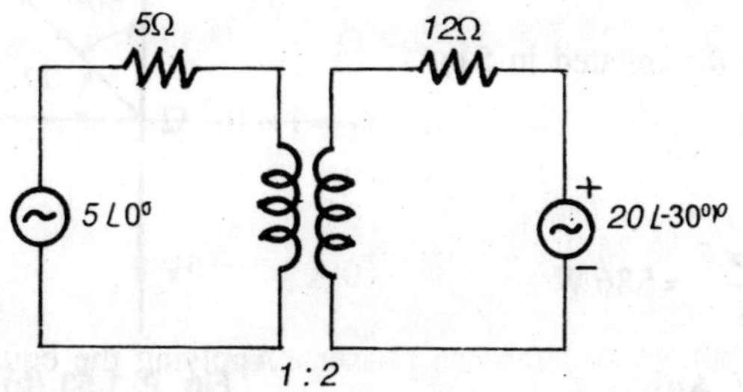

Fig. P. 1.53

Solution :

The secondary impedance reffered to primary side,

$$Z_2' = \left(\frac{1}{2}\right)^2 \times 12 = 3\,\Omega$$

The secondary load voltage reffered to primary side

$$V_2' = \frac{1}{2} \times 20 \angle -60° = 10 \angle -60°$$

Now, the circuit can be drawn as

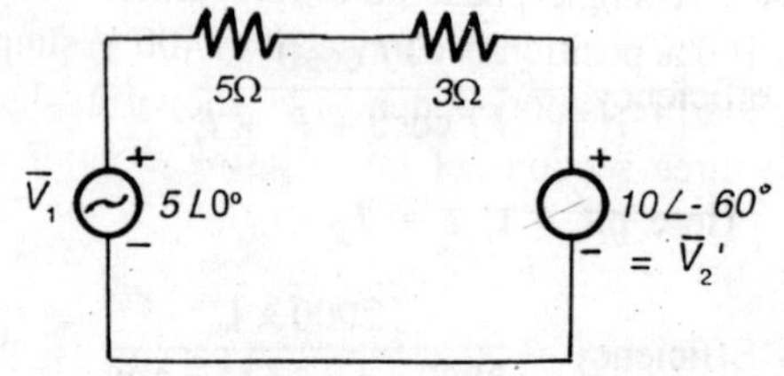

Fig. P. 1.53 (a)

Let us take 5 ∠ 0 V as reference phasor

From circuit the resultant voltage V_R in the circuit is given by

$$V_R = 10 \angle -60° - 5 \angle 0 = 10\left(\frac{1}{2} - \frac{j\sqrt{3}}{2}\right) - j(1 + j.0)$$

$$= 5 - j5\sqrt{3} - 5 = -j5\sqrt{3}$$

Now, the phasor diagram can be drawn as

Magnitude of current I_1 in phase with V_R is given by

$$I = \frac{5\sqrt{3}}{5+3} = \frac{5\sqrt{3}}{8} \text{ A}$$

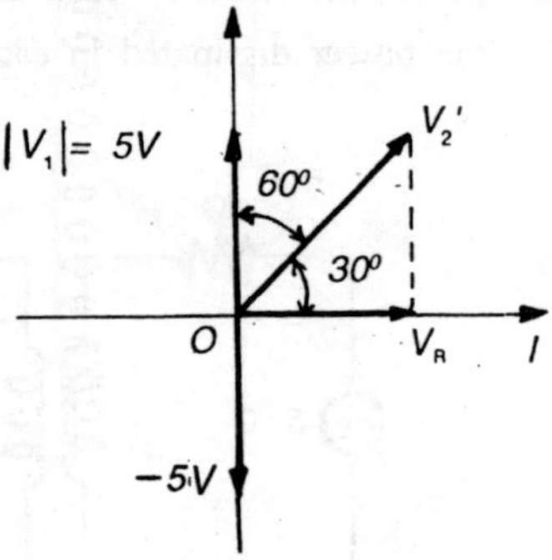

Fig. P. 1.53 (b)

Total power dissipitated in 5 Ω

$$= \left(\frac{5\sqrt{3}}{8}\right)^2 \times 5$$

$$= \frac{25 \times 3 \times 5}{8 \times 8} = 5.86 \text{ W}$$

$= 5.86$ W **Ans.**

Power dissipitated in 3 Ω

$$= \frac{25 \times 3 \times 5}{8 \times 8} = 3.52 \text{ W } \textbf{Ans.}$$

Power delivered by 20 V source

$$= 10 \times \frac{5\sqrt{3}}{8} \times \cos 30° = 10 \times \frac{5\sqrt{3}}{8} \times \frac{\sqrt{3}}{2} = 9.38 \text{ W } \textbf{Ans.}$$

Power delivered by 5 V source

$$= 5 \times \frac{5\sqrt{3}}{8} \times \cos 60° = 25 \times \frac{\sqrt{3}}{8} \times \frac{1}{2} = 2.71 \text{ W } \textbf{Ans.}$$

Problem 1.54 : A single phase auto-transformer with taps at 0%, 25%, 75% and 100% position is connected to 400 V supply and feeds two secondary load resistors connected as shown in figure. Find the currents in the three sections of the winding assuming ideal core.

Solution :

When tapping is 25%, secondary voltage $= \frac{25}{100} \times 400 = 100 \text{ V}$

When tapping is 50% secondary voltage $= \frac{50}{100} \times 400 = 200 \text{ V}$

When tapping is 75% secondary voltage $= \frac{50}{100} \times 400 = 300 \text{ V}$

When tapping is 100% secondary voltage = 400 V

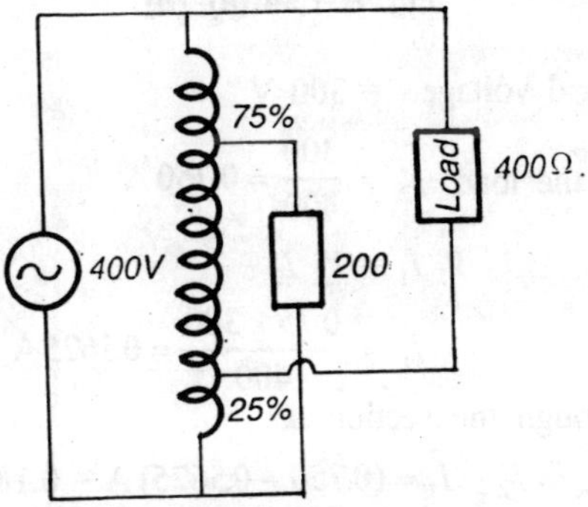

Fig. P. 1.54

This problem can be solved easily by considering are load at a time and then superimposing two results.

when the load 200 Ω is connected between *b* and *d*.

Secondary voltage = 300 V,

Current through the load

$$I_2 = \frac{300}{200} = 1.5 \text{ A}$$

In transformer $V_1 I_1 = V_2 I_2$

$$\therefore \quad I_1 = \frac{300 \times 1.5}{400} = 1.125 \text{ A}$$

$\therefore$ Current through the winding bd

$$I_{bd} = 1.5 - 1.125 = 0.375 \text{ A}$$

Thus, $I_{ab} = 1.125$ A, $I_{bd} = 0.375$ A **Ans.**

When load 400 Ω is connected across the terminals ac, as shown in figure

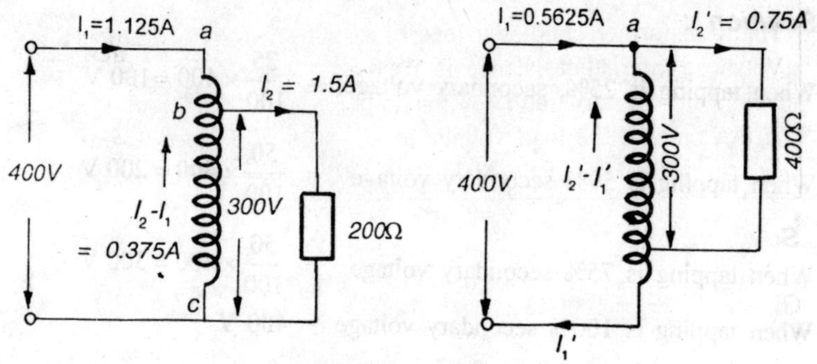

Fig. P. 1.54 (a), (b)

Secondary terminal voltage = 300 V

Current through the load $I_2' = \frac{300}{400} = 0.750$ A

For transformer $V_1 I_1 = V_2 I_2'$

$$\therefore \quad I_1 = \frac{0.75 \times 300}{400} = 0.5625 \text{ A}$$

Now current through the section *ac*.

$$I'_{ac} = I'_2 - I_1 = (0.750 - 0.5625) \text{ A} = 0.1875 \text{ A.}$$

Now combining these two results, we get

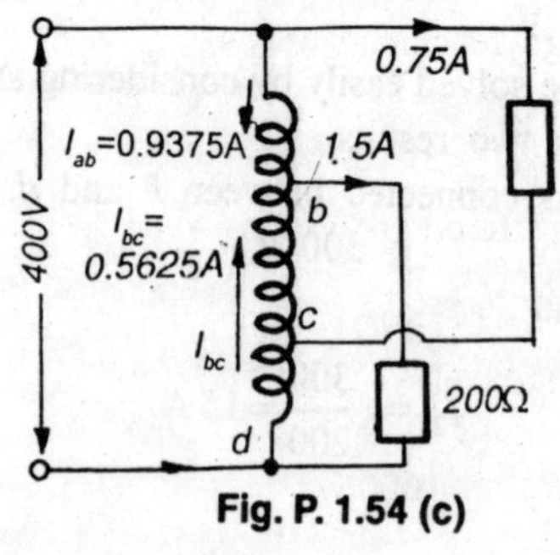

Fig. P. 1.54 (c)

Current through the section *ab* $I_{ab} = 1.125 - 0.1875\text{ A} = 0.9375$

Current through the section *bc* $I_{bc} = (0.375 - 0.1875\text{ A} = 0.5625$

Current through the section *cd* $I_{cd} = 0.5625 - 0.375 = 0.1857$ A **Ans.**

> **Problem 1.55 :** An audio transformer has the following parameters:
>
> $r_1 = 10\,\Omega, \qquad l_1 = 2m\text{H}, \; R_2 = 0.4\;\Omega$
>
> $l_2 = 0.02\; m\text{H}, \qquad M_{21} = M_{12} = 0.15\text{ H}$
>
> This transformer couples a load of 40 Ω to a voltage source of 4 V whose internal resistance is 1600 Ω. Neglect iron loss.
>
> *(i)* Find the turn ratio for maximum power transfer to load.
>
> *(ii)* Compute the load voltage at the following frequencies
> *(a)* 100 Hz (LF) *(b)* 5000 Hz (IF) *(c)* 15,000 Hz (HP)

Solution :

Given, $r_g = 1600\,\Omega, \quad r_l = 40\,\Omega, \quad \dfrac{N_1}{N_2} = ? \quad E_g = 4\text{ V}$

(i) According to Maximum power transfer theorem,
Source impedance = Load impedance

or $$1600 = \left(\frac{N_1}{N_2}\right)^2 \times 40$$

$$\therefore \quad \frac{N_1}{N_2} = \sqrt{40} = 6.32$$

$$\therefore \text{ Turn ratio}\left(\frac{N_1}{N_2}\right) = 6.32 \textbf{ Ans.}$$

(ii) Secondary resistance reffered to primary side is given by

$$r_2' = \left(\frac{N_1}{N_2}\right)^2 \times r_2 = 40 \times 0.4 = 16\;\Omega$$

Secondary leakage reactance reffered to primary is given by

$$l_2' = \left(\frac{N_1}{N_2}\right)^2 \times l_2 = 40 \times 0.02 \times 10^3\text{ H} = 0.8\text{ mH}$$

Load resistance reffered to primary side is given by

$$R_l' = \left(\frac{N_1}{N_2}\right)^2 \times 40 = 1600\,\Omega$$

$$R_g = 1600\,\Omega$$

$$R_s' = R_g + r_1 + r_2' + R_l' = 1600 + 10 + 16 + 1600 = 3226\,\Omega$$

$$R_p' = \frac{1616 \times 1670}{3226} = 806.5\,\Omega$$

Now, $L_1 \cong M_{12} = N_{21} = 0.15,\ l_{eq} = l_1 + l_2' = 2 + 0.8$

$$= 2.8\text{ mH}$$

(i) Load voltage at 100 Hz, Applying, the equation

$$\frac{V_l}{E_g} = \frac{N_2}{N_1} \cdot \frac{R_l'}{R_s'\left[1 + \left(\frac{R_p'}{X_i}\right)^2\right]^{1/2}}$$

$$= \frac{1}{\sqrt{40}} \cdot \frac{1600}{3226\left[1 + \left(\frac{806.5}{2 \times \pi \times 100 \times 2.8 \times 10^{-3}}\right)^2\right]^{1/2}}$$

$$\therefore \quad V_L = \frac{1600 \times 4}{\sqrt{40} \times 3226 \times \left[1 + 2.1 \times 10^{-7}\right]} = 0.31368\text{ V } \textbf{Ans.}$$

(ii) For intermediate frequency,

$$\frac{V_l}{E_g} = \frac{N_2}{N_1} \cdot \frac{R_l'}{R_s'}$$

$$\therefore \quad V_L = E_g \times \frac{N_2}{N_1} \times \frac{R_l'}{R_s'}$$

$$= 4 \times \frac{1}{\sqrt{40}} \times \frac{1600}{3226} = 0.31368\text{ V } \textbf{Ans.}$$

(iii) For high frequency

$$\frac{V_L}{E_g} = \frac{N_2}{N_1} \cdot \frac{R_l'}{R_s'} \times \frac{1}{\left[1 + \left(\frac{X_{eq}}{R_s'}\right)^2\right]^{1/2}}$$

$$\therefore V_L = \frac{R_g \times \frac{N_2}{N_1} \times \frac{R_l'}{R_s'}}{\left[1 + \left(\frac{X_{eq}'}{R_s'}\right)^2\right]^{1/2}} = \frac{0.31368}{\left[1 + \left(\frac{42 \times 15000 \times 0.8 \times 10^{-3}}{3226}\right)^2\right]^{1/2}}$$

$$= \frac{0.31368}{1.00027} = 0.3136 \text{ V } \textbf{Ans.}$$

Problem 1.56 : The self and mutual winding of two winding transformer are

$$L_1 = 2 \text{ mH}, \quad L_2 = 4 \text{ mH}, \quad M_{21} = M_{12} = 1.6 \text{ mH}$$

Calculate the current which would flow in the winding 1 when this winding is connected to a 200 V, $\left(\frac{500}{Z}\right)$ Hz supply and the load of 0.3 mH inductance is connected across the winding 2. Assume power losses in the windings and the magnetic circuit to be Negligible.

Solution :

Given, $L_L = 0.3 \; m\text{H}$

The voltage across the load on secondary side is given by

$$\bar{V}_2 = \bar{I}_2 \bar{X}_L = j\,2\pi\, f\, L_L\, I_2 \qquad (1)$$

$$= j\,2 \times \pi \times \frac{500}{\pi} \times 0.3 \times 10^{-3}\, I_2 = j\,0.3\, I_2$$

Now, the voltage equation for primary winding is

$$\bar{V}_1 = r_1 \bar{I}_1 + j\,w L_1 \bar{I}_1 - j\,w M \bar{I}_2 \qquad (2)$$

The voltage equation for secondary winding is

$$\bar{V}_2 = j\,w\,M\,I_1 - j\,w\,L_2\,I_2 - r_2\,I_2 \qquad (3)$$

Since $r_1 = r_2 = 0$

Let as take $\bar{V}_1$ as reference phasor, Substituting the values in (2) and (3)

$$200 \angle 0° = j2\pi \times \frac{500}{\pi} \times 2 \times 10^{-3}\, I_1 - j\,2\pi \times \frac{500}{\pi} \times 1.6 \times 10^{-3}\, I_2$$

or $200 + j0 = j2I_1 - j1.6I_2$

On equating, this yields $j2I_1 - j1.6I_2 = 200$ (4)

Now from (3)

$$j\,0.3\, I_2 = j2\pi \times \frac{500}{\pi} \times 1.6 \times 10^{-3}\, I_1 - j\,2\pi \times \frac{500}{\pi} \times 4 \times 10^{-3}\, I_2$$

or $0.3\, I_2 = 1.6\, I_1 - 4\, I_2$

or $4.3\,I_2 = 1.6\,I_1 \quad \therefore\ I_1 = \frac{4.3}{1.6} I_2 = 2.7\,I_2, \quad \therefore\ I_1 = \frac{4.3}{1.6} I_2 = 2.7\,I_2$

$I_1 = 2.72\,I_2$ (5)

From (4) and (5) $j\,2 \times 2.7\,I_2 - j\,1.6\,I_2 = 200$

or $j\,5.4\,I_2 - j\,1.6\,I_2 = 200$

or $j\,3.8\,I_2 = 200 \therefore\ I_2 = \frac{200}{3.8} = 52.63$ A

And $I_1 = 52.63 \times 2.7 = 142.11$ **Ans.**

Thus, Primary Current = 142.11 A **Ans.**

Problem 1.57 : A 20 kVA, 4400/440 V, single phase transformer has the following parameters

$$r_1 = 10\,\Omega,\ r_2 = 0.20\,\Omega,\ l_1 = 50\text{ mH}$$

$$l_2 = 5 \times 10^{-4}\text{ H},\ M = 10\text{H}$$

Subscripts 1 and 2 indicate h.v. and l.v. winding respectively.

(a) Find the self impedances of primary and secondary windings.

(b) Find the values of equivalent-circuit parameters referred to *(i)* primary side and *(ii)* the secondary side

(c) The primary of this transformer is energized from 4400 V, 50 hz. If its secondary is connected to a load of impedance $6 + j\,6\ \Omega$. Find the secondary terminal voltage.

Solution :

Given, $r_1 = 10\,\Omega,\ r_2 = 0.20\,\Omega,\ l_1 = 50\text{ mH},\ l_2 = 5 \times 10^{-4}\text{ H},\ M = 10\text{ H}$

(a) Here, turn ratio $= \frac{N_1}{N_2} = \frac{4400}{440} = 10$

Primary self-inductance

$$L_1 = \left(\frac{N_1}{N_2}\right) \times M + l_1 = 10 \times 10 + 50 \times 10^{-3}$$

$$= 100 + 50 \times 10^{-3} = 100.05\text{ H } \textbf{Ans.}$$

Secondary self inductance

$$L_2 = \frac{N_2}{N_1} \times M + l_2 = \frac{1}{10} \times 10 + 5 \times 10^{-4}$$

$$= 1.00005 \text{ H } \textbf{Ans.}$$

(b)(i) Parameters on primary side,

$$r_1 = 10, \Omega$$

$$r_2' = (10)^2 \times r_2 = 100 \times 0.20 = 20\,\Omega$$

$$l_1 = 50 \text{ mH}$$

$$l_2' = (10)^2 \times l_2 = 100 \times 5 \times 10^{-4} = 0.05 \text{ H}$$

Thus, the equivalent circuit parameters on primary side is as below

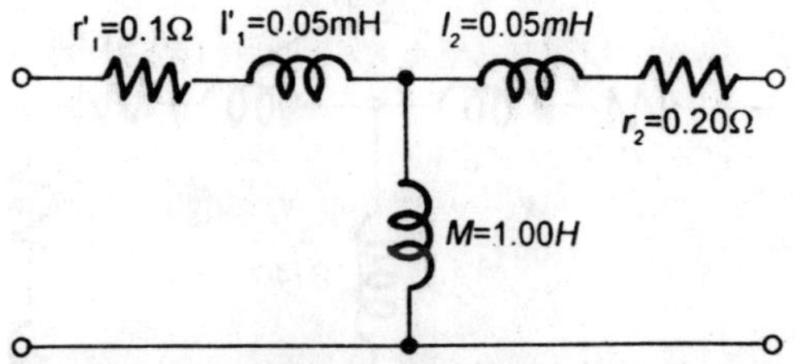

Fig. P. 1.57 (a)

(ii) Parameters on secondary side

$$r_2 = 0.20\,\Omega$$

$$r_1' = \frac{1}{100} \times 10 = 0.1 \text{ W}$$

$$l_2 = 5 \times 10^{-4} \text{ H}$$

$$l_1' = \frac{1}{100} \times 50 \times 10^{-3} = 5 \times 10^{-4} \text{ H}$$

$$l_{m2} = \frac{1}{10} \times 10 = 1.00 \text{ H}$$

The equivalent circuit on secondary side can be drawn as

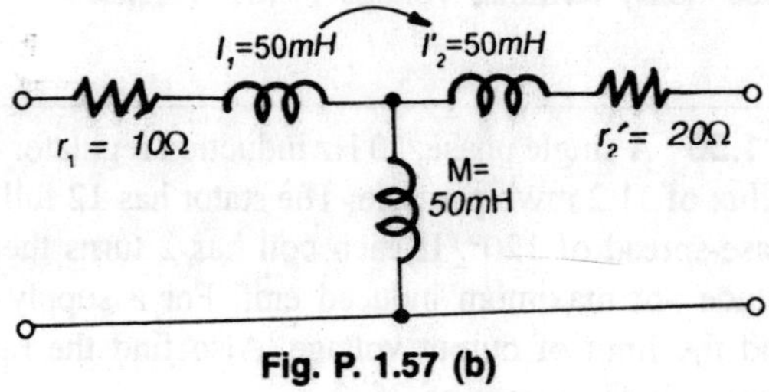

Fig. P. 1.57 (b)

(c) The primary input voltage refered to secondary side is given by

$$V_1' = \frac{1}{10} \times 4400 = 440 \text{ A}$$

$$X_1' = 2\pi f l_1 = 2\pi \times 50 \times 5 \times 10^{-4}\ \Omega = 0.157\ \Omega$$

$$X_2 = 2\pi f l_2 = 0.157$$

and $$X_{m2} = 2\pi f l_{m2} = 2 \times \pi \times 50 \times 1 = 314\ \Omega$$

The equivalent circuit under load condition on secondary side can be drawn as

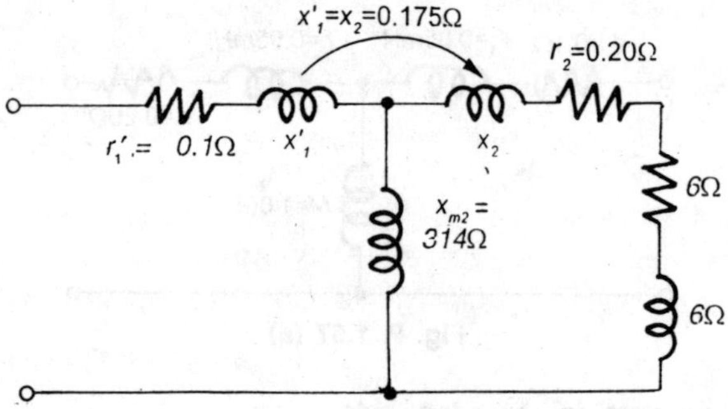

Fig. P. 1.57 (c)

The current flowing through Xm_2 is much smaller as compared to that in $6.2 + j\ 6.157\ \Omega$ connected in parallel with it.

∴ Secondary terminal voltage

$$= \frac{440}{6.2 + j\,6.157} \times (6 + j\,6)$$

$$= \frac{440 \times 6\sqrt{2}\ \angle 45°}{8.74\ \angle 44.80} = 427\ \angle 0.20°$$

∴ Secondary terminal voltage = 427 V **Ans.**

Problem 1.58 : A single phase, 50 Hz induction regulator is designed to have a flux of 11.2 mwb per pole. The stator has 12 full pitch coils with a phase-spread of 120°. If each coil has 2 turns then calculate the magnitude of maximum induced emf. For a supply voltage of 250 V, find the limit of output voltage. Also find the rating of the regulator for a load current of 35 A.

Solution :

The distribution factor is given by $K_w = \dfrac{\sin\frac{\sigma}{2}}{\frac{\sigma}{2}}$

For a phase spread of 120° $\quad K_w = \dfrac{\sin\frac{120°}{2}}{\frac{1}{2}\times 120° \times \frac{2}{180}} = \dfrac{3}{\pi}(0.866)$

Now, Magnitude of maximum induced emf E_2

$$E_2 = j\,2\,\pi\,f\,N\,\phi\,K_w$$

$$= \sqrt{2}\times\pi\times 50\times(12\times 2)\times 0.0112\times\frac{3}{\pi}\times 0.866 = 49.38\text{ V}\quad \textbf{Ans.}$$

Limit of output voltage $= V_t \pm E_2 = 250 \pm 49.38$

Maximum output voltage $= 299.38$ V

Minimum output voltage $= 200.62$ V **Ans.**

Rating of the induction regulation $= E_2 \times I_L$

$= 49.38 \times 35 = 1.73$ kVA **Ans.**

Problem 1.59 : A three phase induction regulator is designed to vary the output voltage V_2 between the limits of $V_1 \pm E_2$. Where V_1 is the supply voltage and E_2 is the magnitude of secondary induced emf. For $E_1 = 500$ V, $E_2 = 60$ V, load = 10 kW, load p.f. = 0.8 and overall frequency of 0.90. Calculate the rating of primary and secondary winding.

Solution :

Given, $\quad E_1 = 500$ V, $E_2 = 60$ V, load = 10 kW

$p.f = 0.8$, $\eta = 0.90$

Input current $= \dfrac{10\times 1000}{\sqrt{3}\times 500\times 0.9\times 0.8} = 16.04$ A

Minimum output voltage $= 500 - 60 = 440$ V

Maximum output current $= \dfrac{10\times 1000}{\sqrt{3}\times 440\times 0.8} = 16.40$ A

Rating of the secondary winding $= \sqrt{3}\times 60\times 16.40 = 1.704$ kVA **Ans.**

Maximum output voltage $= 500 + 60 = 560$ V

$$\text{Minimum output current} = \frac{10 \times 1000}{560 \times \sqrt{3} \times 0.8} = 12.89 \text{ A}$$

As we know that the primary winding current is the sum of difference of input and output current and the current induced in it by the transformer action due to secondary current.

Maximum primary current = primary induced current + (Input current – minimum output current)

$$= \frac{12.89 \times 60}{500} + (16.04 - 12.89)\, 4.7 \text{ A}$$

Rating of primary winding = $\sqrt{3} \times 4.7 \times 500$ = 4.07 kVA **Ans.**

Problem 1.60 : A delta-star 11000/433 V transformer is provided with a tap changer. A three phase star-connected load of 12 kVA at 400 V and at 0.8 lagging, is supplied by the transformer through leads of impedance $0.6 + j1\ \Omega$ pa lead the leakage impedance of l.v. winding is $0.3 + j1\ \Omega$ per phase and that of h.v. winding is $400 + j1600$ per phase. In order to maintain a voltage of 500 V at the load terminals, find tap setting.

Solution :

Given, p.f. = 0.8

Let us calculate on per phase basis.

Primary voltage/phase = 11000 V, being delta

$$\text{Secondary voltage/phase} = \frac{433}{\sqrt{3}} \text{ V}$$

$$\text{Now, Transformation ratio} = \frac{11000}{433/\sqrt{3}} = 44$$

H.V. leakage impedance reffered to l.v. side.

$$Z'_{HV} = (400 + j\,1600) \times \left(\frac{1}{44}\right)^2 = 0.207 + j\,0.826\ \Omega$$

Total impedances on the secondary side of the transformer

$$\overline{Z} = \overline{Z}'_{H.V.} + \overline{Z}'_{L.V.} + \overline{Z}_L$$

$$= (0.207 + j\,0.826 + 0.3 + j\,1 + 0.6 + j\,1)\ \Omega$$

$$= 1.107 + j\,2.826$$

$$\text{Load current} \quad I_L = \frac{12 \times 1000}{\sqrt{3} \times 400} = 17.32 \text{ A}$$

Total impedance drop per phase

$$= 17.32\,(1.107 \times 0.8 + 2.862 \times 0.6) = 45.1 \text{ V}$$

Phase voltage at load terminal $= \dfrac{400}{\sqrt{3}} = 231 \text{ V}$

Per phase voltage that must be maintained at transformer l.v. side

$$= 231 + 45.1 = 276.1 \text{ V}$$

At No-load, l.v. terminal voltage $= \dfrac{433}{\sqrt{3}} = 250 \text{ V}$

∴ The voltage boost that tap changer must proved

$$= 276.1 - 250 = 26.1 \text{ V}$$

∴ Tap setting $= \dfrac{26.1}{250} \times 100\% = 10.44\%$ **Ans.**

Problem 1.61 : A single phase ideal transformer has 1000 and 500 turns for its primary and secondary windings respectively. The secondary winding delivers a resistive load of 500 KW between the terminals A and C and a load of $50 + j\,50\ \Omega$ between B and C, there are 300 turns between A and B and 200 turns between B and C so that there are 500 secondary turns. For a primary voltage of 40 kV, find the primary current and primary input.

Solution :

The circuit diagram can be draw as

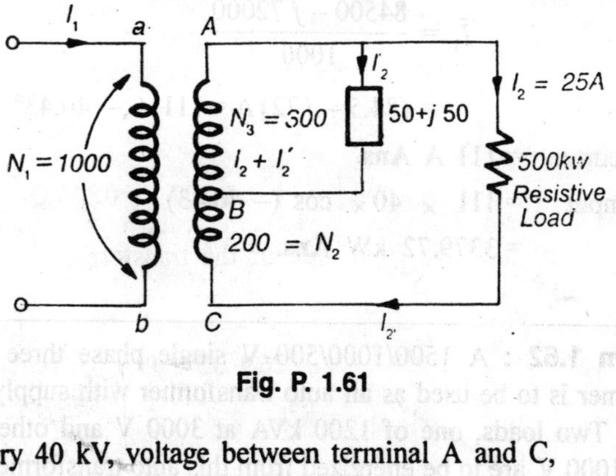

Fig. P. 1.61

For primary 40 kV, voltage between terminal A and C,

$$V_{AC} = \frac{40 \times 1000}{1000} \times 500 = 20 \text{ kV}$$

And load current through 500 kW,

$$I_2 = \frac{500 \times 1000}{20,000} = 25 \text{ A}$$

Voltage between A and B

$$V_{AB} = \frac{40 \times 1000}{1000} \times 300 = 12000 \text{ V}$$

And load current $\quad I_2' = \dfrac{12000}{\bar{Z}_L}$

$$= \frac{12000}{25\sqrt{2} \angle 45^\circ} = 339.4\angle -45^\circ$$

Current through AB winding

$$= I_2' + I_2$$
$$= 25 \angle 0^\circ + 339.4\,(\cos 45^\circ - j \sin 45^\circ)$$
$$= (265 - j\,240)\text{A}$$

Now, secondary mmf $= N_{AB} \times I_{AB} + N_{BC} \times I_{BC}$

$$= 200 \times 25 + 300 \times (265 - j\,240)$$
$$= 5000 + 79500 - j\,72000$$
$$= 84500 - j\,72000$$

In loaded transformer

$$N_1 I_1 = N_2 I_2$$

$$\therefore \quad \bar{I}_1 = \frac{84500 - j\,72000}{1000}$$

$$= (84.5 - j72) \text{ A} = 111 \angle -40.43^\circ$$

Primary current = 111 A **Ans.**

Primary input $= 111 \times 40 \times \cos(-40.43)$

$= 3379.72$ kW **Ans.**

Problem 1.62 : A 1500/1000/500 V single phase three winding transformer is to be used as an auto transformer with supply voltage 2500 V. Two loads, one of 1200 kVA at 3000 V and other of 150 kVA at 1000 V, are to be energized from this auto transformer output. Draw a suitable diagram of connection and find the current in various parts of the circuit. Assume the loads have same power factor.

Solution :

The circuit diagram can be drawn as

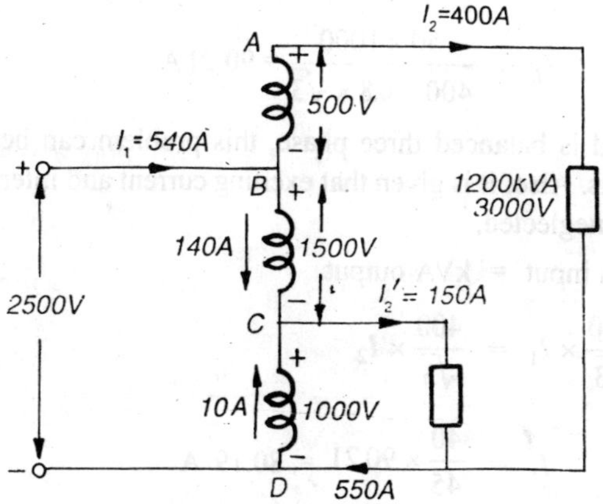

Fig. P. 1.62

Current through the load 1200 kVA at 300 V is

$$I_2 = \frac{12000 \times 1000}{3000} = 400 \text{ A}$$

Current through the load 150 kVA at 1000 V is

$$I_2' = \frac{150 \times 1000}{1000} = 150 \text{ A}$$

Total kVA supplied = 1200 + 150 = 1350 kVA

$$\Rightarrow \quad V_1 \times I_1 = 1350 \times 1000$$

$$\therefore \quad I_1 = \frac{1350 \times 1000}{2500} = 540 \text{ A } \textbf{Ans.}$$

From figure current through winding AB = 400 A

Current through winding BC = 540 − 400 = 140 A

Current through winding CD = 150 − 140 = 10 A **Ans.**

Problem 1.63 : A three phase star-connected auto-transformer supplies a balanced 3-phase load of 50 kW at 400 V and at 0.8 p.f. lagging. If the supply voltage is 450 V, determine the currents in various branches of the auto-transformer. Exciting current and internal voltage drops are neglected.

Solution :

$$V_s = 450 \text{ V, Load} = 50 \text{ kW at } 400 \text{ V at } 0.8 \text{ p.f.}$$

Load current $\quad I_L = \dfrac{50 \times 1000}{400 \times 0.8 \times \sqrt{3}} = 90.21 \text{ A}$

Since the load is balanced three phase, this problem can be solved on per phase basis. Also, it is given that exciting current and internal voltage drop can be neglected.

$\therefore \quad$ kVA input = kVA output

or $\quad \dfrac{450}{\sqrt{3}} \times I_1 = \dfrac{400}{\sqrt{3}} \times I_2$

$\therefore \quad I_1 = \dfrac{40}{45} \times 90.21 = 80.19 \text{ A}$

The circuit diagram of loaded auto transformer can be drawn as

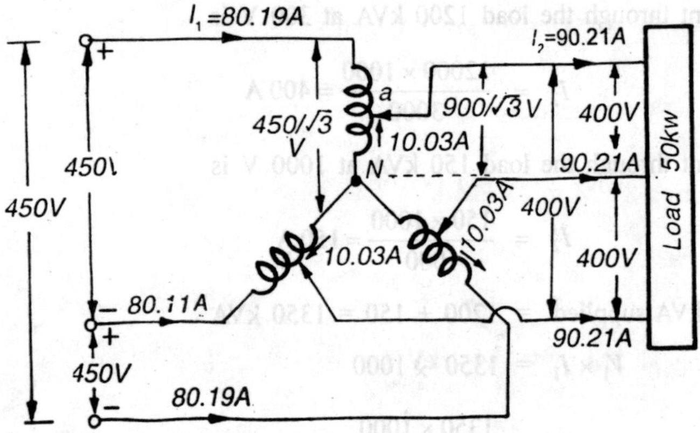

Fig. P. 1.62 (a)

the current flowing through neutral N to tapping as bc is

$$(90.21 - 80.19) = 10.03 \text{ A.}$$

Problem 1.64 : A 6 kVA transformer has a core loss of 50 watts and full load ohmic loss of 120 watts. The daily variation of the load on the transformer is as follows:

8 AM to 1 PM — 3 kW at p.f. 0.8
1 PM to 8 PM — 4 kW at p.f. 0.6

8 PM to 11 PM — 2 kW at p.f. 0.9 11 PM to 4 AM — 1 kW at p.f. 0.1 4 AM to 8 AM — No. load Determine the all day efficiency of the transformer

Solution :

We know, that kVA $= \dfrac{\text{kW}}{\text{p.f.}}$

The core loss remains fixed to 50 watts, whereas ohmic losses vary as the square of the kVA load.

During 8 AM to 1 PM,

$$\text{kVA load} = \frac{3}{0.8} = 3.75 \text{ kVA}$$

$$\therefore \text{ Ohmic losses for 5 hrs} = 120 \times \left(\frac{3.75}{6}\right) = 75 \text{ W}$$

During 1 PM to 8 PM kVA load $= \dfrac{4}{0.6} = 6.67$ kVA

$$\therefore \text{ Ohmic losses for 7 hours} = 120 \times \left(\frac{6.67}{6}\right) = 133.3 \text{ W}$$

During 8 PM to 11 PM kVA load $= \dfrac{2}{0.9} = 2.22$ kVA

$$\therefore \text{ Ohmic losses for 3 hours} = 120 \times \left(\frac{2.22}{6}\right) = 133.3 \text{ W}$$

During 11 PM to 4 AM kVA load $= \dfrac{1 \text{ kW}}{1} = 1$ kVA

$$\therefore \text{ Ohmic losses for 5 hrs} = 120 \times \left(\frac{1}{6}\right) = 20 \text{ W}$$

During 4 AM to 8 AM kVA $= 0$

Ohmic loss $= 0$ W

Daily everage lost as Ohmic losses

$$= \frac{1}{1000}[5 \times 75 + 7 \times 133.3 + 3 \times 44.44 + 5 \times 20] \text{ kWh}$$

$$= 1.54 \text{ kWh}$$

Daily energy lost as core loss $= \dfrac{50 \times 24}{1000}$ kWh $= 1.2$ kWh

$\therefore$ Total kilowatt hour lost $= (1.54 + 1.2) = 1.74$ kWh

Daily kWh output $= [5 \times 3 + 7 \times 4 + 3 \times 2 + 5 \times 1]$ kWh

$= 15 + 28 + 6 + 5 = 54$ kWh

Now, day efficiency $= 1 - \dfrac{\text{Daily losses in kWh}}{\text{Daily output Daily losses in kWh}}$

All day efficiency $= \left(1 - \dfrac{1.74}{1.74 + 54}\right) \times 100\,\% = 96.82\%$ **Ans.**

> **Problem 1.65 :** The maximum efficiency of a 120 kVA, single phase transformer is 96% and occurs at 85% of full load at 0.8 p.f. If the leakage impedance of the transformer is 4%. Find the voltage regulation at rated load of 0.8 p.f. lagging.

Solution :

Given, $\eta_m = 0.96$ at 85% of full load

p.f. $= 0.8$ (lagging)

Output at maximum efficiency $= 120 \times 1000 \times 0.85 \times 0.8 = 81600$ W

Total losses $= \left(\dfrac{1}{\eta_m} - 1\right) \times \text{output} = \left(\dfrac{1}{0.96} - 1\right) \times 81600 = 3400$ W

At maximum efficiency.

Core loss = Ohmic loss $= \dfrac{1}{2} \times$ Total losses $= \dfrac{1}{2} \times 3400 = 1700$ W

Now, Ohmic loss at full load current $= 1700 \times \left(\dfrac{100}{85}\right) = 2000$ W

Equivalent resistance in p.u. $R_{pu} = \dfrac{2000}{120000} = 0.0167$ p.u.

Given, $Z_{pu} = 0.04$ p.u.

Equivalent resistance in p.u. $X_{pu} = \sqrt{(0.04)^2 - (0.0167)^2} = 0.036$ p.u.

We know that Voltage regulation $= R_{pu} \cos\theta + X_{pu} \sin\theta$

when θ = p.f. angle $= 0.016.7 \times 0.8 + 0.036$

$= 0.345 = 3.45\%$ **Ans.**

> **Problem 1.66 :** The full load voltage drop in a single-phase transformer are 3% and 5% due respectively to resistance and leakage reactance. The full load ohmic loss is equal to iron loss. Calculate

(*a*) the efficiency on full load at unity p.f.
(*b*) the full load p.f. at which voltage drop is maximum, and
(*c*) the load p.f. at which voltage drop is zero

Solution :

Given, Percentage full-load voltage drop in resistance $= \frac{I_2 r_{e2}}{E_2} = 3$

Percentage full-load voltage drop in leakage reactance $= \frac{I_2 x_{e2}}{E_2} = 5$

(*a*) Given, $R_{pu} = 0.03$

Full-load Ohmic loss $= R_{pu} \times$ rating in VA

$= 0.03 \times$ rating in VA = iron loss

Now, Full load efficiency at unity p.f.

$$= \frac{\text{kVA} \times 1}{\text{kVA} \times 1 + 0.03 \times \text{kVA} + 0.03 \times \text{kVA}}$$

$$= \left(\frac{1}{1.06}\right) \times 100\,\% = 94.335 \textbf{ Ans.}$$

(*b*) The full load p.f. at which voltage drop is maximum is given by

$$\text{Cos}\,\theta = \frac{r_{e2}}{z_{e2}} = \frac{3}{\sqrt{3^2 + 5^2}} = \frac{3}{\sqrt{9 + 25}} + \frac{3}{\sqrt{34}}$$

$$= 0.5145 \text{ (lagging)}$$

$$= 0.5145 \text{ (lagging)}$$

(*c*) Load p.f. at which voltage drop is zero is

$$\cos\theta = \frac{x_{e2}}{z_{e2}} = \frac{x_{e2} \text{ in p.u.}}{z_{e2} \text{ in p.u.}}$$

$$= \frac{5}{\sqrt{3^2 + 5^2}} = 0.8575 \text{ (leading) } \textbf{Ans.}$$

Problem 1.67 : When 25 kVA, 4000/400 V, 50 Hz transformer is operated at rated voltage on no load, its power input is 200 watts at p.f. 0.2 under rated load conditions, the voltage drop in the total resistance and the total reactance are respective 1 and 4% of rated voltage. Determine input power and p.f. when the transformer delivers 15 kW at 400 V at 0.8 p.f. lagging to a load on the l.v. side

Solution :

Load current at 400 V at 0.8 p.f $I_L = \dfrac{15 \times 1000}{400 \times 0.8} = 46.88\text{ A}$

Rated current on l.v. side $I_{Lr} = \dfrac{25 \times 1000}{400} = 62.5\text{ A}$

Given, Core loss, $Pc = 200\text{W}, \; R_{pu} = 0.01$

or $\dfrac{I_{rated} \times r_{e2}}{E_2} = 0.01$

$\therefore \quad r_{e2} = \dfrac{0.01 \times 400}{62.5} = 0.064\ \Omega$

Similarly, $x_{e2} = \dfrac{0.04 \times 400}{62.5} = 0.256\text{ W}$

Ohmic loss at load $= (46.88)^2 \times 0.064 = 140.66\text{ W}$

Power input $= 15000 + 140.66 + 200 = 15340.6\text{ W}$

Let us take I_2 as reference phasor

$$\bar{I}_2 = 46.88 \angle 0°$$

and $\bar{V}_2 = 400 \angle 36.9°$

For secondary side $\bar{E}_2 = \bar{V}_2 + \bar{I}_2\, \bar{Z}_{e2}$

$$= 400\,(0.8 + j\,0.6) + 46.08\,(0.064 + j\,0.256)$$

$$= 320 + j\,240 + 3 + j\,12 = 323 + j\,252$$

$$\bar{E}_2 = 410 \angle 37.96°$$

For convenience, let us take E_2 as phasor reference

$\therefore \quad \bar{E}_2 = 410 \angle 0°$

$$I_2 = 46.88 \angle -37.96°$$

From No-load $P_c = 200\text{ W}$

Exciting current $I_e = \dfrac{200}{400 \times 0.2} = 2.5\text{ A}$

Core loss component of $I_e = I_e \cos\theta_0 = 2.5 \times 0.2 = 0.5\text{ A}$

$\therefore$ Magnetising current $I_e = \sqrt{(2.5)^2 - (0.3)^2} = 2.45\text{ A}$

Since I_c is in phase with E_2 and I_4 lags E_2 by 90, therefore, no-load current with E_2 as reference phasor, can be written as

$$I_e = 0.5 - j\,2.45$$

Now, primary input current,

$$\begin{aligned} I_1 &= \bar{I}_e + \bar{I}_2 \\ &= (0.5 - j\,2.45) + 46.88 \angle -37.96° \\ &= 0.5 - j\,2.45 + 36.96 - j\,28.84 \\ &= 37.46 - j\,31.29 = 48.81 \angle -39.87° \end{aligned}$$

$\therefore$ Input power factor $= 0.7675$ (lagging) **Ans.**

Input power $= 15.34$ kW **Ans.**

> **Problem 1.68 :** The efficiency of 15 kVA, 2000/200 V, single phase transformer at unity p.f. is 98% at rated load and also at half of rated load. Determine
> *(i)* The transformer core loss and ohmic losses, and
> *(ii)* p.u. value of equivalent resistance of the transformer

Solution :

Let P_c and P_{sc} be the core loss and ohmic loss respectively.

Given, Efficiency at rated load at unity p.f. = 0.98

or $$\frac{15000}{15000 + P_c + P_{sc}} = 0.98 \qquad (1)$$

Efficiency at half of rated load unity p.f. = 0.98

or $$\frac{15000\frac{1}{2}}{\frac{1}{2} \times 15000 + P_c + \frac{1}{4}P_{sc}} = 0.98 \qquad (2)$$

or $$7500 = 0.98\ (7500 + P_c + \frac{1}{4}P_{sc})$$

or $$P_c + \frac{1}{4}P_{sc} = \frac{7500\,(1 - 0.98)}{0.98}$$

or $$P_c + \frac{1}{4}P_{sc} = 153.1 \qquad (3)$$

From (1) $$P_c + P_{sc} = \frac{15000\,(1 - 0.98)}{0.98}$$

$$P_c + P_{sc} = 306.1 \qquad (4)$$

From (3) and (4) $P_{sc} + \frac{1}{4} P_{sc} = 306.1 - 153.1$

or $\frac{3}{4} P_{sc} = 153$

$\therefore \quad P_{sc} = 153 \times \frac{4}{3} = 204$ W **Ans.**

From (4) $Pc = 306.1 - 204 = 102.1$ W **Ans.**

(2) p.u. value of equivalent resistance of the transformer

$$= \frac{204}{15 \times 1000} = 0.0136 \text{ p.u. } \textbf{Ans.}$$

> **Problem 1.69 :** A 200 V, 50 Hz, single phase transformer has hystersis loss of 320 watts and eddy current loss of 100 watts. If the transformer is operated at 220 V, 25 Hz supply mains, then compute its total core loss. Take the value of Steineutz's constant as 1.6.

Solution :

Given, at $V_1 = 200$ V, $f_1 = 50$ Hz

$P_{h1} = 320$ Watts, $P_{e1} = 100$ Watts

At $V_2 = 220$ V, $f_2 = 25$ Hz

$P_{h2} + P_{e2} = ?$

Applying the relation, $V = \sqrt{2}\, \pi f\, N B_m A_i$

Now, $$\frac{V_1}{V_2} = \frac{f_1}{f_2} \times \frac{B_{m1}}{B_{m2}}$$

or $$\frac{B_{m1}}{B_{m2}} = \frac{V_1}{V_2} \times \frac{f_2}{f_1} = \frac{200}{220} \times \frac{25}{50} = \frac{5}{11}$$

$$\therefore \quad B_{m2} = \frac{11}{5} B_{m1}$$

We know that hystersis loss, $P_h = k_h f B_{m1}^x$

$$\therefore \quad \frac{P_{h2}}{P_{h1}} = \frac{f_2}{f_1} \times \left(\frac{B_{m2}}{B_{m1}}\right)^x$$

$$\therefore \quad P_{h2} = \frac{f_2}{f_1} \times \left(\frac{B_{m2}}{B_{m1}}\right)^x \times P_{h1}$$

$$= \frac{25}{50^2} \times \left(\frac{11}{5}\right)^{1.6} \times 320 = 564.93 \text{ W}$$

Also, we have $P_e = k_e f^2 B_m^2$

$$\therefore \quad \frac{P_{c2}}{P_{e1}} = \frac{f_2^{\,2}}{f_1^{\,2}} \times \left(\frac{B_{m2}}{B_{m1}}\right)^2$$

$$P_{e2} = \left(\frac{25}{50}\right)^2 \times \left(\frac{11}{5}\right)^2 \times 100 = 121 \text{ W}$$

Thus, Total core loss, $P_c = P_h + P_e = 121 + 564.93$
$= 685.93$ W **Ans.**

Problem 1.70 : A 15 kVA, 2500/250 V, 50Hz, single phase transformer when working on rated voltage at no-load, takes an input 200 W at 0.2 p.f. Its percentage leakage impedance based on its own kVA rating is (0.5 + *j* 1) of the transformer delivers 12 kW at 250 V at p.f. 0.8 lagging on its h.v. side, determine input power and power factor.

Solution :

Let us analysis this problem on h.v. side

$$\text{Turn ratio} = \frac{2500}{250} = 10$$

$$\text{Secondary load current, } I_2 = \frac{12 \times 1000}{250 \times .8} = 60 \text{ A}$$

Secondary load current reffered to h.v. side

$$I_2' = \frac{60}{10} = 6 \text{ A}$$

$$\text{Load p.f.} = 0.8 = \cos\theta_2$$

$$\therefore \quad \bar{I}_2' = 6\angle -36.9°$$

Now, Base impedance on h.v. side,

$$Z_{BH} = \frac{(V_1)^2}{\text{Rating}} = \frac{(2500)^2}{15 \times 1000} = \frac{2500 \times 2500}{15 \times 1000} = 416.67\ \Omega$$

Leakage impedance on h.v. side

$$= (0.5 + j\,1) \times \frac{416.67}{100} = (2.08 + j\,4.17)\ \Omega$$

Given, Input power = core loss = P_c = 200 W

$$\text{Core loss resistance} = \frac{V_1^2}{P_c} = \frac{(2500)^2}{200} = 312.50\ \Omega$$

$$\text{No-load current,} \quad I_e = \frac{200}{2500 \times 0.2} = 0.4\ \text{A}$$

$$\cos\theta_0 = 0.2\,,\ \sin\theta_0 = 0.98$$

$$\text{Magnetising reactance} \quad X_\phi = X_\phi = \frac{V_1}{I_e \sin\theta_0} = \frac{2500}{0.4 \times 0.98} - 6377.6\ \Omega$$

The equivalent circuit reffered to h.v. side can be drawn as

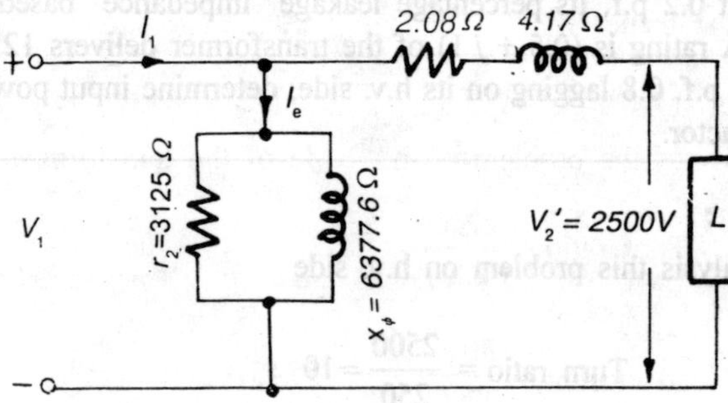

Fig. P. 1.70

Now, Primary input voltage,

$$V_1 = V_2' + I_1\,(r_{e1} \cos\theta_1 + x_{e1} \sin\theta_1)$$
$$= 2500 + 6\ (2.08 \times 0.8 + 4.17 \times 0.6)$$
$$= 2524.996\ \text{V} \cong 2525\ \text{V}$$

Active power loss is series resistance $= (6)^2 \times 2.08 = 74.89$ W

Reactive power loss is series reactance $= (6)^2 \times 4.17 = 150.12$ VA

$$\text{Active power loss in core resistance} = \frac{(2525)^2}{31250} = 204.02\ \text{W}$$

$$\text{Reactive power loss in magnetising reactance} = \frac{(2525)^2}{6377.6} = 999.7 \text{ W}$$

$$\text{Complex power at load end} = 12\,(1 + j \tan(\cos^{-1} 0.8))$$

$$= 12\,(1 + j\,0.75) = (12 + j\,9) \text{ kVA}$$

Complex power input to transformer.

$$\text{Complex power input} = (0.075 + .204 + 12) + j\,(0.150 + 0.999 + 9)$$

$$= (12.279 + j\;10.149)$$

Input power = 12.279 W **Ans.**

$$\text{Input p.f.} = \cos\left[\tan^{-1}\left(\frac{10.149}{12.279}\right)\right] = 0.771 \text{ (lagging) } \textbf{Ans.}$$

Problem 1.71 : A single phase load is fed through a 22 kV feeder whose is $20 + j\;100\;\Omega$ and a 22 kV/2.2 kV transformer of equivalent leakage impedance $0.2 + j\;1.2\;\Omega$ referred to l.v. side. For a load of 80 kW at p.f. 0.8 leading at 2.2 kV terminals, compute

(i) Voltage at the sending end of the feeder

(ii) Voltage at the primary terminals of the transformer

(iii) Active and reactive powers and also the p.f. at the sending end of the feeder

Solution :

Let us analyses the problem on h.v. side of the transformer

$$\text{Turn ratio} = \frac{22000}{2200} = 10$$

$$\text{Secondary load current} = \frac{80 \times 1000}{2200 \times 0.8} = 45.45 \text{ A}$$

Secondary load current referred to primary side

$$I_2' = \frac{45.45}{10} = 4.55 \text{ A}$$

Secondary terminal voltage referred to primary side

$$V_2' = 10 \times 22000 = 22000 \text{ V}$$

Transformer impedances referred to h.v. side

$$\overline{Z}_{HV} = 100 \times (0.2 + j\,1.2)\,\Omega = 20 + j\,120\,\Omega$$

The equivalent circuit diagram can be drawn as

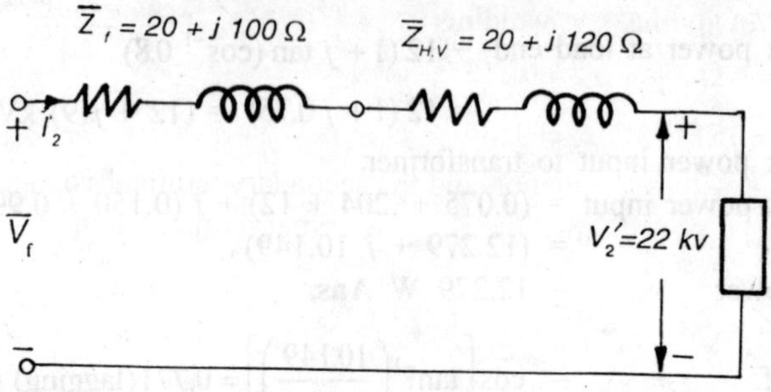

Fig. P. 1.71

Then, Total impedance on feeder side

$$\bar{Z} = \bar{Z}_f + \bar{Z}_{H.V.} = (40 + j\,220)\,\Omega = 223.6\angle 79.7°$$

$$\bar{I}_2' = \bar{I}_1 = 4.55\angle 36.9°$$

Let us take V_2' as reference phasor.

$$\bar{V}_f = \bar{V}_2' + \bar{I}_2'\,\bar{Z} = 22000 + 4.55\angle 36.9° \times 223.6\angle 79.7°$$

$$= 22000 + 1017.4\angle 116.6° = 21544.4 + j909.7$$

$$= 21563.6\angle 2.42°$$

(i) Voltage at the sending end of the feeder = 21.563 kV **Ans.**

(ii) Voltage at the sending end of the transformer.

$$\bar{V}_s = 22000 + 4.55\angle 36.9^c \times 121.66\angle 80.54°$$

$$= 22000 + 553.6\angle 117.44°$$

$$= 21744.9 + j491.32 = 21750\angle 1.3°$$

Voltage at primary terminals = 21.75 kV **Ans.**

(iii) Power input at the sending end of the feeder

$$= \bar{V}_f I_1$$

$$= 21.56\angle 2.42 \times 4.55\angle -36.9°$$

$$= 98.1\angle -34.48°$$

Active power = 80.87 kW **Ans.**

Reactive power = 55.54 kVA **Ans.**

p.f. = (– 34.48) = 0.824 3 (lagging) **Ans.**

Problem 1.72 : A 25 kVA, 3000/300 V_1 single phase transformer has the following parameters:

H.V. winding	l.v. winding
$r_1 = 6\,\Omega$	$r_2 = 0.2\,\Omega$
$x_1 = 15\,\Omega$	$x_2 = 0.5\,\Omega$

Find the voltage regulation and the secondary terminal voltage at full load for a p.f. of 0.8 lagging. The primary voltage is held constant at 3000 V.

Solution :

Here turn ratio $= \dfrac{3000}{300} = 10$

Equivalent resistance reffered to secondary side

$$= 0.2 + 6 \times \frac{1}{100} = 0.2 + 0.06 = 0.26\,\Omega$$

Equivalent reactance reffered to secondary side,

$$= 0.5 + 15\frac{1}{100} = 0.65\,\Omega$$

Full load secondary current, $I_2 = \dfrac{25 \times 1000}{300} = 83.33$ A

Full load voltage drop in the secondary terminal voltage at 0.8 lagging

$$= I_2\, r_{e2} \cos\theta + I_2\, x_{e2} \sin\theta$$

$$= 83.33 \times 0.26 \times 0.8 + 83.33 \times 0.65 \times 0.6 = 49.83 \text{ V}$$

$\therefore$ Secondary terminal voltage = 300 − 49.83 = 250.17 V **Ans.**

$$\text{Voltage regulation} = \frac{49.83}{300} = 0.1661 = 16.61\%$$ **Ans.**

Problem 1.73 : A 15 kVA, 2500/250 V, 50 Hz, single phase transformer has following parameters.

$r_1 = 3.4\,\Omega$	$x_1 = 5.4\,\Omega$
$r_2 = 0.03\,\Omega$	$r_2 = 0.03\,\Omega$

Script 1 and 2 denote high and low voltage windings respectively. Find the p.u. values of r_{e1}, r_{e2}, x_{e1} and x_{e2}

Solution :

Given, V_{BH} = 2500 V V_{BL} = 250 V, Rating = 15 kVA

Base impedance on h.v. side $= \dfrac{2500 \times 2500}{15000} = 416.67\ \Omega$

Base impedance or l.v. side $= \dfrac{250 \times 250}{15000} = 4.17\ \Omega$

Now, $r_{1pu} = \dfrac{r_1}{Z_{BH}} = \dfrac{3.4}{416.67} = 0.008$ p.u.

$$x_{1pu} = \frac{x_1}{Z_{BH}} = \frac{5.4}{416.67} = 0.013 \text{ p.u.}$$

$$r_{e1} = 3.4 + (10)^2 \times 0.03 = 6.4\ \Omega$$

$$r_{e1pu} = \frac{6.4}{416.67} = 0.015 \text{ p.u} \ \textbf{Ans.}$$

$$x_{e1} = x_1 + \left(\frac{N_1}{N_2}\right)^2 \times x_2$$

$$= 5.4 + 100 \times 0.052 = 5.4 + 5.2 = 10.6\ \Omega$$

$$x_{e1} \text{ (p.u.)} = \frac{10.6}{416.67} = 0.025 \text{ p.u.} \ \textbf{Ans.}$$

Calculation on l.v. side

$$r_{e2} = r_2 + \left(\frac{N_2}{N_1}\right)^2 \times r_1$$

$$= 0.03 + \frac{1}{100} \times 3.4 = 0.064$$

$$r_{e2} \text{ (p.u.)} = \frac{0.064}{4.17} = 0.015 \ \textbf{Ans.}$$

$$x_{e2} = x_2 + \left(\frac{N_2}{N_1}\right)^2 x_1 = 0.052 + \frac{1}{100} \times 5.4$$

$$= 0.052 + 0.054 = 0.106\ \Omega$$

$$x_{e2} \text{ (p.u.)} = \frac{0.106}{4.17} = 0.025 \text{ p.u.} \ \textbf{Ans.}$$

Problem 1.74 : A 100 kVA, 10,000/500 V, delta-star distribution transformer gave the following test results: open circuit test – 500 V, 10 A, 1.06 kW, short circuit test – 400 V, rated current, 2kW. Calculate the equivalent circuit parameters reffered to h.v. side and efficiency at half full load of unity p.f.

Solution :

As we know that OC test yields shunt parameters and core loss, where as SC test yields equivalent series parameters and full load copper loss.

Shunt Parameters on l.v. Side (l.v. side is star connected)

$$\text{Applied voltage/phase} = \frac{500}{\sqrt{3}} = 288.7 \text{ V}$$

Per phase exciting current, $I_e = 10$ A

$$\text{Per phase core loss, } P_c = \frac{1600}{3} = 533.33 \text{ W}$$

or $$V_t I_e \cos\theta_0 = V_t I_c = 533.33$$

$$\therefore \quad I_c = \frac{533.33}{500/\sqrt{3}} = 1.85 \text{ A}$$

$$\text{Magnetising current } I_\phi = \sqrt{(I_e)^2 - (I_c)^2}$$

$$= \sqrt{(10)^2 - (1.85)^2} = 9.83 \text{ A}$$

$$\text{Core resistance, } r_c = \frac{288.7}{1.85} = 156.1\ \Omega$$

$$\text{Magnetising reactance } x_\phi = \frac{288.7}{9.83} = 26.37\ \Omega$$

Series Parameters on h.v. Side

$$\text{Applied voltage/phase} = \frac{400}{\sqrt{3}} \text{ A}$$

$$\text{Rated current/phase} = \frac{100{,}000}{10000 \times 3} = 3.33 \text{ A}$$

$$\text{Ohmic loss/phase} = \frac{2000}{3} = 666.67 \text{ W}$$

$$Z_{eH} = \frac{V_{sc}}{I_{sc}} = \frac{400}{\sqrt{3} \times 3.33} = 69.35\ \Omega$$

$$r_{eH} = \frac{P_{sc}}{I_{sc}^{\ 2}} = \frac{666.67}{(3.33)^2} = 60.12\ \Omega$$

$$\therefore \quad x_{eH} = \sqrt{(69.35)^2 - (60.12)^2} = 34.57\ \Omega$$

Shunt parameters referred to h.v. side

$$r_c' = 156.1 \times \left(\frac{11000 \times \sqrt{3}}{400} \right)^2 = 3541.18 \text{ k}\Omega$$

$$x_\phi' = 26.37 \times \left(\frac{11000 \times \sqrt{3}}{400} \right)^2 = 59.83 \text{ k}\Omega$$

Efficiency at half full load

$$= 1 - \frac{\text{per phase losses}}{\text{per phase losses} + \text{per phase output}} = 1 - \frac{P_c + P_{sc}}{P_c + P_{sc} + S_r}$$

$$= 1 - \frac{533.33 + 666.67 \times \frac{1}{4}}{\frac{1}{2}\left(\frac{100}{3} \times 1000 \times 1 \right) + 533.3 + 666.7 \times \frac{1}{4}}$$

$$= \left(1 - \frac{699.9975}{17366.64} \right) = 0.9597 \text{ or } 95.75\% \quad \textbf{Ans.}$$

Problem 1.75 : A 20 kVA, 5000/500 V, single phase transformer has resistances and leakage reactances as below:

$$r_1 = 4.6\,\Omega, \quad x_1 = 10.2\,\Omega, \quad r_2 = 0.04\,\Omega, \quad x_2 = 0.104\,\Omega$$

subscripts 1 and 2 denote high voltage and low voltage winding respectively.

With primary supply voltage held constant at 5000 V, calculate secondary terminal voltage when

(a) The l.v. winding is connected to a load impedance of $6 + j\,4\,\Omega$.

(b) The transformer delivers its rated current at 0.8 p.f. lagging on l.v. side.

Solution :

$$\text{Here, Turn ratio} = \frac{5000}{500} = 10$$

(a) Let us analyse this problem on l.v. side by reffering h.v. parameters to l.v. side.

$$\left(\overline{Z}_{eq} \right)_{L.V.} = \overline{Z}_2 + Z_1' = \overline{Z}_2 + Z_1' \times \left(\frac{1}{\text{Turn ratio}} \right)^2$$

$$= (0.046 + j\,0.014) + (4.6 + j\,10.2) \times \frac{1}{100}$$

$$= 0.046 + j\,0.014 + 0.046 + j\,0.102$$

$$= 0.092 + j\,0.206 = r_{e2} + j\,x_{e2}$$

Total impedance on l.v. side,

$$\bar{Z} = (\bar{Z}_{eq})_{LV} + \bar{Z}_{load} = 0.092 + j\,0.206 + 6 + j\,4$$

$$= 6.092 + j\,4.206 = 7.40 \angle 34.62°$$

Primary input voltage reffered to h.v. side

$$V_1' = 5000 \times \frac{1}{10} = 500 \text{ V}$$

Hence, load current $(I_2) = \dfrac{500}{7.40} = 67.57$ **Ans.**

Secondary terminal voltage $= 67.57 \times \left| 6 \quad j\,4 \right|$ V

$$= 67.57 \times \sqrt{36 + 16}$$

$$= 67.57 \times \sqrt{52} = 487.25 \text{ V } \textbf{Ans.}$$

(b) Rated current on l.v. side $(I_2) = \dfrac{20 \times 1000}{500} = 40$ A

The phasor diagram can be drawn as

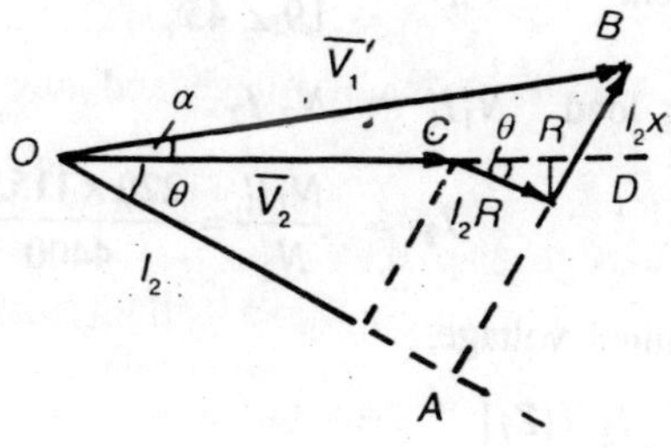

Fig. P. 1.75

From phasor diagram.

$$CD = 40 \times 0.092 \times 0.8 + 40 \times 0.206 \times 0.6 = 7.89 \text{ V}$$

Secondary terminal voltage $= 500 - 7.89$ V $= 492.11$ V **Ans.**

Problem 1.76 : The equivalent circuit parameters reffered to low voltage side of a 220/4400 V single phase transformer is as below

$$r_{e1} = 0.4\,\Omega, \quad x_{e1} = 0.8\,\Omega \quad r_c = 600\,\Omega, \quad x_\phi = 300\,\Omega$$

The load connected to the high voltage side is $400 + j\,200\,\Omega$. For a primary voltage of 220 V, compute

(a) The secondary terminal voltage
(b) Primary current and power factor.
(c) Power output and efficiency

Solution :

This problem can be easily analysed by reffering all quantites to l.v. side.

$$\text{Here turn ratio} = \frac{220}{4400} = \frac{1}{20}$$

Load impedance on the transformer reffered to l.v. side

$$Z_L' = (400 + j\,200) \times \frac{1}{400} = (1 + j\,0.5)\,\Omega$$

$$\begin{aligned}\text{Total impedance on primary side } \overline{Z} &= Z_L + \overline{Z}_{eq} \\ &= 1 + j\,0.5 + 0.4 + j\,0.8 \\ &= 1.4 + j\,1.4 = 1.98 \angle 45^\circ\end{aligned}$$

Let us take input voltage as reference phasor.

$$\text{Primary load current} \quad (I_1') = \frac{220 \angle 0}{1.9 \angle 45^\circ} = 115.79 \angle -45^\circ$$

$$\text{For transformer at load} \quad N_1 I_1' = N_2 I_2$$

$$\therefore \quad I_2 = \frac{N_1 I_1'}{N_2} = \frac{220 \times 115.79}{4400} = 5.79 \text{ A}$$

$\therefore$ Secondary terminal voltage,

$$\begin{aligned}V_2 &= I_2 \times |Z_L| \\ &= 5.79 \times \sqrt{(400)^2 + (200)^2} = 2589.37 \text{ V } \textbf{Ans.}\end{aligned}$$

(b) The core loss current $\overline{I}_c = \dfrac{V_1 \angle 0^\circ}{r_c \angle 0^\circ} = \dfrac{220}{600} = 0.367 \angle 0^\circ \text{ A}$

The magnetising Current $\overline{I}_\phi = \dfrac{220 \angle 0^\circ}{300 \angle 90^\circ} = 0.733 \angle -90^\circ \text{ A}$

$$\begin{aligned}\text{Exciting current, } \overline{I}_e &= \overline{I}_c + \overline{I}_\phi \\ &= 0.307 \angle 0^\circ + 0.733 \angle -90^\circ \\ &= 0.367 - j\,733\end{aligned}$$

Total primary current

$$\bar{I}_1 = \bar{I}_e + \bar{I}_1'$$
$$= 0.307 - j\,0.733 + 115.79 \angle -45°$$
$$= 82.23 - j\,82.61 - 116.57 \angle -45.13°$$

Primary current = 116.57 A **Ans.**

Power factor = cos 45.13 = 0.706 (lagging) **Ans.**

(c) Load power factor $= \dfrac{400}{\sqrt{400^2 + 200^2}} = 0.894$

Power output = $2589.37 \times 5.79 \times 0.894$ W = 13.41 kW

Power input = $V_1 I_1 \cos \phi_1$

$= 220 \times 116.57 \times .706 = 18.11$ kW

% Efficiency $= \left(\dfrac{\text{output}}{\text{input}}\right) \times 100\% = \left(\dfrac{13.41}{18.11}\right) \times 100\%$

= 74.05 % **Ans.**

Problem 1.77 : In figure shown, there are two single phase transformers T_1 and T_2 of turn ratio 4 : 1 and 2 : 1 connected with their primaries in series across a 120 V, 50 Hz source. The dot points indicate terminals of the corresponding polarity on the primary and secondary sides. Assume that the transformers are ideal and take $R = 10\ \Omega$. Determine the current drawn from the source.

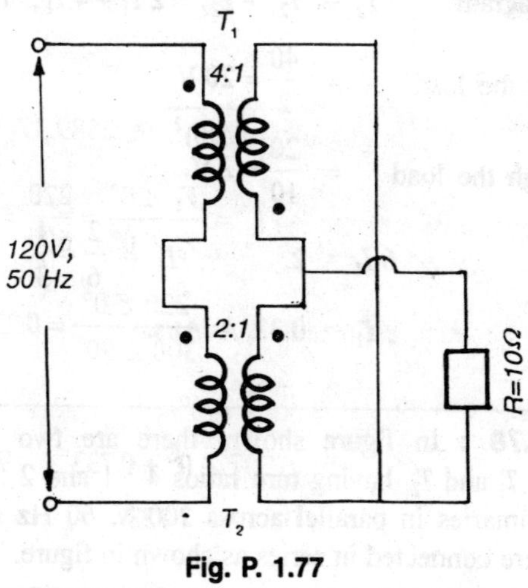

Fig. P. 1.77

Solution :

Let I_1 be the current in two series connected primaries and I_1 is also the source current.

The circuit diagram can be drawn as

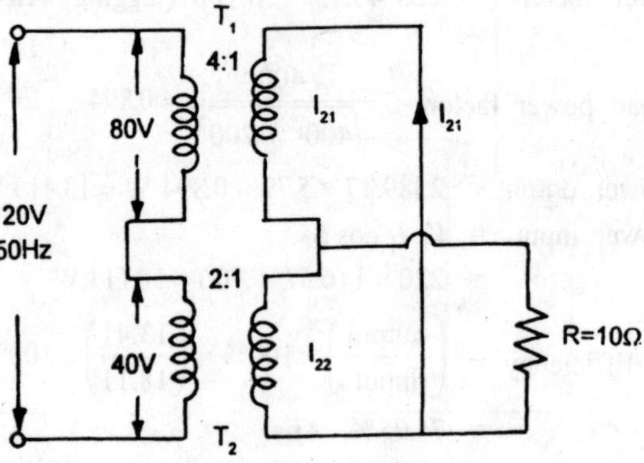

Fig. P. 1.77 (a)

For transformer, T_1, Applying mmf balance equation

$$4 \times I_1 = I_{21} \times 1, \quad \therefore I_{21} = 4 I_1 \qquad (1)$$

Where I_{21} = secondary current of T_1, for transformer T_2.

$$2 \times I_1 = I_{22} \times 1 \quad \therefore \quad I_{22} = 2 I_1 \qquad (2)$$

is the secondary current of T_{21}

Now, from diagram $\quad I_2 = I_{21} + I_{22} = 2 I_1 + 4 I_1, \; I_2 = 6 I_1$

Voltage across the load $= \dfrac{40}{2} = 20$ V

Current through the load $= \dfrac{20}{10} = 2$ V

or $\quad 6 I_1 = 2, \quad \therefore I_1 = \dfrac{2}{6} = \dfrac{1}{3}$

$$I_1 = 0.33 \text{ A } \textbf{Ans.}$$

Problem 1.78 : In figure shown, there are two single phase transformers T_1 and T_2 having turn ratios 4 : 1 and 2 : 1 connected with their primaries in parallel across 200 V, 60 Hz source. Their secondaries are connected in series as shown in figure. The dot point indicates the terminals of the corresponding polarity on the primary and secondary sides. The value of load R is 20 Ω. Determine the current taken from the source.

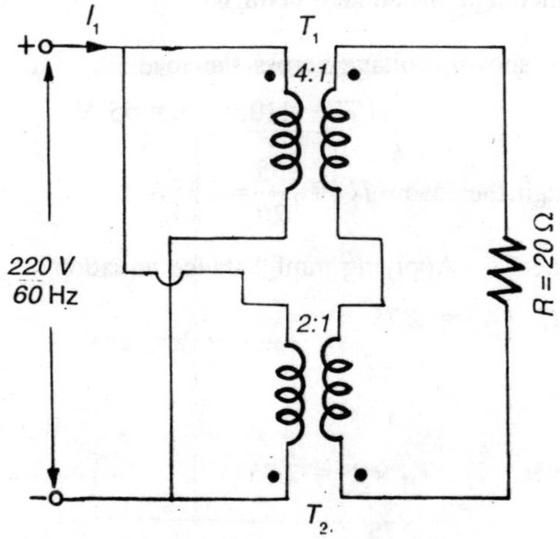

Fig. P. 1.78

Solution :

Let I_1 be the current drawn from the source

I_{11} be the current in the primary of transformer T_1

I_{12} be the current in the primary of transformers T_2

Now, the circuit can be drawn as

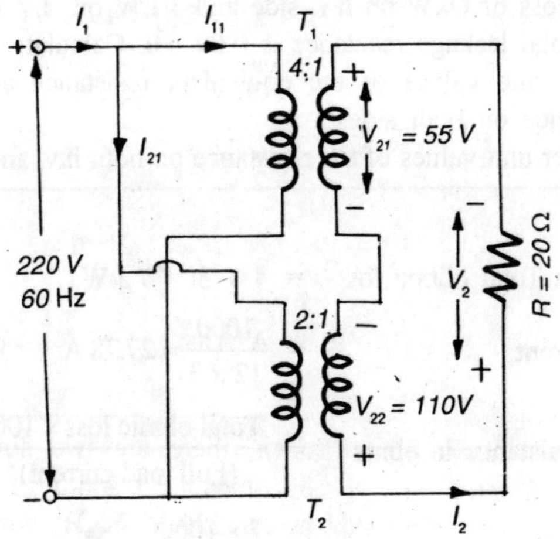

Fig. P. 1.78 (a)

For transformer T_1, Secondary voltage $V_{21} = \frac{220}{4} = 55\text{ V}$

For transformer T_2, Secondary voltage $V_{22} = \frac{220}{2} = 110\text{ V}$

From polarity shown, voltage across the load

$$V_2 = 110 - 55 = 55\text{ V}$$

Current through the load $I_2 = \frac{55}{20} = 2.75\text{ A}$

For transformer T_1, Applying mmf balance equation

$$I_{11} \times 4 = 2.75$$

$$\therefore \quad I_{11} = \frac{2.75}{4}$$

For transformer T_2, $I_{21} \times 2 = 2.75$

$$\therefore \quad I_{21} = \frac{2.75}{2}$$

$$\therefore \text{ Source current } I_1 = I_{11} + I_{21} = \frac{2.75}{2} + \frac{2.75}{4} = 2.75\,(0.5 + 0.25)$$

$$= 2.75 \times 0.75 = 2.06\text{ A} \quad \textbf{Ans.}$$

Problem 1.79 : A 1000 kVA, 12/.6 kV, 3 phase delta-star transformer has ohmic loss of QkW on h.v. side and 3 kW on l.v. under rated load. The total leakage reactance is 0.09 p.u. Calculate:

(a) The ohmic values of the equivalent resistance and leakage reactance on both sides.

(b) The per unit values of the resistance on both h.v. and l.v. sides.

Solution :

For H.V. side Total ohmic loss $= 4 + 3 = 7\text{ kW}$

Full load current $= \frac{1000}{12 \times 3} = 27.78\text{ A}$

Equivalent resistance in ohms $= \frac{\text{Total ohmic loss} \times 1000}{(\text{Full load current})^2}$

$$= \frac{7 \times 1000}{(27.78)^2} = 9.07\ \Omega \quad \textbf{Ans.}$$

Since primary is delta connected $Z_{p.u.} = \dfrac{\left(\dfrac{Z}{3} \text{ in ohms}\right) \times (\text{MVA})B}{(\text{kV})^2 B}$

Equivalent resistance in p.u. $= \dfrac{9.07 \times 1}{3 \times (12)^2} = 0.0211\ p.u$ **Ans.**

Equivalent leakage reaction $= 3 \times \dfrac{0.09 \times (12)^2}{1} = 38.88\ \Omega$ **Ans.**

On l.v. side, Total ohmic loss $= 4 + 3 = 7$ kW

Full load current $= \dfrac{1000 \times \sqrt{3}}{3 \times .6} = 962.3$ A

Equivalent resistance in ohms $= \dfrac{7 \times 1000}{(\text{Full load current})^2}$

$= \dfrac{7000}{(962.3)^2} = 0.0076\ \Omega$ **Ans.**

For star connected impedances $Z_{pu} = \dfrac{(Z_{\text{in ohms}}) \times (\text{MVA})B}{(\text{kW})^2 B}$

Equivalent resistance in p.u. $= \dfrac{0.0076 \times 1}{(.6)^2} = 0.0211\ p.u$ **Ans.**

Equivalent leakage reactance in ohms $= \dfrac{0.09 \times (12)^2}{1} = 12.96\ \Omega$ **Ans.**

Problem 1.80 : Three identical single phase transformers are connected with primaries in delta and secondaries in star. Each of the transformer is rated at 100 kVA, 6600/230 V and has an equivalent impedance reffered to h.v. side of $(20 + \sqrt{36})$ ohms. The primary side of the transformer is connected to a 3-ϕ supply through a feeder having resistance and leakage reactance $(4 + j6)$ ohms per phase.

Calculate the line to line voltage at the secondary end of the feeder when the transformer secondary is delivering rated load at 400 V (L – L) at 0.8 lagging.

Solution :

When the single phase transformers are connected into delta-star transformer, primary side line voltage = 6600 V

Secondary side line voltage = $230 \times \sqrt{3} = 400$ V.

Now, transformation ratio $= \left(\dfrac{6600}{400}\right) = \dfrac{33}{2}$.

The connection under loaded condition can be drawn as

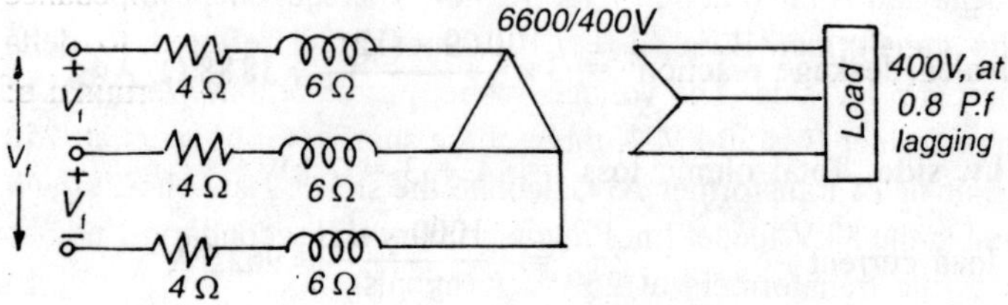

Fig. P. 1.80

H.V. feeder reffered to l.v. side of the transformer

$$Z'_f = (4 + j\,6) \times \left(\frac{2}{33}\right)^2 = 0.0147 + j\,0.0220$$

Transformer impedance reffered to l.v. side is

$$\overline{Z'}_1 = (20 + j\,36) \times \left(\frac{230}{6600}\right)^2 = 0.0243 + j\,0.044$$

Total impedance on l.v. side

$$Z = 0.0243 + j\,0.044 + 0.0147 + j\,0.0220$$

$$= 0.0390 + j0.660 = r_2 + j\,x_2$$

Rated current on l.v. side $I_2 = \dfrac{100000}{230} = 434.78\ \Omega$

Per phase load voltage $= \dfrac{400}{\sqrt{3}} = 230.9$ V

Applying the formula

$$V_s = \sqrt{(V_r \cos\theta + I_2\,r_2)^2 + (V_2 \sin\theta + I_2\,x_2)^2}$$

$$= \sqrt{(230.9 \times 0.8 + 434.78 \times 0.0390)^2 + (230.9 \times 0.6 + 434.78 \times .066)^2}$$

$= 261.99$ V

Line to line voltage at the h.v. side sending end of the feeder

$= 261.99 \times \dfrac{6600}{230} = 7518.1$ V **Ans.**

Problem 1.81 : Three identical single phase-transformers are connected delta-delta. Each of the transformer is rated to 100 kVA, 5000/250 V and has an equivalent impedance reffered to h.v. side of $(1 + j\,9)\ \Omega$ The primary side of this transformer A is fed from a second transformer B through a three phase feeder of leakage impedance $(0.2 + j\ 1)$ ohms/phase. The transformer B is connected star-delta and is rated at 500 kVA 50/5kV. The equivalent impedance of the transformer B is $(2 + j\,10)$ ohms/phase reffered to delta connected 5 kV side. The voltage applied to the primary terminal of the transformer B is 50 kV. A three phase short circuit occurs at 250 V terminals of transformer A. Calculate the steady state short circuit current in the 5 kV feeder lines in the primary and secondary windings of both the transformers at 250 V terminals.

Solution :

This problem can be easily analyzed by drawing circuit diagram according to given situation as below:

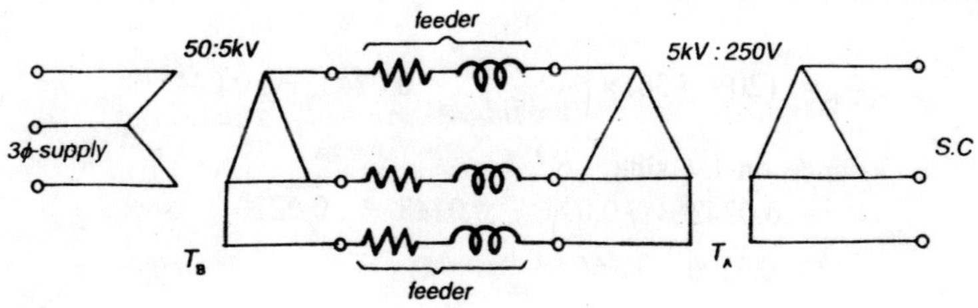

Fig. P. 1.81

For the feeder, Source voltage/phase $= \dfrac{5000}{\sqrt{3}}$

To solve this problem, reffering all the quantities to 5 kV feeder on an equivalent star connection per phase.

For transformer A $\quad \bar{Z}_{eq\,1} = \dfrac{1}{3}(1 + j\,9) = (0.333 + j\,3)\ \Omega$

For transformer B, $\quad \bar{Z}_{eq\,2} = \dfrac{1}{3}(2 + j\,10) = (.667 + j\,3.33)\ \Omega$

Feeder leakage impedance $\bar{Z}_f = (0.2 + j\ 1)\ \Omega\,/\,\text{phase}$

The total impedance between source voltage and the fault point

$$= (0.333 + 0.667 + 0.2) + j(3 + 3.333 + 1)$$
$$= 1.200 + j\,7.333 = (1.200 + j\,7.333)\ \Omega$$
$$= 7.431\angle 80.71°$$

Current through the feeder $= \dfrac{5000}{\sqrt{3}} \times \dfrac{1}{3.431} = 388.47$ A **Ans.**

Current in the 5 kV delta-connected winding on both the transformers A and B.

$$= \frac{388.47}{\sqrt{3}} = 224.29 \text{ A} \quad \textbf{Ans.}$$

Current in the 250 V winding of transformer A

$$= 224.29 \times \frac{5000}{250} = 4485.7 \text{ A} \quad \textbf{Ans.}$$

Current in the 50 kV winding of transformer B

$$= \frac{224.29 \times \sqrt{3}}{10} = 38.85 \text{ A} \quad \textbf{Ans.}$$

Chapter 2

Principles of Electromechanical Energy Conversion

SINGLY-EXCITED MAGNETIC SYSTEM

In this system, there exists only an excitation source as shown in figure.

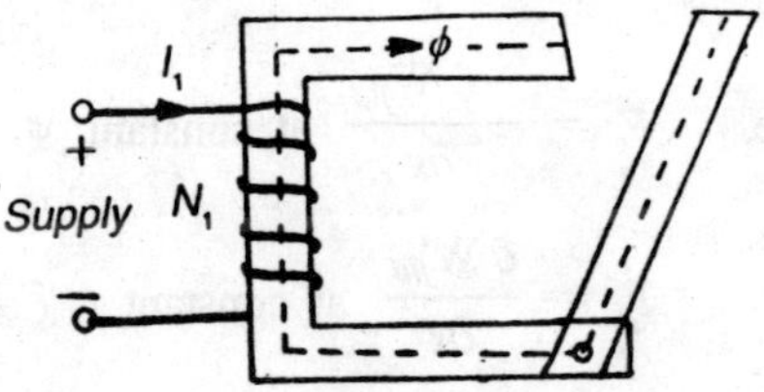

Fig. 2.1

From figure, Field energy,

$$W_{fld} = \int_{\phi_1}^{\phi_2} (\phi)\, d\phi$$

$$= \int_{\lambda_1}^{\lambda_2} i(\lambda)\, d\lambda$$

Co-energy

$$W'_{fld} = \int_0^i i(\lambda)\, d\lambda$$

$$= \int_0^i \psi\, di$$

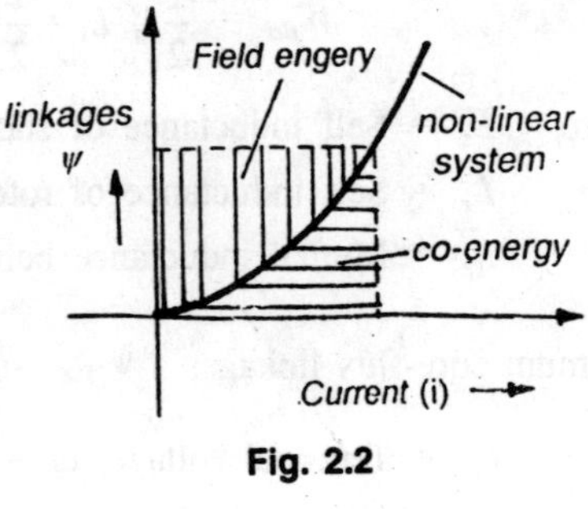

Fig. 2.2

For linear magnetic system.

$$W_{fld} = W'_{fld} = \frac{1}{2}\psi i = \frac{1}{2}F\phi = \frac{1}{2}Li^2 = \frac{1}{2}\phi^2 \Re$$

Where, F = mmf of the coil

$\Re$ = Reluctance.

L = Self-inductance of the coil

Magnetic stored energy density,

$$W_{fld} = \frac{1}{2}HB = \frac{1}{2}\mu H^2 = \frac{1}{2}\frac{B^2}{\mu}$$

Where, H = Magnetic field intensity.

B = Magnetic field

In linear system, Co-energy density,

$$W'_{fld} = W_{fld} = \frac{1}{2}\mu H^2 = \frac{1}{2}\frac{B^2}{\mu} = \frac{1}{2}\text{BHJ/m}^2.$$

Electrical input,

$$W_e = \int_{\psi 1}^{\psi 2} i_1\, d\psi = i_1(\psi_2 - \psi_1) \text{ for constant current,}$$

Mechanical force, $$F_e = \frac{-\partial \text{W}_{fld}}{\partial x} \text{ at constant } \psi,$$

$$F_e = \frac{\partial \text{W}'_{fld}}{\partial x} \text{ at constant } i,$$

Electromagnetic torque,

$$T_e = \frac{\partial W_{fld}(\psi_1, \theta)}{\partial \theta} = \frac{\partial\{W'_{fld}(i_1, \theta)\}}{\partial \theta}$$

For double excited magnetic system,

$$W_{fld} = \frac{1}{2}i_s^2 L_s + \frac{1}{2}i_r^2 \text{L}_r + M_{rs}\, i_s\, i_r$$

Where, L_s = Self inductance of stator

L_r = Self inductance of rotor

M_{sr} = Mutual inductance between rotor and stator

Maximum core-flux linkage, $$\psi_{\max} = \frac{\sqrt{2}\,V_t}{\omega}$$

Where, V_t = Terminal voltage, ω = speed in rad/sec.

SINGLE PHASE MOTORS

Single Phase Induction Motor

The Schematic diagram of a single phase induction motor is shown below

As we know that whenever a single phase ac supply is given to main-winding, a stator mmf is produced which is stationary in space but pulsating in magnitude.

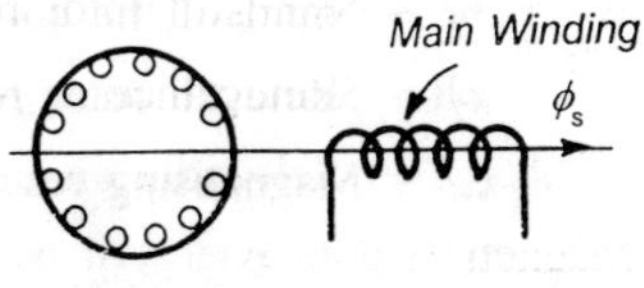

Fig. 2.3

Slip of the rotor current due to backward field,

$$S_b = 2 - s = \frac{n_s + n_r}{n_s}$$

Slip of the rotor current due to forward field,

$$S_f = s = \frac{n_s - n_r}{n_s}, \text{ where } n_s = \frac{120 \times f}{p}$$

n_r = rotor speed (in rpm)

Forward slip frequency, $f_f = s \times f$

Backward slip frequency, $f_b = (2 - s) f,$

Where, f = supply frequency,

Equivalent Circuit of Single Phase Induction Motor

The equivalent circuit of single phase induction motor can be drawn as:

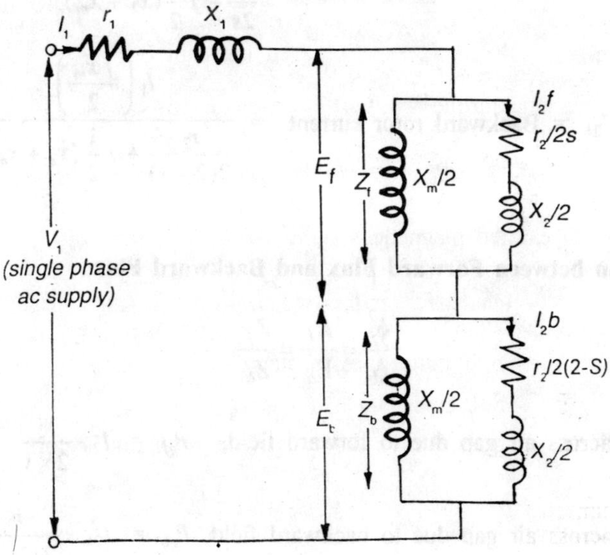

Fig. 2.4

In figure,

r_1 = Stator resistance.

x_1 = Stator leakage reactance.

r_2 = Standstill rotor resistance reffered to stator.

x_2 = Standstill rotor reactance reffered to stator side

x_m = Magnetising reactance.

$$z_f = \text{Forward impedance} = \frac{j\dfrac{x_m}{2}\left[\dfrac{r_2}{2\times s}+\dfrac{j\,x_2}{2}\right]}{\dfrac{r_2}{2s}+j\left[\dfrac{x_2}{2}+\dfrac{x_m}{2}\right]}$$

$$z_b = \text{Backward impedance} = \frac{j\dfrac{x_m}{2}\left[\dfrac{r_2}{2\,(2-s)}+\dfrac{j\,x_2}{2}\right]}{\dfrac{r_2}{2\,(2-s)}+j\left[\dfrac{x_2}{2}+\dfrac{x_m}{2}\right]}$$

V_I = 1-ϕ ac supply.

f_I = Forward induced emf

F_b = Backward induced emf

$$I_{2f} = \text{Forward rotor current} = \frac{I_1\left(j\dfrac{x_m}{2}\right)}{\dfrac{r_2}{2s}+j\dfrac{1}{2}(x_2+x_m)}$$

$$I_{2b} = \text{Backward rotor current} = \frac{I_1\left(\dfrac{j\,x_m}{2}\right)}{\dfrac{r_2}{2(2-s)}+j\dfrac{1}{2}(x_2+x_m)}$$

Relation between Forward Flux and Backward Flux

$$\frac{\phi_f}{\phi_b} = \frac{E_f}{E_b} = \frac{Z_f}{Z_b}$$

Power across air gap due to forward field, $P_{gf} = I_{2f}^2\,\dfrac{r_2}{2\times s}$

Power across air gap due to backward field, $P_{gb} = I_{2b}^2 \times \dfrac{r_2}{2\,(2-s)}$

Forward field torque, $T_f = \frac{P_{gf}}{\omega_s}$

Backward field torque, $T_b = \frac{P_{gb}}{\omega_s}$

Internal torque developed, $T_e = T_f - T_b = \frac{1}{\omega_s}\left[I_{2f}^2 \frac{r_2}{2s} - I_{2b}^2 \frac{r_2}{2(2-5)} \right]$

Net power developed, $P = \omega_r \times T_e$

$$= (1-s) \times \omega_s \times (I_f - T_b)$$

$$P = (1-s)(P_{gf} - P_{gb})$$

Forward field rotor ohmic loss $= sP_{gf}$

Backward field rotor ohmic loss $= (2-s)\, P_{gb}$

Resultant air-gap power $= P_{gf} - P_{gb} = I_1^2 (R_f - R_b)$

Block Rotor Test of Single Phase Induction Motor

The block rotor test yields stator and rotor impedances. The equivalent circuit under block rotor test is given by.

Watt meter reading

$= W_{sc}$

Total impedance,

$$Z_{sc} = \frac{V_{sc}}{I_{sc}}$$

Equivalent resistance,

$$R_{sc} = \frac{W_{sc}}{(I_{sc})^2} = r_1 + r_2$$

$$X_{sc} = \sqrt{Z_{sc}^2 - R_{sc}^2}$$

$$= x_1 + x_2$$

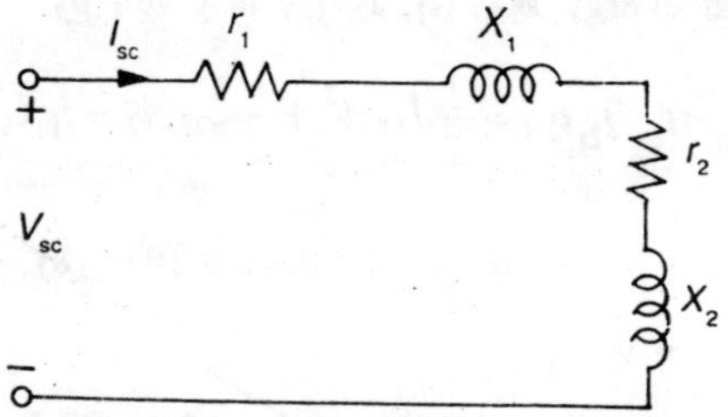

Fig. 2.5

NO-LOAD TEST

Let W_{nl} be watt meter reading

$V_l = V_{nl}$ be voltmeter reading and I_{nl} be ammeter reading

$$\therefore \quad Z_{nl} = \frac{V_1}{I_{ne}}$$

no-load power factor $\cos \theta_{nl} = \dfrac{W_{ul}}{V_1 I_{nl}}$

$$X_{nl} = \left(\sqrt{1-\cos^2 \theta_{nl}}\right) \times X_{nl} = x_1 + \frac{1}{2}(x_2 + x_m)$$

Problem 2.1 : In a doubly-excited rotary machine, the inductance co-efficients are

$L_{11} = (1.1 + 0.4 \cos 2\theta)$;

$L_{22} = (0.03 + 0.005 \cos 2\theta)$

$L_{12} = 0.2 \cos \theta$.

The exciting current are $i_1 = 8$ A and $i_2 = 50$ A.

Obtain the torque/angular displacement relation. Derive the expression used, if any. *[I.A.S., 1994]*

Solution :

Given $L_{11} = (1.1 + 3.4 \cos 2\theta)$

$L_{22} = (0.03 + 0.005 \cos 2\theta)$

$L_{12} = 0.2 \cos\theta$

$i_1 = 8$ A, $i_2 = 50$ A.

Field energy $W_{fld}(i_1, i_2, \theta)$ is given by

$$W_{fld}(i_1, i_2, \theta) = \frac{1}{2} L_{11}, i_1^2 + \frac{1}{2} L_{22}, i_2^2 + i_1 L_{12} \times i_2$$

$$= \frac{1}{2}(1.1 + 0.4 \cos 2\theta) \times (8)^2 + \frac{1}{2}(0.03 + 0.005 \cos 2\theta) \times (50)^2 + 50 \times 8 \times 0.2 \cos\theta$$

$$= 35.2 + 12.8 \cos 2\theta + 37.5 + 6.25 \cos 2\theta + 80 \cos\theta$$

$$= 72.7 + 12.8 \cos 2\theta + 6.25 \cos 2\theta + 80 \cos\theta$$

$$= 72.7 + 19.05 \cos 2\theta + 80 \cos\theta.$$

Torque is given by $T_e = \dfrac{\partial W_{fld}(i_1, i_2, \theta)}{\partial \theta}$

$$= \frac{\partial (72.7 + 19.05 \cos 2\theta + 80 \cos\theta)}{\partial \theta}$$

$$= -38.1 \sin 2\theta - 80 \sin\theta$$

$T_e = -38.1 \sin 2\theta - 80 \sin\theta$ Nm **Ans.**

Problem 2.2 : P. 2.2 shows an electromagnet used to left on iron piece. If N = 1000, i = 1 Amp, x = 0.5 mm A = 1 cm^2, Calculate *(a)* coil inductance *(b)* stored magnetic field energy and *(c)* the force F on the iron piece. ($\mu_0 = 4\pi \times 10^{-7}$ H/m)

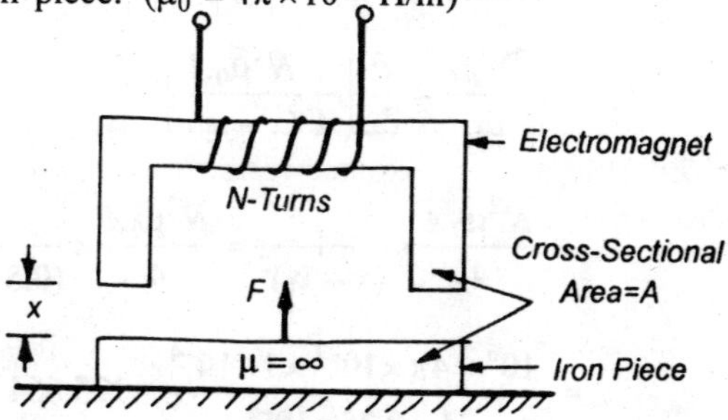

Fig. P. 2.2

Solution :

Given, $N = 1000$,

$$i = 1\ A,\quad x = 0.5\ \text{mm} = 0.5 \times 10^{-3}\ m$$

$$A = 1\ \text{cm}^2 = 1 \times 10^{-4}\ \text{m}^2$$

(a) Reluctance offered by the flux $= \dfrac{2x}{\mu_0 A}$

$$\text{coil inductance } (L) = \frac{(\text{No. of coil turns})^2}{\text{Reluctance}}$$

$$= \frac{(1000)^2}{2 \times 0.5 \times 10^{-3}} \times 4\pi \times 10^{-7} \times 1 \times 10^{-4}$$

$$= 0.125664\ \text{H} \quad \textbf{Ans.}$$

(b) We know that magnetic energy stored $= \dfrac{1}{2} L i^2$

$$= \frac{1}{2} \times 0.125664 \times (1)^2$$

$$= 0.062832\ \text{W-sec} \quad \textbf{Ans.}$$

(c) Let the electromagnet be displaced vertically downward by small distance x_1,

Then length between the air gap $= x - x_1$

Inductance in this situation is given by, $L_x = \dfrac{N^2 \mu_0 A}{2(x - x_1)}$

Field energy stored, $W_{fld} = \dfrac{1}{2} L_x i^2$

Since $i = 1\text{ A}$

$$W_{fld} = \frac{1}{2} L_x = \frac{1}{2} \frac{N^2 \mu_0 A}{2(x - x_1)}$$

Force on the armature is given by

$$F_e = \frac{\partial W_{fld}}{\partial x} = \frac{\partial}{\partial x}\left(\frac{1}{4}\frac{N^2\mu_0 A}{(x - x_1)}\right)$$

$$= \frac{N^2\mu_0 A}{4} \times \frac{1}{(x-x_1)^2} = \frac{N^2\mu_0 A}{4} \times \frac{1}{(0.5\times10^{-3})^2}$$

$$= \frac{10^6 \times 4\pi \times 10^{-7} \times 1 \times 10^{-4}}{4 \times 0.25 \times 10^{-6}} = 125.664 \text{ N} \quad \textbf{Ans.}$$

Problem 2.3 : A double excited rotating machine has following self and mutual inductances.

$$r_s = 40\,\Omega. \qquad L_a = 0.10 \text{ H}$$

$$r_r = 2\,\Omega. \qquad L_r = 0.04 + 0.02 \cos 2\theta_r$$

$$M_{sr} = 0.08 \cos \theta_r$$

where θ_r is the space angle between coil axes. The rotor is revolving at a speed of 100 rad/sec.

For $i_s = 10$ A d.c. and $i_r = 2$ A d.c

(a) derive expressions for the instantaneous voltages applied to the stator and rotor windings.

(b) obtain expressions for magnetic torque and the corresponding electrical power.

For both the parts *(a)* and *(b)*, find their values for $\theta_r = 90°$

Solution :

Given, $r_s = 40\,\Omega$, $L_s = 0.16$ H,

$r_r = 2\,\Omega$, $L_r = 0.04 + 0.02 \cos 2\theta_r$,

$M_{sr} = M_{rs} = 0.08 \cos \theta_r$, $i_s = 10$ A d.c.,

$\omega_r = 100$ rad/sec, $i_r = 2$ A d.c.

(a) We have to calculate v_s and v_r

Now, $$v_s = i_s r_s + \frac{d\psi_s}{dt} \qquad (1)$$

and $\psi_s = L_s\, i_s + M_{sr} \times i_r$

$$= 0.16 \times 10 + 0.08 \times z \times \cos\theta_r$$

$$= 0.6 + 0.16 \times \cos\omega_r \times t$$

$$\therefore \quad \frac{d\Psi_s}{dt} = -0.16 \times \omega_r \sin\omega_r$$

$$= -0.16 \times 100 \sin\theta_r - 16 \sin\theta_r$$

Putting $\frac{d\Psi_s}{dt} = -0.16 \sin\theta_r$ in (1)

$$V_s = 10 \times 40 - 16 \sin\theta_r$$

$$V_s = 400 - 16 \sin\theta_r \quad \textbf{Ans.}$$

Again,

We have $V_r = i_r\, r_r + \frac{d\varphi_r}{dt}$

and $\Psi_r = L_r\, i_r + M_{sr} \times i_s$

$$= (0.04 + 0.02 \cos 2\theta_r) \times 2 + 0.08 \cos 2\theta_r \times 10$$

$$= 0.08 + 0.04 \cos 2\,\omega_r\, t + 0.8 \cos\omega_r\, t$$

$$\therefore \quad \frac{d\Psi_s}{dt} = -0.08 \times \omega_r \sin\, 2\,\omega_r\, t - 0.8 \times \omega_r \sin\omega_r\, t$$

$$= -8 \sin\theta_r - 80 \sin\theta_r$$

$$\therefore \quad V_r = 2 \times 2 - 8 \sin 2\theta_r - 80 \sin\theta_r$$

$$V_r = 4 - 8 \sin 2\theta_r - 80 \sin\theta_r \quad \textbf{Ans.}$$

Where $\theta_r = 90°$

$$V_s = 384 \text{ V}, \quad \textbf{Ans.}$$

$$V_r = 76 \text{ V} \quad \textbf{Ans.}$$

The magnetic torque is given by the relation,

$$T_m = \frac{1}{2} i_s^2 \frac{dL_s}{d\theta_r} + \frac{1}{2} i_r^2 \frac{dL_r}{d\theta_r} + i_s\, i_r \frac{dM_{sr}}{d\theta_r}$$

$$= \frac{1}{2} \times (10)^2 \times \frac{d\,(0.16)}{d\theta_r} + \frac{1}{2} \times (2)^2 \times \frac{d\,(0.04 + 0.02 \cos 2\theta_r}{d\theta_r}$$

$$+ 10 \times 2 \frac{d\,(0.08 \cos\theta_r)}{d\theta_r}$$

$$= \frac{1}{2} \times 100 \times 0 + \frac{1}{2} \times 4\,[0 - 0.04 \sin 2\theta_r] - 20 \times 0.08 \times \sin\theta_r$$

$$T_m = -0.08 \sin 2\theta_r - 1.6 \sin\theta_r \text{ Nm} \quad \textbf{Ans.}$$

When $\theta_r = 90°$, $T_m = 1.6$ Nm **Ans.**

We know that, electric power $= E_a I_a = T_e \times w_r$

$\therefore$ Electric power $= T_m \times 100$

$$= -8 \sin 2\theta_r - 160 \sin\theta_r \quad \textbf{Ans.}$$

When $\theta_r = 90°$

Electric power $= -160$ W **Ans.**

Problem 2.4 : Two air-cored coils have their magnetic axes co-incident. Their parameters are:

$$L_1 = 0.4 \text{ H}, \; L_2 = 0.2 \text{ H},$$
$$M_{12} = M_{21} = 0.1e^{-4x} \text{ H}, \; I_l = 10\,A, \; I_2 = 5\,A$$

where x is the distance between the two coils.

For steady current of $I_1 = 10$ A and $I_2 = 5$ A, calculate the magnetic force developed. Does this force tend to increase or decrease x ? Calculate the magnitude of force for $x = 2.5$ cm.

Solution :

Given, $L_1 = 0.4$ H, $L_2 = 0.2$ H

$M_{21} = M_{12} = 0.1e^{-4x}$ H, $I_1 = 10\,A$, $I_2 = 5$ A,

Since magnetic axis of the coil concides, M_{12} and M_{21} have positive sign

Field energy is given by

$$W_{fld}(i_1, i_2, x) = \frac{1}{2} L_1 i_1^2 + \frac{1}{2} L_2 i_2^2 + M_{12}\, i_1 \times i_2$$

Since i_1 and i_2 are steady currents

$$= \frac{1}{2} L_1 I_1^2 + \frac{1}{2} L_2 I_2^2 + I_1 \times I_2 M_{12}$$

Magnetic force acting between the coil is given by

$$F_e = \frac{\partial W_{fld}}{\partial x} = \frac{1}{2} I_1^2 \frac{\partial L_1}{\partial x} + \frac{1}{2} I_2^2 \frac{\partial L_2}{\partial x} + I_1 I_2 \frac{\partial M_{12}}{\partial x}$$

Since L_1 and L_2 are constants $= 0 + 0 + 10 \times 5 \times 0.1 \dfrac{\partial\left(e^{-4x}\right)}{\partial x}$

$$= 5 \times (-4)\, e^{-4x}$$

$$F_e = -20\, e^{-4x} \text{ N} \quad \textbf{Ans.}$$

Since coupling is positive, i.e. fluxes along the axis of coils are additive; they attract each other, hence x decreases.

At $x = 2.5 \text{ cm} = 2.5 \times 10^{-2}$ m

$$F_e = -20\, e^{-4 \times 2.5 \times 10^{-2}} = -20\, e^{-0.1}$$

$$= \frac{-20}{e^{0.1}} = -18.097 \text{ N} \quad \textbf{Ans.}$$

> **Problem 2.5 :** In a 4 pole, 50 Hz sctorage motor, the emf available across the brushes one pole pitch apart is 64. The secondary emf at standstill is 80 V. Find the no-load speed range.

Solution :

Given, Maximum voltage available across the brushes,

$$E_{j\max} = 64 \text{ V}$$

Secondary emf at standstill,

$$E_2 = 80 \text{ V}, \; k = \frac{E_{j\max}}{E_2} = \frac{64}{80} = \frac{4}{5}$$

The no-load speed range is given by

$$n_{ro} = n_s \left(1 \pm k \sin \frac{\theta}{2}\right)$$

Where $n_s = \dfrac{120 \times 50}{4} = 1500$ rpm

For one pitch coil, $\theta = 180°$

Minimum no-load speed $= 1500\left(1 - \dfrac{4}{5} \sin \dfrac{180°}{2}\right)$

$$= 1500 \times \frac{1}{5} \times 1 = 300 \text{ rpm}$$

Maximum no-load speed $= 1500\left(1 + \dfrac{4}{5} \sin \dfrac{180°}{2}\right)$

$$= 1500 \times \frac{9}{5} = 2700 \text{ rpm}$$

$\therefore$ Speed range is 2700 : 300 or 9 : 1 **Ans.**

Problem 2.6 : A 6 pole, 440 V, 50 Hz, 3-phase star connected shage motor has following data:

Effective secondary turns per phase = 60

Total lap-connected tertiary turns = 200

Calculate the minimum and maximum no-load speeds of this motor for a brush separation 20° mechanical.

Solution :

Given, $\theta_m = 20°$, $Z = 200 \times 2 = 400$, $N_2' = 60$, $A = P = 6$, for lap

In electrical degree, $\theta_e = \frac{P}{2} \times \theta_m = \frac{6}{2} \times 20° = 60°$

Now $k = \frac{Z}{A \pi N_2} = \frac{400}{60 \times \pi \times 6} = 0.354$

Synchronous speed, $n_s = \frac{120 \times f}{P} = \frac{120 \times 50}{6} = 1000$ rpm

The no load speed range of this motor is given by

$$n_{ro} = n_s \left(1 \pm k \sin \frac{60°}{2}\right)$$

Minimum speed $= 1000\left(1 - 0.354 \times \frac{1}{2}\right) = 823$ rpm

Maximum no-load speed $= 1000\left(1 + 0.354 \times \frac{1}{2}\right) = 1177$ rpm

For brush separation of 20° mechanical,
Minimum no-load speed = 823 rpm **Ans.**
Maximum no-load speed = 1177 rpm **Ans.**

Problem 2.7 : A 6 pole, 50 Hz schrage motor, running at no-load given minimum and maximum speeds of 600 rpm and 3000 rpm respectively, for brush separation of 180° electrical. Find the brush separation in mechanical and electrical degrees for no-load speeds of *(a)* 500 rpm, and *(b)* 1500 rpm

Solution :

Given, $p = 6$, $f = 50$ Hz, $\eta_{ro \cdot \min} = 600$ rpm (1)

$\eta_{ro \cdot \max} = 3000$ rpm, (2)

$\theta_e = 180°$

Synchronous speed, $n_s = \dfrac{120 \times f}{P} = \dfrac{120 \times 50}{6}$ 1000 rpm

Now, From (1) and (2) $= \dfrac{n_s\left(1 - k\sin\dfrac{180°}{2}\right)}{n_s\left(1 + k\sin\dfrac{180°}{2}\right)} = \dfrac{600}{3000} = \dfrac{1}{5}$

$$\Rightarrow \frac{1-k}{1+k} = \frac{1}{5} \text{ or}$$

$$5 - 5k = 1 + k \quad \text{or} \quad 6k = 4$$

$$\therefore \quad k = \frac{2}{3}$$

(a) when rotor speed is 500 rpm, i.e. less than synchronous speed 1000 rpm, then no load speed in equal to

$$500 = 1000\left(1 - \frac{2}{3}\sin\frac{\theta}{2}\right) \text{ or } \frac{1}{2} = 1 - \frac{2}{3}\sin\frac{\theta}{2}$$

$$\frac{2}{3}\sin\frac{\theta}{2} = \frac{1}{2} \text{ or } \sin\frac{\theta}{2} = \frac{3}{4} \text{ or } \frac{\theta}{2} = \sin^{-1}\left(\frac{3}{4}\right) = 48.59°$$

$$\therefore \quad \theta_e = 97.18° \textbf{ Ans.}$$

In mechanical degree,

$$\theta_m = \frac{2}{p} \times \theta_e = \frac{2}{6} \times 97.18° = 32.39° \textbf{ Ans.}$$

(b) When the speed is 1500 rpm alove synchronous speed,

we have $1500 = 1000\left(1 + k\sin\dfrac{\theta}{2}\right)$ or $\dfrac{1500}{1000} = 1 + \dfrac{2}{3}\sin\dfrac{\theta}{2}$

or $\dfrac{2}{3}\sin\dfrac{\theta}{2} = \dfrac{3}{2} - 1 = \dfrac{1}{2}$ or $\sin\dfrac{\theta}{2} = \dfrac{3}{4}$

$$\therefore \quad \theta_e = 97.18° \text{(electrical)} \textbf{ Ans.}$$

$$\theta_m = 32.39° \text{(mechanical)} \textbf{ Ans.}$$

Problem 2.8 : A single phase 240 V, 50 Hz, 2 pole inductively compensated series motor has the following constants series field winding,

$$r_f = 4\,\Omega,\ L_f = 0.04\text{ H}$$

Armature winding $r_a = 4\,\Omega,\ L_a = 0.04\text{ H}$

Compensating winding, $r_a = 0\,\Omega,\ L_a = 3\text{ H},\ M_q = 1.2\text{ H}$

At full load speed of 4500 rpm, the motor takes 3 A. Find the power factor, power output and torque developed at the same speed,

Solution :

Given, $f = 50\text{ Hz},\ P = 2\text{ pole},\ N_{fl} = 4500\text{ rpm}$

$$L_q = 3\text{ H},\ M_q = 1.2\text{ H},\ r_q = 0,\ r_a = 2\,\Omega$$

$$L_a = 1.00\text{ H},\ r_f = 4\,\Omega,\ L_f = 0.04\text{ H}$$

Synchronous speed $= \dfrac{120 \times 50}{2} = 3000\text{ rpm}$

Here $v = \dfrac{4500}{3000} = 1.5$

Armature transient inductance,

$$L'_a = \left(L_a - \frac{M_q^{\,2}}{L_q}\right) = 1 - \frac{(1.2)^2}{3} = 0.52\text{ H}$$

To Calculate M_d, Applying the relation

$$\text{v} = \left[r_a + r_f + M_d\,\omega_r + j\left(L_f + L'_a\right)2\,\pi f\right] \times 3$$

$$240 = \left[2 + 4 + M_d\,\omega_r + j\,(0.04 + 0.52) \times 2 \times 3.14 \times 50\right] \times 3$$

or $3\,M_d\,\omega_r = 222 - j175.84 \times 3$

or $M_d\,\omega_r = 74 - j175.84$

$$M_d\,\omega_r = \sqrt{(74)^2 + (175.84)^2} = 190.78$$

$\therefore$ $M_d = \dfrac{190.78 \times 60}{2\,\pi \times 4500} = 0.405\text{ H}$

The power factor is given by

$$\text{P.f.} = \frac{r_a + r_f + \text{v}\omega M_d}{\sqrt{\left(r_a + r_f + \text{v}\,\omega\,M_d\right)^2 + \left[314 \times \left(L_a + L_f\ 5\right)\right]^2}}$$

$$= \frac{2 + 4 + 1.5 \times 2\,\pi \times 50 \times 0.405}{\sqrt{(2 + 4 + 1.5 \times 2\,\pi \times 50 \times 0.405)^2 + \left[314 \times (0.52 + 0.04)\right]^2}}$$

$$= \frac{196.76}{\sqrt{(196.76)^2 + (175.84)^2}} = 0.7456 \quad \text{p.f} = 0.7456 \quad \textbf{Ans.}$$

Torque developed, $= M_d I^2 = 9 \times (.405)^2 = 1.476$ Nm **Ans.**

Power output $= T_e \,.\, \omega_r = T_e \times \nu \times \omega = 1.476 \times 1.5 \times 314 = 695.3$ W **Ans.**

> **Problem 2.9 :** The constants of an universal motor obtained from the tests with rotor standstill are
> Total resistance = $15\,\Omega$, Total reactance = $90\,\Omega$
> when 220 V d.c. is given to motor, it runs at $7200/\pi$ rpm and draws a no-load current of 3 A. Find the speed, power factor and output when the motor is connected to 220 V a.c. and takes 3 A again.

Solution :

Given, $R = 15\,\Omega, \quad X = 90\,\Omega,$

$$n_r = \frac{7200}{\pi} \text{ rpm}, I_0 = 3 \text{ A (d.c.)}, I_0 = 3 \text{ A (a.c.)}$$

Speed of rotor $= \dfrac{7200}{\pi} \text{ rpm} = \dfrac{7200 \times \pi}{60 \times \pi} = 240$ rad/sec

To Calculate M_d, Applying the relation, $V = I\left(R + M_d\,\omega_r\right)$

$$\therefore \quad M_d = \left[\frac{V}{I} - R\right] \times \frac{1}{\omega_r} = \left[\frac{220}{3} - 15\right] \times \frac{1}{240} = 0.243 \text{ H}$$

For AC operation $\bar{V} = \left(R + M_d\,\omega_r + j\,X\right)\bar{I}$

or $220 = (15 + 0.243\,\omega_r + j\,90) \times 3$

or $\dfrac{220}{3} = (15 + 0.243\,\omega_r) + j\,20$

or $\left(\dfrac{220}{3}\right)^2 = (15 + 0.243\,\omega_r)^2 + (20)^2$

or $(15 + 0.243\,\omega_r)^2 = \left(\dfrac{220}{3}\right)^2 - 20^2$

or $(15 + 0.243\,\omega_r)^2 = 4977.78$

or $$15 + 0.243\,\omega_r = 70.55$$

or $$0.243\,\omega_r = 55.55$$

$$\therefore \quad \omega_r = \frac{55.55}{0.243} = 228.6 \text{ rad/sec}$$

$$\text{speed} = 228.6 = \frac{228.6 \times 60}{2\pi} \text{ rpm} = 2184.1 \text{ rpm}$$

Here, $\nu = 2184.1$

$$\text{Power factor} = \frac{R + VX_{md}}{(R + \text{V}\,X_{md})^2 + X^2}$$

Here, $$\text{V}\,X_{md} = \frac{\omega_r}{\omega} \times 2\pi f \times M_d$$

$$= \frac{\omega_r}{\omega} \times \omega \times M_d = 228.6 \times .243 = 55.55$$

$$\text{Now p.f.} = \frac{55.55 + 15}{\sqrt{(15 + 55.55)^2 + (90)^2}} = \frac{70.15}{\sqrt{(70.15)^2 + (90)^2}} = 0.615 \quad \textbf{Ans.}$$

Torque developed, $T_d = M_d \times I^2 = 9 \times .243 = 2.187$ Nm

Power output = Torque $\times\, \omega_r = 2.187 \times 228.6 = 499.95$ W **Ans.**

Problem 2.10 : A 220 V, 400 W, 3000 rpm, 50 Hz single-phase series motor has total resistance $2\,\pi$ and total reactance $16\,\Omega$. For a stray power loss of 30 W, determine the current and p.f. when working under rated conditions.

Solution :

Given, $r = 2\,\Omega,\ X = 16\,\Omega,\ P_s = 20$ W

Now, Internal mechanical power developed in armature.

$$P_a = \text{Shaft power output + stray losses}$$

$$P_a = 400 + 20 = 420 \text{ W}$$

Since $$P_a = E_{rad} \times I$$

$$\therefore \quad E_{rad} = \frac{P_a}{I}$$

To calculate current, Applying

$$V_t^2 = (I\,X)^2 + (r\,I + E_{rad})^2$$

or $$V_t^2 = (I\,X)^2 + \left(r\,I + \frac{P_a}{I}\right)^2$$

or $$I_a^2\,V_t^2 = I^4\,X^2 + \left(r\,I^2 + P_a\right)^2$$

or $$V_t^2 \times I^2 = I^4\,X^2 + r^2\,I^4 + 2\,P_a r\,I^2 + P_a^2$$

or $$\left(X^2 + r^2\right)I^4 + \left(2P_a \times r - V_t^2\right)I^2 + P_a^2 = 0$$

$$(256+4)\,I^4 + \left(2\times 420\times 2 - (220)^2\right)I^2 + (220)^2 = 0$$

or $$260\,I^4 - 46720\,I^2 + 48400 = 0$$

or $$I^4 - 179.7\,I^2 + 186.15 = 0$$

This yields, $$I^2 = \frac{179.7 \pm \sqrt{(179.7)^2 - 4\times 186.15}}{2} = \frac{179.7 \pm 177.82}{2}$$

Neglecting higher values, $I^2 = 0.941$

$\therefore$ $$I = 0.97\text{ A}$$

Now $$\therefore \sin\theta = \frac{IX}{V_t} = \frac{0.97\times 16}{220} = 0.071$$

$\therefore$ $$\text{p.f.} = \sqrt{1-\sin^2\theta} = \sqrt{1-(0.071)^2}$$

$$= 0.997 \text{ lagging } \textbf{Ans.}$$

Problem 2.11 : A universal series motor when operating on 240 V dc draws 12 A and runs at 1500 rpm. Find the new speed and p.f. when connected to 240 V, 50 Hz supply, the motor current remaining same. The motor has total resistance of 1 Ω and total inductance 0.02 H.

Solution :

Give, $r = 1\,\Omega,\ L = 0.2\text{ H},\ I = 12\text{ A},\ N_r = 1500\text{ rpm},\ f = 50\text{ Hz}$

When working on d.c. $V_t = I_r + CIn$ or $240 = 12\times 1 + CI\times 1500$

or $$CI = \frac{228}{1500} = \frac{19}{125} \text{ V/rpm}$$

Total reactance of the motor $X = 2\pi fl = 314 \times 0.2 = 6.28$

When working on ac, Current remain same

$$V_t = I(r + jx) + CIn$$

or $$240 = 12(1 + j\,6.2 + 8) + CIn$$

or $$240 = (CIn + 12) + 6.28 \times 12$$

or $$240 = (CIn + 12) + j\,75.36$$

or $$\therefore (240)^2 = (CIn + 12)^2 + (75.36)^2$$

or $$CIn + 12 = 227.86$$

or $$CIn = 215.86$$

$\therefore$ $$n = \frac{215.86}{CI} = \frac{215.86 \times 125}{19} = 1499 \text{ rpm}$$

New speed = 1499 rpm **Ans.**

$$\text{p.f.} = \frac{CIn + I_r}{V_t} = \frac{215.86 + 12}{240} = \frac{227.86}{240}$$

$$= 0.9494\,(\text{lagging})$$ **Ans.**

Problem 2.12 : A universal motor gave the following test results

	Input current	*Voltage across field winding*	*Voltage across armature winding*	E_{rad} / ω_r
DC Test	3 A	16 V	8V	0.30 V-s/rad
AC Test	3 A	111 V	18 V	0.26 V-s/rad

In case motor current is 3 A at 240 V, estimate motor speed and torque when it is energized from *(a)* dc source, and *(b)* ac source

Solution :

For D.C. test $I = 3$ A, $V_t = I\left(r_a + r_f\right) + E_{rad}$

$$240 = 16 + 8 + 0.30 \times \omega_r$$

or $$0.30\,\omega_r = 218$$

$\therefore$ $$\omega_r = \frac{218}{0.30} = 726.67 \text{ rad/sec}$$

$$\text{Speed in r.p.m.} = \frac{726.67 \times 60}{2\pi} = 6942.7 \text{ rpm} \quad \textbf{Ans.}$$

$$\text{Torque} = \frac{T_{rad}}{\omega_r} \times I = 0.30 \times 3 = 0.90 \text{ Nm} \quad \textbf{Ans.}$$

(b) For A.C. operation, $I_{ra} = 8$, $\therefore r_a = \frac{8}{3} = 2.67\ \Omega$

$$I_{rf} = 16, \quad \therefore r_f = \frac{16}{3} = 5.34\ \Omega$$

Now, Voltage across armature winding = 18

or $$I\sqrt{r_a^2 + X_a^2} = 18$$

or $$\sqrt{r_a^2 + X_a^2} = 6$$

or $$r_a^2 + X_a^2 = 36$$

$\therefore$ $$x_a = \sqrt{36 - (2.67)^2} = 5.37\ \Omega$$

And, Voltage across field winding = 111 V

or $$I \times \sqrt{R_f^2 + X_f^2} = 111$$

or $$\sqrt{R_f^2 + X_f^2} = \frac{111}{3} = 37$$

or $$R_f^2 + X_f^2 = (37)^2$$

or $$X_f = \sqrt{(37)^2 - (R_f)^2} = \sqrt{(37)^2 - (5.34)^2} = 36.61\ \Omega$$

To Calculate speed, Applying the relation,

$$V_t^2 = \left[E_{rad} + I\left(r_a + r_f\right)\right]^2 + \left[I\left(X_a + r_f\right)\right]^2$$

or $$(240)^2 = \left[0.26 \times \omega_r + 3\,(8.01)\right]^2 + \left[3 \times (5.37 + 36.61)\right]^2$$

or $$(0.26 \times \omega_r + 24.03)^2 = 41739.12$$

or $$0.26 \times \omega_r + 24.03 = 204.3$$

or $$0.26 \times \omega_r = 204.3 - 24.03 = 180.27$$

$$\therefore \quad \omega_r = \frac{180.27}{0.26} = 693.4 \text{ rad/sec}$$

$$\text{Speed} = \frac{693.4 \times 60}{2 X Z} \text{ rpm} = 6624.38 \text{ rpm} \quad \textbf{Ans.}$$

$$\text{Torque} = \frac{E_{rad}}{\omega_r} \times I = 0.26 \times 3 = 0.78 \text{ Nm} \quad \textbf{Ans.}$$

Problem 2.13 : A 2-pole a.c. commutator machine has the following data:

Armature turns = 200

Armature speed = 1500 rpm

Frequency and maximum value of pubating flux = 50 Hz and 0.01 wb respectively.

Find the magnitude and frequency of the emf appearing at brushes if the brush axis is

(a) 90° away from the field axis

(b) Aligned along the field axis

Solution :

Given, $Z = 200 \times 2 = 400$, $N_r = 1500$ rpm, $A = P = 2$

(a) As we know that when the brush axis is 90° away from the field, transformer emf is zero while rotational emf appears.

The magnitude of rotational emf is given by

$$E_r = \sqrt{2}\, f_r\, \phi_m \frac{Z}{A} \text{ volts} = \sqrt{2} \times \frac{np}{2} \times \phi_m \times \frac{Z}{A}$$

$$= \sqrt{2} \times \frac{1500 \times 2}{2 \times 60} \times 0.01 \times \frac{400}{2} = 70.71 \text{ V} \quad \textbf{Ans.}$$

Frequency of emf = frequency of field–flux frequency

= 50 Hz **Ans.**

(b) When the brush axis is aligned along the field axis rotational emf is zero while transformer emf appears.

$$\text{Magnitude of transformer emf} = \sqrt{2}\, f\, \phi_m \frac{Z}{A}$$

$$= \sqrt{2} \times 50 \times 0.01 \times \frac{400}{2} = 141.42 \text{ V } \textbf{Ans.}$$

Frequency of the emf = 50 Hz **Ans.**

> **Problem 2.14 :** A 4-pole rotor carries a lap-connected winding with full-pitched coils. The armature, with 340 turns, rotates in a constant sine-distributed d-axis flux of 0.02 wb. One end of the winding is tapped at 4 equidistant points and connected to 4 slip rings. At the other end, a pair of commutator brushes is placed in the q-axis. Compute the magnitude and frequency of emfs appearing across the q-axis brushes and between adjacent slip rings if the armature rotates at *(i)* 1800 rpm *(ii)* 1200 rpm.

Solution :

Here, $Z = 340 \times 2 = 680$ Conductors

(i) Since flux is constant, the brush emf is given by

$$E_r = \frac{\phi n Z P}{60 A} = \frac{0.02 \times 1800 \times 680 \times 4}{60 \times 4} = 40.8 \text{ V at zero Hz}$$

For 4 equi distant taps, the winding is equivalent to 4-phase winding with a phase speed of $\sigma = \frac{2\pi}{4} = \frac{2 \times 180}{4} = 90°$

Now, Distribution factor $(k_d) = \dfrac{\sin \frac{90°}{2}}{\frac{\pi}{4}} = \dfrac{4}{\sqrt{2}\,\pi} = \dfrac{2\sqrt{2}}{\pi}$

For full pitch winding, the winding factor is $k_w = \frac{2\sqrt{2}}{\pi} \times 1 = \frac{2\sqrt{2}}{\pi}$

Emf appearing between adjacent slip rings, $E_{ph} = \sqrt{2}\, Z\, f_r\, N_{ph} \times \phi \times k_w$

$$= \sqrt{2} \times 3.14 \times \frac{4 \times 1800}{2 \times 60} \times \frac{340}{4} \times 0.02 \times \frac{2\sqrt{2}}{\pi} = 408 \text{ V}$$

Frequency of the slip ring end $\frac{P_h}{2} = \frac{4 \times 1800}{2 \times 60} = 60$ Hz **Ans.**

(ii) When the armature rotates at 1200 rpm

Brush emf $E_r = 40.8 \times \frac{1200}{1800} = 27.2$ V at zero Hz and $$E_{ph} = 408 \times \frac{1200}{1800} = 272 \text{ V at } \left(60 \times \frac{2}{3}\right) \text{Hz}$$

$$= 272 \text{ V at } 40 \text{ Hz } \textbf{Ans.}$$

Problem 2.15 : An induction motor, with negligible stator resistance develops a maximum torque of 12 Nm at a negative slip of 0.5 when balanced 2-phase 210 V are applied to its symmetrical 2-phase stator windings.
This motor is now used as a 2-phase control motor with reference voltage of 210 V in time quadrature with control voltage of *(a)* 210 V, *(b)* 105 V. Find the torque at slip 0.4.

Solution :

Given, $s_{\max T} = 0.5,\ T_m = 12 \text{ Nm}$

For balanced two phase motor during running, we may apply the relation

$$\frac{T_e}{T_m} = \frac{2}{\dfrac{s_{\max T}}{s} + \dfrac{s}{s_{\max T}}}$$

(a) When control voltage is 210 V, the control motor runs as a balanced 2-phase motor.

For a slip 0.4, $$\frac{T_e}{T_m} = \frac{2}{\dfrac{s_{\max T}}{s} + \dfrac{s}{s_{\max T}}}$$

or $$\frac{T_e}{12} = \frac{2}{\dfrac{0.5}{0.4} + \dfrac{0.4}{0.5}} = \frac{2}{1.25 + .8} = \frac{2}{2.05}$$

$$\therefore \quad T_e = \frac{2 \times 12}{2.05} = 5.85 \text{ Nm } \textbf{Ans.}$$

(b) When control voltage is 105 V, then positive sequence voltage

$$V_{r1} = \frac{1}{2}(210 + 105) = \frac{315}{2} = 157.5 \text{ V}$$

Negative sequence voltage $V_{r2} = \frac{1}{2}(210 - 105) = \frac{105}{2} = 52.5 \text{ V}$

For a positive sequence of voltage 157.5 V, torque at slip 0.4 is

given by $T_1 = 5.85 \times \left(\frac{157.5}{210}\right)^2 = 3.29 \text{ Nm}$

The slip for positive sequence flux is 0.4. The slip for negative sequence flux is $2 - s = 2 - 0.4 = 1.6$. Negative sequence torque T_3 at slip 1.6 for 210 V as the control voltage is given by

$$\frac{T_3}{T_m} = \frac{2}{\frac{s_{\max T}}{1.6} + \frac{1.6}{s_{\max T}}} = \frac{2}{\frac{0.5}{1.6} + \frac{1.6}{0.5}} = 0.57$$

$$\therefore \quad T_3 = 0.57 \times 12 = 6.83 \text{ Nm}$$

Let T_2 be the torque at negative sequence voltage 52.5 V.

$$T_2 = 6.83 \times \left(\frac{52.5}{210}\right)^2 = 0.427 \text{ Nm}$$

$\therefore$ Net torque at slip $0.4 = T_1 - T_2 = 3.29 - 0.427 = 2.863$ Nm **Ans**

> **Problem 2.16 :** A 2-phase, 4-pole servo-motor has the following parameters at 50 Hz $r_1 = 350\ \Omega$, $r_2 = 200\ \Omega$
> $x_1 = x_2 = 40\ \Omega$, $r_m = 1000\ \Omega$
> This motor is operated with 110 V across the reference winding and 70 V across the control winding, the latter leading by former by 90°,
> Calculate the stator current, output power and motor torque at a slip of 0.4. The rotational losses are 0.6 W. Both the windings have equal number of turns.

Solution :

Given, $r_1 = 350\ \Omega, r_2 = 200\ \Omega,\ x_1 = x_2 = 40\ \Omega, x_m = 1000\ \Omega$
$V_r = 110$ V, $V_c = 70$ V, $\theta = 90°$, $P_{wf} = 0.6$, slip = 0.4

Now, synchronous speed

$$N_s = \frac{120 \times 50}{4} = 1500 \text{ rpm} = \frac{1500 \times 2}{60} = 50\,\pi \text{ rad/sec}$$

Positive sequence component of applied voltage

$$\bar{V}_{r1} = \frac{1}{2}(110 + 70) + j\,0 = 90 + j\,0 \text{ volts}$$

Negative sequence component of applied voltage

$$\bar{V}_{r2} = \frac{1}{2}(110 - 70) + j\,0 = 20 + j\,0 \text{ volts}$$

Impedance offered to positive sequence voltage

$$Z_1 = r_1 + j x_1 + \frac{j X_m\left(\frac{r_2}{s} + x_2\right)}{\frac{r_2}{s} + j\left(x_m + x_2\right)}$$

$$= 350 + j40 + \frac{j \times 1000\left(\frac{200}{0.4} + j40\right)}{\frac{200}{0.4} + j\left(1000 + 40\right)}$$

$$= 350 + j40 + \frac{1000\angle 90° \times 501.6\angle 4.57°}{1153.95\angle 64.32°}$$

$$= 350 + j40 + 438.68\angle 30.25° = 728.95 + j\,261$$

$$= 774.27\angle 19.7°$$

∴ Positive sequence current

$$\bar{I}_{r1} = \frac{V_{r1}\angle 0°}{Z_1} = \frac{90\angle 0°}{774.27\angle 19.7°} = 0.116\angle -19.7°$$

For negative sequence, slip = 1.6
Impedance offered to negative sequence voltage,

$$\bar{Z}_2 = 350 + j40 + \frac{j\,1000\left(\frac{200}{1.6} + j40\right)}{\frac{200}{1.6} + j\left(1040\right)}$$

$$= 350 + j40 + \frac{1000\angle 90° \times 131.24\angle 17.74°}{1047.49\angle 83.12°}$$

$$= 350 + j40 + 125.29\angle 24.62° = 463.9 + j\,92.2$$

$$= 472.97\angle 11.03°$$

Negative sequence current

$$\bar{I}_{r2} = \frac{20\angle 0°}{472.97\angle 11.03} = 0.042\angle -11.03°$$

Phase current is reference winding, $\bar{I}_r = \bar{I}_{r1} - \bar{I}_{r2}$

$$= 0.1092 - j\,0.039 + 0.0412 - j\,0.008$$

$$= 0.1504 - j\,0.047 + 0.1576\angle -17.35°$$

Phase current in control winding, $\bar{I}_c = \bar{I}_{r1} - \bar{I}_{r2}$

$$= 0.1092 - j\,0.039 - 0.0412 + j\,0.008$$

$$= 0.068 - j\,0.031 = 0.075 \angle 24.51°$$

Now, phase sequence gap power

$$= 2 \times \bar{I}_{r1}^{\ 2} \times R_{f1} = 2 \times (0.116)^2 \times 318.9 = 10.2 \text{ W } \textbf{Ans.}$$

Negative sequence gap power

$$P_{g2} = 2 \times I_{r2}^{\ 2} \times r_{f2} = 2 \times (0.042)^2 \times 113.9 = 0.402 \text{ W}$$

Resultant air gap power,

$$P_g = P_{g1} - P_{g2} = 10.2 - 0.402 = 9.798 \text{ W}$$

Internal mechanical power developed

$$P = (1 - s)(P_{g1} - P_{g2})$$

$$= (1 - 0.4) \times 9.798 = 0.6 \times 9.796 = 5.879 \text{ W}$$

$\therefore$ Power output $= P - P_{wf} = 5.879 - 0.6 = 5.279$ W

$\therefore$ Shaft torque $= \dfrac{P_{sh}}{\omega_r} = \dfrac{P_{sh}}{\omega_s (1 - S)}$

$$= \frac{5.279}{0.6 \times 2\,\pi \times 25} = 0.0624 \text{ Nm } \textbf{Ans.}$$

Problem 2.17 : The following data pertains to a 220 V, 50 Hz capacitor-start single phase induction motor at standstill.

Main winding excited alone: 100 V, 2 A, 30 W

Auxiliary-winding excitede alone: 80 V, 1 A, 40 W

Determine the volue of capacitance for obtaining the maximum starting torque.

Solution :

Given, For main windings,

$$V_m = 100 \text{ V}, I_m = 2 \text{ A}, P_m = 30 \text{ W}$$

$$Z_m = \frac{100}{2} = 50\,\Omega, r_m = \frac{30}{(2)^2} = \frac{30}{4} = 7.5\,\Omega$$

$$\therefore \quad X_m = \sqrt{Z_m^{\ 2} - r_m^{\ 2}} = \sqrt{(50)^2 - (7.5)^2} = 49.43\,\Omega$$

For auxiliary winding, $V_a = 80\text{ V}, I_a = 1\text{ A}, P_a = 40\text{ W}$

$$\therefore \quad Z_a = \frac{V_a}{I_a} = \frac{80}{1} = 80\,\Omega,\ r_a = \frac{40}{I_a^2} = 40\,\Omega$$

$$\therefore \quad X_a = \sqrt{Z_a^2 - r_a^2} = \sqrt{(80)^2 - (40)^2} = 69.28\,\Omega$$

To obtain the maximum torque,

$$X_c = X_a + \frac{r_a \times r_m}{X_m + Z_m} = 69.28 + \frac{40 \times 7.5}{49.43 + 50} = 72.30\,\Omega$$

We know that, $X_c = \dfrac{1}{2\,\pi f c}$

$$\therefore \quad c = \frac{1}{2\pi f\, X_c} = \frac{1}{2 \times 3.14 \times 50 \times 72.30} = 44.05\,\mu\text{F}$$

$\therefore$ Capacitance $= 44.05\ \mu$ F **Ans.**

Problem 2.18 : A 220 V, 2 pole, 50 Hz, single phase induction motor has the following constants and losses:

Stator resistance	$= 2\,\Omega$
Stator leakage reactance	$= 3\,\Omega$
Rotor resistance reffered to stator side	$= 4\,\Omega$
Rotor leakage reactance reffered to stator side	$= 3\,\Omega$
Magnetising reactance	$= 70\,\Omega$
Core loss = 10 W, Friction and windage loss	= 20 W

The following data relates to tests on a 120 volts, 160 W, 50 Hz, 4-pole single phase induction motor:

No-load test: 120 V, 60 W, 2.5 ampere

Blocked rotor test: 60 V, 120 W, 5 A

The stator winding resistance is 2.4 ohms and during the block rotor test, the stating winding is open.

Determine the equivalent circuit parameter. Also, find the core, windage and friction losses.

Solution :

Given, For Blocked rotor test,

$$V_{sc} = 60\text{ V},\ P_{sc} = 120\text{ W},\ I_{sc} = 5\text{ A}$$

$$\therefore \qquad Z_{sc} = \frac{V_{sc}}{I_{sc}} = \frac{60}{5} = 12\ \Omega$$

$$\therefore \qquad R_{sc} = r_1 + 2\left(\frac{r_2}{2}\right) = \frac{P_{sc}}{(I_{sc})^2} = \frac{120}{25} = 4.8\ \Omega$$

This yields $\qquad r_2 = 4.8 - r_1 = 4.8 - 2.4 = 2.4\ \Omega$

and $\qquad X_{sc} = x_1 + x_2 = \sqrt{(12)^2 - (4.8)^2} = 10.998\ \Omega$

$$\therefore \qquad x_1 = x_2 = \frac{X_{sc}}{2} = 5.498\ \Omega$$

From no-load test, $\cos\theta_{nl} = \frac{60}{120 \times 2.5} = \frac{1}{5}$

$$\therefore \qquad \sin\theta_{nl} = \sqrt{1 - \left(\frac{1}{5}\right)^2} = \frac{\sqrt{24}}{5}$$

$$\because \qquad X_{nl} = Z_{nl}\sin\theta_{nl} = \frac{60}{2.5} \times \frac{\sqrt{24}}{5} = 23.52\ \Omega$$

Since $\qquad X_{nl} = x_1 + \frac{1}{2}(x_2 + x_m) = 23.52$

or $\qquad \frac{1}{2}(x_2 + x_m) = 23.52 - 5.498$

$$\therefore \qquad X_m = 30.54\ \Omega$$

The equivalent circuit can be drawn as

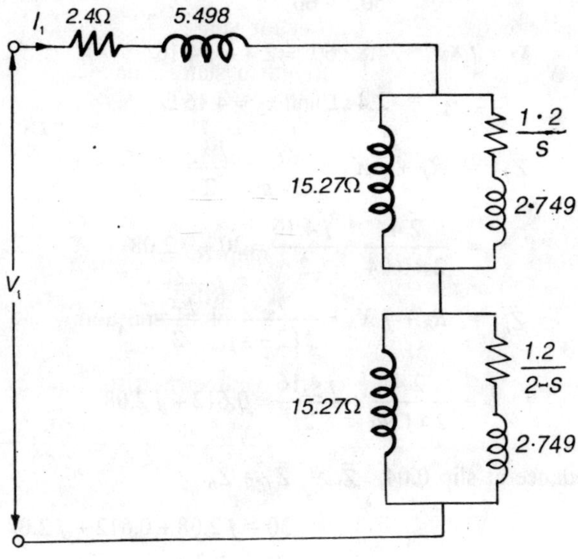

Fig. P. 2.18

The core loss, windage loss and friction is given by

$$= W_{nl} - I_{nl}^{2}\left(r_1 + \frac{r_2}{4}\right) = 60 - (2.5)^2\left(2.4 + \frac{2.4}{4}\right)$$

$$= 60 - (2.5)^2 \times 3 = 41.25 \text{ W } \textbf{Ans.}$$

> **Problem 2.19 :** A single phase, 6-pole, induction motor, takes a line current $50\angle -60°$ at standstill with its main winding excited from 240 V, 50 Hz, source. Neglecting stator impedance, magnetising current and rotational loss. Compute the torque at a slip of 0.04.

Solution :

It is given that stator leakage impedance and magnetising reactance are negligible. Hence, single-phase induction motor equivalent circuit at standstill gives

$$\bar{Z}_2 = \frac{\bar{V}_t}{\bar{I}_L},$$

and $\omega_s = \dfrac{120 \times f}{P} = \dfrac{120 \times 50}{6} = \dfrac{1000 \times 2\pi}{60} = \dfrac{100\pi}{3}$ rad/sec

or $x_2 + j\,x_2 = \dfrac{240\angle 0°}{50\angle -60°}$

or $x_2 + j\,x_2 = 4.8\angle 60° = 2.4 + j\,4.16$

$\therefore$ $r_2 = 2.4\ \Omega$, and $x_2 = 4.16\ \Omega$

Now $\bar{Z}_f = R_f + j\,X_f = \dfrac{r_2}{2 \times s} + \dfrac{j\,x_2}{2}$

$$= \frac{2.4}{2 \times 0.04} + \frac{j\,4.16}{2} = 30 + j\,2.08$$

and $\bar{Z}_b = R_b + j\,X_b = \dfrac{r_2}{2\,(2-s)} + \dfrac{j\,x_2}{2}$

$$= \frac{2.4}{2 \times 1.96} + \frac{j\,4.16}{2} = 0.612 + j\,2.08$$

Total impedance at slip 0.04, $\bar{Z}_1 = \bar{Z}_f + \bar{Z}_b$

$$= 30 + j\,2.08 + 0.612 + j\,2.08$$

$$= 30.612 + j\,4.16\ \Omega = 30.89\ \angle 7.7°$$

$\therefore$ Stator current $(I_1) = \dfrac{V_1}{Z_1} = \dfrac{240}{30.89\angle 7.7°} = 7.77\angle -7.7°$

Mechanical power developed $= (1-s)\, I_1^2\, (R_f - R_b)$

$$\text{Torque} = \frac{P_m}{\omega_r} = \frac{(1-S)\, I_1^2\, (R_f - R_b)}{(1-s)\,\omega_s} = \frac{I_1^2\,(R_f - R_b)}{\omega_s}$$

$$= \frac{(7.77)^2\,(30-.612)}{\dfrac{100\pi}{3}} = 16.95 \text{ Nm} \quad \textbf{Ans.}$$

> **Problem 2.20 :** A 240 V, 6 pole, 50 Hz, Single-phase motor has the following constants and losses:
>
> $$r_1 = 2\,\Omega,\, x_1 = 3\,\Omega,\, r_2 = 4\,\Omega,\, x_2 = 3\,\Omega,\, X_m = 60\,\Omega$$
>
> Core loss = 80 W,
>
> Friction and windage loss = 20 W
>
> If this motor is running with a slip of 0.04 at rated voltage and frequency, then compute the stator current, pf, power output, torque and efficiency with its auxiliary winding open.

Solution :

Given, $P = 6,\; f = 50$ Hz

$$N = \frac{120 \times 50}{6} = 1000 \text{ rpm} = \frac{1000 \times 2\pi}{60} = \frac{100\pi}{3} \text{ rad/sec}$$

$$r_1 = 2\,\Omega,\, x_1 = 3\,\Omega,\, r_2 = 4\,\Omega,\, x_2 = 3\,\Omega,\, X_m = 60\,\Omega$$

Coreloss = 80 W, P_{wf} = 20 W

Forward field impedance,

$$\bar{Z}_f = \frac{j\dfrac{X_m}{2}\left(\dfrac{r_2}{2\times s} + j\dfrac{x_2}{2}\right)}{\dfrac{r_2}{2\times s} + j\dfrac{1}{2}(x_2 + X_m)} = \frac{30j\left(\dfrac{4}{2\times 0.04} + j\dfrac{3}{2}\right)}{\dfrac{4}{2\times 0.04} + j\dfrac{1}{2}(60+3)}$$

$$= \frac{30j\,\angle 90° \times 50.02\angle 1.72°}{59.1\angle 32.21°} = 25.39\angle 59.51° = 12.89 + j\,21.88$$

Here, $R_f = 12.89\,\Omega, X_f = 21.88\,\Omega$

Backward field impedance,

$$\bar{Z}_b = \frac{j\dfrac{X_m}{2}\left(\dfrac{r_2}{2-s}+j\dfrac{x_2}{2}\right)}{\dfrac{r_2}{2-s}+j\left(\dfrac{x_2}{2}+\dfrac{X_m}{2}\right)} = \frac{j\,30\left(\dfrac{4}{2\times(2-0.04)}+j\,1.5\right)}{\dfrac{4}{2\times(2-0.04)}+j\,(1.5+30)}$$

$$= \frac{j\,30\,[1.02+j\,.15]}{1.02+j\,31.5} = \frac{30\angle 90^\circ \times 1.81\angle 55.78^\circ}{31.52\angle 88.15^\circ}$$

$$= 1.72\angle 57.63^\circ = 0.921 + j\,1.45\,\Omega$$

Here, $R_b = 0.921\,\Omega, X_b = 1.45\,\Omega$

Total motor impedance $= \bar{Z}_1 + \bar{Z}_f + \bar{Z}_b$

$$= 2 + j\,3 + 12.89 + j\,21.88 + 0.921 + j\,1.45$$

$$= 15.811 + j\,26.33 = 30.71\angle 59.02^\circ$$

$\therefore$ Stator current, $I_1 = \dfrac{V}{Z} = \dfrac{240}{30.71} = 7.82$ A

p.f. = cos 59.02 = 0.5147 (lag) **Ans.**

Power delivered to forward field

$$P_{gf} = I_1^2\,R_f = (7.82)^2 \times 12.89 = 788.25\text{ W}$$

Power delivered to backward field

$$P_{gb} = I_1^2\,R_b = (7.82)^2 \times 0.921 = 56.32\text{ W}$$

Total electrical power delivered = 788.25 – 56.32 = 731.93 W

Rotational loss = Core loss + windage and friction loss

= 80 + 20 = 100 W

Total mechanical power delivered $= P_m = (1-s)\times 731.93$

$$= 0.96 \times 731.93 = 702.65\text{ W}$$

Power output $= P_m - (P_c + P_{wf}) = 702.56 - 100 = 602.50$ W

$\therefore$ Torque $= \dfrac{\text{Power output}}{\text{Rotor speed}} = \dfrac{602.56\times 3}{0.96\times 100\times\pi} = 5.998$ Nm **Ans.**

Power input $= V_1\,I_1\cos\theta = 240\times 7.82\times 0.5147 = 965.99$ W

$$\therefore \text{ Efficiency} = \frac{\text{Output}}{\text{Input}} = \left(\frac{602.56}{965.99}\right) \times 100\% = 62.38\% \textbf{ Ans.}$$

Problem 2.21 : An inductor is made from magnetic core with one air gap g as shown in figure. The exciting coil has 500 turns, $A = 5 \text{ cm} \times 5 \text{ cm}$ and $g = 1$ cm.
Core is assumed to have infinite permeability and fringing is neglected.

(a) Find the coil inductance.

(b) Find the magnetic energy stored in the coil if current of 10 A flows though the coil.

(c) With the coil current held fixed at 10 A, find the electrical energy supplied by the source if the gap length decreases from 1 cm to 0.5 cm. Also calculate the mechanical work done.

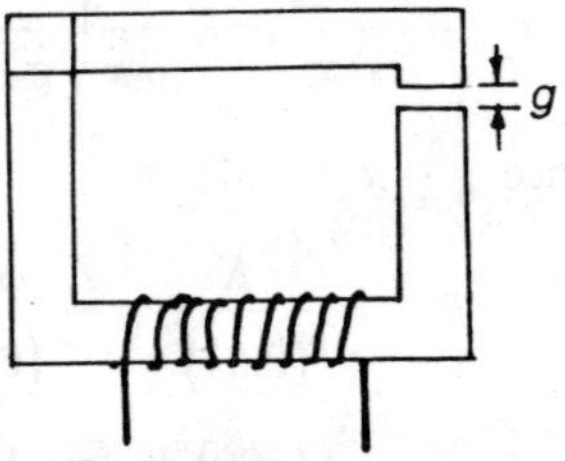

Fig. P. 2.21

Solution :

Given, $N = 500,\ A = 25$ sq. cm., $g = 1$ cm, $I = 10$ A

$$\text{Reluctance of the air gap} = \frac{g}{\mu_o \times A}$$

$$\text{Coil inductance} = \frac{N^2}{\text{reluctance}}$$

or
$$L = \frac{N^2 \times \mu_o A}{g} = \frac{a}{g} \text{ where } a = N^2 \mu_o A$$

$$= \frac{(500)^2 \times 4\pi \times 10^{-7} \times 25 \times 10^{-4}}{1 \times 10^{-2}} = 78.5 \text{ mH } \textbf{Ans.}$$

$$\text{Magnetic energy stored} = \frac{1}{2} LI^2 = \frac{1}{2} \times 78.5 \times 10^{-3} \times 100$$

$$= 43.925 = 3.93 \text{ W-sec } \textbf{Ans.}$$

To calculate the force on the armature, let the armature be displaced though a distance x so that air gap length becomes $(g - x)$

$$\therefore \qquad \text{Inductance} = \frac{N^2 \mu_o A}{g - x}$$

Field energy stored, $W_{fld} = \frac{1}{2} L_x \times I^2 = \frac{1}{2} \frac{N^2 \mu_o A}{g - x} \times I^2$

$$= \frac{K}{g - x}, \text{ where } K = \frac{N^2 \mu_o A I^2}{2}$$

We know that, force on the armature,

$$F_a = \frac{\partial W_{fld}}{\partial x} = K \frac{\partial}{\partial x}\left(\frac{1}{g - x}\right) = -K \times \frac{1}{(g-x)^2} \times -1 = \frac{K}{(g-x)^2}$$

Since $g - x = 0.01$ m.

$$F_a = \frac{K}{(0.01)^2} = \frac{N^2 \mu_o A \times I^2}{(0.01)^2}$$

$$= \frac{(500)^2 \times 4\pi \times 10^{-7} \times 25 \times 10^{-4} \times 100}{(0.01)^2} \text{ N}$$

$$= \frac{25 \times 10^4 \times 4\pi \times 10^{-7} \times 25 \times 10^{-4} \times 100}{1 \times 10^{-4}} = 785 \text{ N} \textbf{ Ans.}$$

(b) Electrical energy input at constant current movement.

$$W_{ele} = (L_{x2}, \text{ when air gap is } 0.05) \times I^2 - (L_{x1}, \text{ when air gap is } 0.01) \times I^2$$

$$= N^2 \mu_{oA} \left[\frac{1}{0.005} - \frac{1}{0.01}\right] \times 100 = \frac{N^2 \mu_{o \times A}}{2}\left[\frac{1000}{5} - \frac{100}{1}\right] \times 100$$

$$= \frac{(500)^2 \times 4\pi \times 10^{-7} \times 25 \times 10^{-4} \times 100 \times 100}{2} \text{ watt-sec}$$

$$= \frac{7.85}{2} \text{ watt-sec} = 3.93 \text{ watt-sec } \textbf{Ans.}$$

Mechanical work done

$$= \int_0^{0.5} f_e . dx = \int_0^{0.5} \frac{K}{(g-x)^2} dx$$

$$= K \int_0^{0.5} \frac{1}{(g-x)^2} dx = K \left[\frac{1}{(g-x)^2}\right]_0^{0.5}$$

$$= \frac{N^2 \mu o A I^2}{2}\left[\frac{1}{1-0.5} - \frac{1}{1}\right] = \frac{N^2 \mu o A I^2}{2} \times 1 \times 10^{-2}$$

$$= \frac{(500)^2 \times 4\pi \times 10^{-7} \times 25 \times 10^{-4} \times 100}{2}$$

$$= 0.0393 \text{ Watt-sec } \textbf{Ans.}$$

> **Problem 2.22 :** Two parallel plates, each of area $A = 1\,\text{m}^2$ are separated by a distance g. The electric field intensity between the plates is 30×10^4 V/m. Find the force acting between the plates.

Solution :

Given, $A = 1\,\text{m}^2, E = 30 \times 10^4$ V/m, distance $= g$

We know that electric field energy stored is

$$W_{fld}(q, x) = \frac{1}{2}\frac{q^2}{c} = \frac{1}{2}\frac{q^2(g-x)}{A\varepsilon_o}$$

We know that, force $= \dfrac{\partial W_{fld}(q,x)}{\partial x} = \dfrac{1}{2}\dfrac{q^2}{A\varepsilon_o}\dfrac{\partial(g-x)}{\partial x} = \dfrac{1}{2}\dfrac{q^2}{A\varepsilon_o}$ (1)

Since $q = DA$ and $D = \varepsilon_o E$ $\therefore$ $q = \varepsilon_o EA$

Putting this in (1),

$$\text{force} = \frac{1}{2}A\varepsilon_o \times \varepsilon_o^2 E^2 A^2 = \frac{1}{2}E^2 \varepsilon_o \text{A} = \frac{1}{2}E^2 \varepsilon_o \text{A}$$

$$= \frac{1}{2} \times (30 \times 10^4)^2 \times \left(\frac{10^{-9}}{36\pi}\right) \times 1$$

$$= \frac{1}{2} \times 9 \times 10^{10} \frac{10^{-9}}{36\pi} = \frac{10}{8\pi} = 0.398 \text{ N } \textbf{Ans.}$$

> **Problem 2.23 :** Two magnetic surfaces separated by distance g and have a flux density of 1.4 T in between them. This value is usually the saturation level for ferromagnetic materials. Find the force between these two surfaces for area $A = 1\,\text{m}^2$.

Solution :

Let the distance between magnetic surfaces be reduced by x_1 then length between the surfaces $= g - x$

Field energy stored, $W_{fld}(B, x) = \frac{1}{2}\frac{B^2}{\mu_o}(g-x)A$

We know that, force $= \frac{\partial W_{fld}(B, x)}{\partial x} = \frac{-\partial}{\partial x}\left[\frac{1}{2}\frac{B^2}{\mu_o}(g-x)A\right] = \frac{1}{2}\frac{B^2 A}{\mu_o}$

$$= \frac{1}{2} \times \frac{(1.4)^2 \times 1}{4\pi \times 10^{-7}} = \frac{1.96}{8\pi} \times 10^7 \text{ N} \quad \textbf{Ans.}$$

Problem 2.24 : Self and mutual inductances in henrys of two coupled coils are

$$L_1 = 4 + \frac{1}{4x}, \quad L_2 = 6 + \frac{1}{2x}, \quad M_{12} = M_{21} = \frac{1}{4x}$$

Over a certain displacement x in metres. The coil resistances are negligible.

For constant currents of $I_1 = 6$ A and $I_2 = -2$ A,

Compute mechanical work done in increasing x from 0.2 to 1m and energy supplied by each electrical source.

Also calculate change in field energy.

Solution :

Given, $I_1 = 6$ A and $I_2 = -2$ A

$$L_1 = 4 + \frac{1}{4x}, \quad L_2 = 6 + \frac{1}{2x},$$

$$M_{12} = M_{21} = \frac{1}{4x}$$

Assuming linear case, Field energy stored,

$$W_{fld}(i_1, i_2, x) = \frac{1}{2}L_1 I^2 + \frac{1}{2}L_2 I_2^{\,2} + I_1 I_2 M$$

$$= \frac{1}{2}\left(4 + \frac{1}{4x}\right) \times 36 + \frac{1}{2} \times \left(6 + \frac{1}{2x}\right) \times 4 - 12 \times \frac{1}{4x}$$

$$= 18\left(4 + \frac{1}{4x}\right) + 2\left(6 + \frac{1}{2x}\right) - \frac{3}{x}$$

$$= 72 + \frac{9}{2x} + 12 + \frac{1}{x} - \frac{3}{x} = 84 + \frac{9}{2x} + \frac{1}{x} - \frac{3}{x}$$

$$= 84 + \frac{9+2-6}{2x} = 84 + \frac{5}{2x}$$

Now, force is given by, $f_e = \dfrac{\partial W_{fld}\left(i_1, i_2, x\right)}{\partial x} = -\dfrac{5}{2x^2}$

Mechanical work done during displacement of 0.2 to 1

$$= \int_{0.2}^{1} \left(\frac{-5}{2x^2}\right) dx = \frac{-5}{2} \int_{0.2}^{1} \frac{1}{x^2}\, dx = \frac{5}{2}\left[\frac{1}{x}\right]_{0.2}^{1}$$

$$= \frac{5}{2}\left[1 - \frac{1}{0.2}\right] = \frac{5}{2}(1-5) = -10 \text{ Watt-sec} \quad \textbf{Ans.}$$

Now, $\psi_1 = L_1\, i_1 + M\, i_2$

$$= \left(4 + \frac{1}{4x}\right) \times 6 + \frac{1}{4x} \times (-2) = 24 + \frac{3}{2x} - \frac{1}{2x}$$

$$= 24 + \frac{3-1}{2x} = 24 + \frac{1}{x}$$

$$\psi_2 = L_2\, i_2 + M\, i_2$$

$$= \left(6 + \frac{1}{2x}\right) \times (-2) + \frac{1}{4x} \times 6 = -12 - \frac{1}{x} + \frac{3}{2x}$$

$$= -12 + \frac{3}{2x} - \frac{1}{x} = -12 + \frac{3-2}{2x} = -12 + \frac{1}{2x}$$

$$W_{ele1} = \int_{\lambda_1 (x=0.2)}^{\lambda_1 (x=1)} i_1\, d\lambda_1 = 6 \int_{(x=0.2)}^{(x=1)} d\psi_1$$

$$= 6\left[24 + \frac{1}{x}\right]_{0.2}^{1} = 6\left[24 + \frac{1}{x} - 24 - \frac{1}{0.2}\right]$$

$$= 6[1-5] = -24 \text{ Watt-sec} \quad \textbf{Ans.}$$

$$W_{ele2} = \int_{\lambda_2 (x=0.2)}^{\lambda_2 (x=1)} i_2\, d\lambda_2 = i_2\left[-12 + \frac{1}{2x}\right]_{0.2}^{1}$$

$$= (-2)\left[-12 + \frac{1}{2} + 12 - \frac{1}{2 \times 0.2}\right]$$

$$= (-2)\left[\frac{-1}{2} - 2.5\right] = 4 \text{ Watt-sec } \textbf{Ans.}$$

Net electrical energy input $W_{ele} = W_{ele1} + W_{ele2}$

$$= 4 - 24 = -20 \text{ Watt-sec } \textbf{Ans.}$$

Now, we have $W_{fld}(i_1, i_2, x) = 84 + \frac{5}{2x}$

Change in field energy = field energy at $x = 1$ – field energy at $x = 0.2$

$$84 + \frac{5}{2} - 84 - \frac{5}{2 \times 0.2} = \frac{5}{2} - \frac{25}{2} = -10 \text{ Watt-sec } \textbf{Ans.}$$

Chapter 3

Basic Concepts in Rotating Electrical Machines

BASIC CONCEPTS IN AC MACHINES

(*i*) Slots per pole per phase (q), $= \dfrac{\text{slots}\,(s)}{\text{pole}\,(p) \times \text{phase}\,(m)}$

(*ii*) Poles per phase $= \dfrac{\text{no. of poles}\,(p)}{\text{no. of phase}\,(m)}$

(*iii*) Angular slot angle (γ)

$$\gamma = \frac{180 \times \text{no. of poles}\,(p)}{\text{Total no. of slots}\,(s)} \text{ electrical degree}$$

$$= \frac{180}{\text{slots per pole}} \text{ electrical degree}$$

(*iv*) Distribution factor $k_d = \dfrac{\sin \dfrac{q\gamma}{2}}{q \sin \dfrac{\gamma}{2}}$

ηth harmonic distribution factor, $k_{dn} = \dfrac{\sin \dfrac{qn\gamma}{2}}{q \sin \dfrac{n\gamma}{2}}$

For $Y < 15°$, $k_d = \dfrac{\sin \dfrac{q\gamma}{2}}{\dfrac{q\gamma}{2}} = \dfrac{\sin \dfrac{\sigma}{2}}{\dfrac{\sigma}{2}}$

Where $\sigma = q\gamma$ phase speed

If q is not integer, $q = \frac{s_k}{p_k}$, $k_d = \frac{\sin \frac{q\gamma}{2}}{s_k \sin \frac{q\gamma}{2s_k}}$

(*v*) Chording angle $\varepsilon = 180° - \text{coil span}$

(*vi*) Pitch factor, $k_p = \cos \frac{\varepsilon}{2}$

η th harmonic pitch factor, $k_{pn} = \frac{\cos n\varepsilon}{2}$

Where n = no. of harmonics.

Relation between Electrical Angle and Mechanical Angle

$$\theta_e = \frac{p}{2}\theta_m$$

Where θ_e = Electrical angle
p = No. of poles
θ_m = Mechanical angle.

Per pole area for a *p*-pole machine.

$$A = \frac{2\pi r l}{p}$$

Where r = Radius of the armature core
l = Axial lenght of the armature core.

Flux per pole for p pole machines,

$$\phi = \frac{4}{p} B_{peak} \times l \times r$$

Where B_{peak} = Peak value of flux density

Generated emf in a full pitched coil is given by

$$E = 4.44 f_r N \phi$$

Where f_r = Speed frequency = $\frac{pn_r}{120}$ Hz, N_r = (in rpm)

Generated emf in a short pitched coil,

$$E = \sqrt{2}\pi f_r N \phi k_p,$$

Where $k_p = \cos \frac{\varepsilon}{2}$

Generated emf in a short-pitched distributed winding is

$$E = \sqrt{2}\,\pi\, f_r\, N\,\phi\, k_p\, k_d$$

$$= \sqrt{2}\,\pi\, f_r\, k_w\, N\,\phi \text{, where } k_w = k_d \times k_p = \text{ winding factor.}$$

Synchronous Machine

Synchronous speed of rotor, $N_s = \dfrac{120\,f}{p}$

Where f = Frequency is Hz, and p = no. of poles

The per phase emf generated in a synchronous machine is given by

$$E_{ph} = \sqrt{2}\, f\, N_{ph}\,\phi\, k_w$$

Where N_{ph} = Total no. of series turns per phase.

Induction Machines

Speed of rotating magnetic field, $(n_s) = \dfrac{120f}{p}$

Where f = Supply frequency is Hz

Slip = $\dfrac{f_r}{f}$ where f_r = Frequency of emf generated in rotor circuit

f = Supply frequency.

Rotor speed, $N_r = (1 - s)\,N_s$

The magnitude of per phase emf generated in stator winding is given by

$$E_s = \sqrt{2}\,\pi\, f\, k_w\, N_{ph}\,\phi$$

Where N_{phs} = No. of stator series turns

The per phase emf generated in rotor circuit of induction machines is given by

$$E_{phr} = \sqrt{2}\,\pi\, f_r\, N_{phr}\,\phi\, k_w$$

or $$E_{phr} = \sqrt{2}\,\pi\, s\, f\, N_{ph}\,\phi\, k_w$$

Where s = Slip.

It the rotor rotates is opposite direction of rotating field.

$$E_{ph} = \sqrt{2}\,\pi f\,(2 - s)\, N_{phr}\, k_w\,\phi$$

D.C. Machines

The back emf developed in d.c. machines is given by

$$E_b = \frac{\phi N Z p}{60\,A}$$

where ϕ = Flux per pole, Z = Total no. of conductors,
N = Speed of rotor in rpm
P = No. of poles
A = No. of parallel paths

Relation between Fundamental emf and η th Harmonic emf

We have, $E_{ph1} = \sqrt{2}\,\pi f k_{w1} N_{ph}\,\phi_1$

and $E_{phn} = \sqrt{2}\,\pi nf k_{wn} N_{ph}\,\phi_n$

$$\frac{E_{phn}}{E_{ph1}} = n \times \frac{k_{wn}}{k_{w1}} \times \frac{\phi_n}{\phi_1}$$

Where ϕ_1 = Total value of fundamantal flux per pole = $\frac{2DL}{P} B_1$

ϕ_n = n^{th} harmonic flux pole, $\phi_n = \frac{2DL}{np} \times B_n$

Where D = Air-gap diameter of the armature

L = Core-length of armature, $\frac{E_{phn}}{E_{ph1}} = \frac{k_{wn}}{k_{w1}} \times \frac{B_n}{B_1}$

- The peak value of mmf for p pole ac machine is given by

$$f_{1\,peak} = \frac{4}{\pi} \times \frac{N\,I_{max}}{p}\ \text{ATs/pole} = \frac{4}{\pi} \times \frac{N\sqrt{2}\,I}{p}\ \text{ATs/pole}$$

Where N = No. of turns of the coils.
I = r.m.s. value of alternating curent

- The peak value of fundamental mmf for p pole, short-pitched distributed winding is given by

$$f_{1\,peak} = \frac{4}{\pi}\frac{N_{ph} \times \sqrt{2}\,I \times k_w}{p}\ \text{ATs/pole}$$

For full-pitched coil, $K_p = 1$, and $k_w = k_d$

$$f_{1\,peak} = \frac{4}{\pi}\frac{N_{ph} \times \sqrt{2}\,I \times k_d}{p}\ \text{ATs/pole}$$

Where N_{ph} = series turns per phase.

- The peak value of fundamental field mmf produced by field winding in p-pole machine is given by

$$F_{1f} = \frac{4}{\pi} k_f \frac{N_f \times I_f}{p}\ \text{ATs/pole}$$

Where N_f = Total no. of field turns

I_f = Field winding current

K_f = Field winding factor.

- Peak value of the fundamental mmf is d.c. machine,

$$f_a = \frac{8}{\pi^2}\left(\frac{N}{P}\ \frac{I_a}{A}\right) \text{AT / pole}$$

Where p = No. of poles, I_a = Armature curent

A = No. of parallel paths

N = No. of armature turns

- Peak value of the fundamental mmf in single phase commutator,

$$F_a = \frac{8\sqrt{2}}{\pi\ 2} \times \frac{N}{P} \times \frac{I}{A} \text{AT / pole}$$

Where N = Total no. of armature turns

A = No. of parallel paths

P = No. of poles

- Peak value of magnetic flux density in air gap of a rotating ac machine is given by

$$B_{peak} = \frac{\mu_o F_R}{g} = \frac{3\mu_o F_m}{2g}$$

Where F_R = Resultant mmf crosses one air gap length g

F_m = Maximum value of resultant mmf.

g = Air gap lenght

- Flux per pole at rotating at synchronous speed.

$$\phi = \frac{6\mu_0 rl F_m}{pg}$$

$$\phi = \frac{24\sqrt{2}}{2\pi g p} \times \mu_o rl k_w N_{ph} I$$

- Amplitude of rotating mmf in terms of series turns per phase and r.m.s. value of current.

$$F_R = 2.7\, k_w \frac{N_{ph} I}{p} \text{ATs / pole}$$

- For *p.f.* pole AC machine,

Total flux per pole (ϕ) = $\frac{4}{p} B_{peak}\, lr = \frac{4}{p} B_R lr$

In dc machines, $\varphi = \frac{4}{p}\frac{\mu_o}{g} F_f\, lr$

- Torque Developed in p pole non-salient machine,

$$T_e = \frac{p}{2} \pi rl\, B_r\, F_r \sin\delta_s$$

or $$T_e = \frac{p}{2}\frac{\mu_o \pi rl}{g} F_s\, F_r \sin\lambda$$

Where F_s = Stator mmf, F_r = rotor mmf

λ = Angle between stator and rotor mmf

- Peak value of fundamental flux density wave, in terms of field mmf

$$B_{peak} = \frac{\mu_o f_{1f}}{g}$$

- Rating of continuous rated motor = $\left[\frac{\sum \{(k_w)^2 \times \text{time}\}}{\sum \text{time}}\right]^{1/2}$

- The temperature rise during loading of a machine is given by

$$\theta = \theta_f\left(1 - e^{-\frac{t_x}{t_n}}\right)$$

Where θ = Temperature rise, θ_f = Final temperature

t_x = Time for loading, t_h = Heating time constant.

- The short time load is given by

$$p_x = p_r\left[\frac{1+\alpha}{1-e^{-\left(t_x/t_h\right)}} - \alpha\right]^{1/2}$$

Where p_r = Rating of the machine

$$\alpha = \frac{\text{core loss}\,(w_c)}{\text{ohmic loss}\,(w_o)},$$

t_h = Heating time constant

Relation between losses at different rating.

$$\frac{w_x}{w_r} = \frac{\alpha + \left(\frac{p_x}{p_r}\right)}{1+\alpha}$$

Where w_x = Total losses at rating, P_x

w_r = Total losses at rating P_r

Problem 3.1 : Calculate the phase and line voltage of a three phase star-connected 4-pole small alternator having a total flux per pole of 0.30 Wb. Each phase has a concentrated rated winding of 10 full-pitched truns and the relative velocity between armature winding and field flux is 1500 r.p.m.

Solution

Given, $\phi = 0.\ 30$ wb/pole, $N_r = 1500$ rpm

$p = 4$, $N_{ph} = 10$

Rotor frequency, $f_r = \dfrac{N_r \times p}{120} = \dfrac{1500 \times 4}{120} = 50$ Hz

Applying the relation, phase voltage,

$$E_{ph} = \sqrt{2}\,\pi\, f_r\, N_{ph}\,\phi$$

$$= \sqrt{2} \times \pi \times 50 \times 10 \times 0.30 = 666.43 \text{ V } \textbf{Ans.}$$

Line voltage, $E_L = \sqrt{3} \times E_{ph} = \sqrt{3} \times 666.43 = 1154.3$ V **Ans.**

Problem 3.2 : The stator of a 4-pole, 3-phase, 50 Hz induction machine is supplied from 3-phase, 50 Hz supply. The slip rings of the machine are open circuited. It is desired that frequency of the voltage across slip rings is (*a*) 50 Hz (*b*) 75 Hz (*c*) 100 Hz (*d*) 25 Hz. Find the speed of the rotor to obtain these frequencies.

Solution:

Given, $p = 4$, $m = 3$, $f = 50$ Hz

Synchronous speed of the rotating field.

$$N_s = \frac{120 \times 50}{4} = 15000 \text{ r.p.m.}$$

(*a*) Here, $f_r = 50$ Hz, slip $= \dfrac{f_r}{f} = \dfrac{50}{50} = 1$

$\therefore$ Rotor speed $(N_r) = N_5\ (1 - s)$

$= 1500\ (1-1) = 0$ r.p.m. **Ans.**

(*b*) Here, $f_r = 75$ Hz, slip $= \frac{f_r}{f} = \frac{75}{50} = \frac{3}{2}$

$\therefore$ Rotor speed, $(N_r) = N_s \; (1 - s)$

$$= 1500\left(1 - \frac{3}{2}\right) = -750 \text{ rpm.}$$

This negative sign shows that the rotor rotates with speed 750 rpm against the direction of rotating field. **Ans.**

(*c*) Here, $f_r = 100$ Hz, slip $= \frac{f_r}{f} = \frac{100}{50} = 2$

$\therefore$ Rotor speed $(N_r) = N_s\,(1-s)$

$$= 1500\;(1 - 2) = -1500 \text{ rpm.}$$

This negative sign shows that the rotor rotates with speed 1500 rpm against the dirextion of rotating field. **Ans.**

(*d*) Here, $f_r = 25$ Hz, slip $= \frac{f_r}{f}\frac{25}{50} = \frac{1}{2}$

$\therefore$ Rotor speed $(N_r) = N_s\,(1-s)$

$$= 1500\left(1 - \frac{1}{2}\right) = +750 \text{ r.p.m.}$$

This positive sign shows that the rotor rotates with speed 750 rpm in the direction of synchronously rotating magnetie field. **Ans.**

Problem 3.3 : The frequency of the e.m.f. in the circuit of a 6-pole induction motor is 50 Hz. circuit e.m.f. has a frequency of 2 Hz, then find:

(*i*) the direction in which the rotor is revolving with respect to the rotating field.

(*ii*) the slip and the rotor speed.

Solution :

Given, $p = 6,\; f = 50$ Hz

Synchronous speed of rotating magnetie field,

$$(N_s)t_h = \frac{120 \times f}{p} = \frac{120 \times 50}{6} = 1000 \text{ r.p.m.}$$

(*a*) Here, $f_r = 2$ Hz, We know that, slip (s) $= \frac{f_r}{f} \Rightarrow \frac{2}{50} = \frac{1}{25}$

Rotor speed $(N_r) = N_s\left(1-\frac{1}{25}\right) > 0$

Since the rotor speed is positive, the rotor rotates in the direction of rotating field. **Ans.**

(b) Here, $f_r = 2$ Hz, Slip $= \frac{f_r}{f} = \frac{2}{50} = 0.04$ **Ans.**

Rotor speed $(N_r) = N_s(1-s) = 1000\ (1 - 0.04)$

$= 960$ r.p.m. **Ans.**

Problem 3.4: (*a*) A 10 kW, 250 V, 8-pole, 600 r.p.m. lap-connected d.c. generator has 400 armature conductors. At rated voltage and current, armature ohmic losses are 150 watts. Compute the useful flux per pole.

(*b*) A 6-pole d.c. generator has a rated speed of 800 r.p.m. Calculate frequency of the e.m.f.

(*i*) Generated in the armature conductors, and

(*ii*) At the brushes.

Solution :

Given, $V_t = 250$ V, $p = 8$, $N = 600$ r.p.m.,
$A = p = 8$, $Z = 400$

Rated armature current $I_a = \frac{10 \times 1000}{250} = 40$ A

Also given, $I_a^2\, r_a = 150$, $\therefore\ r_a = \frac{150}{(40)^2}$

(a) Now, for d.c. generator,

$$E_b = V_t + I_a r_a$$

$$E_b = 250 + 40 \times \frac{150}{(40)^2} = 250 + \frac{15}{4} = \frac{1015}{4}\ \text{V}$$

To calculate flux per pole, Applying the relation,

$$E_b = \frac{\phi NZP}{60\,A}$$

$$\therefore\ \phi = \frac{E_b \times A \times 60}{NZp} = \frac{1015 \times 8 \times 60}{400 \times 4 \times 8 \times 600} = 0.06344\ \text{Wb}\ \ \textbf{Ans.}$$

(*b*) Here, $p = 6$, $N_r = 800$ r.p.m.

(*i*) The frequency of the emf generated in the armature conductors is same as that of rotor.

$$\therefore \text{ Required frequency} = \frac{N_r \times p}{120} = \frac{800 \times 6}{120} = 40 \text{ Hz} \quad \textbf{Ans.}$$

(*ii*) At the brushes, the emf is d.c. in nature.
Hence, Frequency of the e.m.f. at brushes = 0 Hz **Ans.**

Problem 3.5 : A 3-phase star-connected alternator has a rating of 810 kVA at 2.5 kV and has 12 poles. Rated speed is 500 r.p.m. There are three slots per pole per phase and number of conductors per slot is 5. For a full-pitched winding, compute the flux per pole to give rated terminal voltage at no load.

Solution :

Given, $p = 12, \; N_s = 500$ rpm

Slots/pole/phase $q = 3$, conductors/slot = 5

$$E_{ph} = \frac{2.5 \times 1000}{\sqrt{3}}$$

Angular slot pitch $$(\gamma) = \frac{180 \times p}{s} \text{ (electrical degree)}$$

$$= \frac{180 \times 12}{108} = 20°$$

Synchronous frequency, $$f_s = \frac{500 \times 12}{120} = 50 \text{ Hz}$$

Slots per pole per phase $$= \frac{\text{slots}}{\text{pole} \times \text{no. of phase}}$$

$$\therefore \quad \text{Slots (s)} = 3 \times 12 \times 3 = 108$$

Total No. of conductors = no. of slots × 5 = 108 × 5

Total No. of conductors/phase $$= \frac{108 \times 5}{3}$$

No. of turns per phase $$N_{phs} = \frac{108 \times 5}{3 \times 2} = 90$$

For full-pitched winding, $k_p = 1$

$$\text{Distribution factor,} \qquad k_d = \frac{\sin\frac{q\gamma}{2}}{q\sin\frac{\gamma}{2}} = \frac{\sin 30^\circ}{3\sin 10^\circ}$$

$$\text{To calculate flux per pole,} \qquad k_w = k_p \times k_d$$

$$\text{Applying the relation,} \qquad E_{ph} = \sqrt{2}\,\pi f_s \times N_{phs}\, k_w \times \phi$$

$$\therefore \qquad \varphi = \frac{E_{ph}}{\sqrt{2}\,\pi f_s \times N_{phs} \times k_w} = \frac{2 \times 2.5 \times 1000 \times 3 \times \sin 10^\circ}{\sqrt{3} \times \sqrt{2} \times \pi \times 50 \times 90 \times 1}$$

$$= 0.07522 \text{ Wb } \textbf{Ans.}$$

> **Problem 3.6 :** A 3-phase, 50 Hz, 8-pole, star-connected alternator has 120 stator slots and each slot has 10 conductors. The coil span is 12 slots. If the flux per pole is 0.12 Wb, calculate the induced emf between lines. Which of the emf harmonics is eliminated?

Solution :

Given, $m = 3, \quad p = 8, \quad f = 50$ Hz,

Coil span = 12 slots

Total No. of stator slots (s) = 120

No. of conductors per slots = 10

$\therefore$ Total No. of conductors = 120 × 10 = 1200

$$\text{No. of turns per phase, } N_{phs} = \frac{1200}{2 \times 3} = 200$$

Flux per pole (ϕ) = 0.12 Wb.

$$\text{Slots/pole/phase} \qquad (q) = \frac{1200}{8 \times 3} = 5$$

$$\text{Angular slot pitch,} \qquad \gamma = \frac{180 \times p}{\text{Total no. of slots}} = \frac{180 \times 8}{120} = 12^\circ$$

$$\text{Coil span angle} \qquad = 12\,\gamma = 12 \times 12 = 144^\circ$$

$$\text{Chording angle} \qquad (\in) = 180^\circ - 144^\circ = 36^\circ$$

$$\text{Distribution factor,} \quad k_d = \frac{\sin\frac{5 \times 12}{2}}{5\sin\frac{12^\circ}{2}} = \frac{\sin 30^\circ}{5\sin 6^\circ}$$

Pitch factor, $k_p = \cos\frac{\varepsilon}{2} = \cos\frac{36°}{2} = \cos 18°$

To calculate induced emf between lines, Apply the relation,

$$E_l = \sqrt{3} \times E_{ph} = \sqrt{3} \times \sqrt{2} \times \pi \times f \times N_{phs} \times k_p \times k_d \times \phi$$

$$= \sqrt{3} \times \sqrt{2} \times \pi \times 50 \times 200 \times \cos 18° \times \frac{\sin 30°}{5 \sin 6°} \times 0.12$$

$= 8401.92$ V **Ans.**

Now, Third harmonic pitch factor, $k_{p3} = \cos\frac{3 \times 36°}{2} = \cos 54°$

Fifth harmonic pitch factor, $k_{p5} = \cos\frac{5 \times 36°}{2} = \cos 90° = 0$

Since fifth harmonic pitch factor is 0, $k_w = 0$

$\therefore$ Fifth harmonic emf per phase $= \sqrt{2} \times \pi \times 5 \times f \times N_{phs} \times k_w \times \phi = 0$

Thus, fifth harmonic emf is absent **Ans.**

> **Problem 3.7 :** A 3-phase, delta-connected, 4-pole, 50 Hz, 415 V induction motor has 36 stator slots and 10-turn coils, each short-pitched by one slot. If stator leakage impedance is neglected, find the flux per pole of the rotating flux-density wave in case the stator of induction motor has (*a*) single-layer winding (*b*) double-layer winding.

Solution :

Given, $p = 4$, $f = 50$ Hz,

$V_t = 415$ V $(L - L)$, m (phase) $= 3$

Total No. of stator slots $(s) = 36$

Slots per pole per phase, $q = \frac{s}{m \times p} = \frac{36}{3 \times 4} = 3$

Angular slot pitch, $\gamma = \frac{180 \times p}{s} = \frac{180 \times 4}{36} = 20°$

Chording angle (ε) = one slot pitch $= 20°$

Distribution factor, $k_d = \frac{\sin\frac{q\gamma}{2}}{q \sin\frac{\gamma}{2}} = \frac{\sin\frac{3 \times 20°}{2}}{3 \sin\frac{20°}{2}} = \frac{\sin 30°}{3 \sin 10°}$

Pitch factor, $k_p = \cos\frac{\varepsilon}{2} = \cos\frac{20^\circ}{2} = \cos 10^\circ$

No. of series turns/phase $N_{phs} = \frac{10 \times 36}{3} = 120^\circ$

For single layer winding, $N_{ph1} = \frac{N_{phs}}{2} = \frac{120}{2}$

For delta connection, $E_{ph} = E_L = 415 \text{ V}$

Now, Applying the relation, $E_{ph} = \sqrt{2}\,\pi\, f\, N_{ph} \times k_p \times k_d \times \phi$

or $$\phi_1 = \frac{E_{ph}}{\sqrt{2}\,\pi\, f\, N_{ph} \times k_p \times k_d}$$

$$= \frac{415 \times 2 \times 3 \times \sin 10^\circ}{\sqrt{2} \times \pi \times 50 \times 120 \times \cos 10^\circ \times \sin 30^\circ} = 0.032941 \text{ wb } \textbf{Ans.}$$

(*b*) For double layer winding,

$$N_{ph2} = 120 = 2 \times N_{ph1}$$

$$\phi_2 = \frac{E_{ph}}{\sqrt{2}\,\pi\, f\, N_{ph} \times k_p \times k_d}$$

$$= \frac{415 \times 3 \times \sin 30^\circ}{\sqrt{2} \times \pi \times 50 \times 120 \times \cos 10^\circ \times \sin 30^\circ} = 0.016471 \text{ Wb } \textbf{Ans.}$$

Problem 3.8 : A 4-pole, %50 Hz, delta-connected induction motor is supplied from 415 V, 3-phase souce. It has 36 stator slots with double-layer winding and 8-turn coil chorded by one slot. Its stator lenght is 30 cm, mean air-gap radius 10 cm and air-gap lenght 1 mm.

(*a*) For negligible stator impedance, find the flux per pole of the rotating flux density wave.

(*b*) Find the line current drawn by the motor in case no-load ilosses are 1400 watts.

Solution :

Given, $m = 3, \quad p = 4, \quad f = 50 \text{ Hz}$

Total no. of slots $(s) = 36, \; l = 30 \text{ cm} = 30 \times 10^{-2} \text{ m}$

$r = 10 \text{ cm} = 10 \times 10^{-2} \text{ m}, \; g = 1 \text{ mm}$

$= 1 \times 10^{-3} \text{ mm}$

Now, slots per pole per phase $(q) = \dfrac{s}{m \times p} = \dfrac{36}{3 \times 4} = 3$

Angular slot pitch, $\gamma = \dfrac{180 \times p}{\text{No. of slots}} = \dfrac{180 \times 4}{36} = 20$

Chording angle (ε) = one slot pitch = 20

No. of series turns/phase $= \dfrac{\text{No. of coils} \times \text{No. of slots}}{\text{No. of phases}} = \dfrac{8 \times 36}{3} = 96$

$$\text{Distribution factor, } k_d = \frac{\sin\dfrac{q\gamma}{2}}{q \sin\dfrac{\gamma}{2}} = \frac{\sin\dfrac{3 \times 20^\circ}{2}}{3 \sin\dfrac{20^\circ}{2}} = \frac{\sin 30^\circ}{3 \sin 10^\circ}$$

$$\text{Pitch factor, } k_p = \cos\frac{\varepsilon}{2} = \cos\frac{20^\circ}{2} = \cos 10^\circ.$$

For delta connected induction motor

$$E_{ph} = E_L = 415 \text{ V}$$

(*a*) To calculate flux per pole, Apply the relation

$$E_{ph} = \sqrt{2}\, \pi f N_{phs} \times k_p \times k_d \times \phi$$

$$\therefore \quad \phi = \frac{E_{ph}}{\sqrt{2}\, \pi f N_{phs} \times k_p \times k_d}$$

$$= \frac{415 \times 3 \times \sin 10}{\sqrt{2} \times \pi \times 50 \times 96 \times \cos 10 \times \sin 30^\circ}$$

$= 0.02059$ Wb **Ans.**

(*b*) We have, flux per pole

$$(\phi) = \frac{4}{p} B_{peak} \times l \times r$$

$$\therefore \quad B_{peak} = \frac{p\,\phi}{4 \times l \times r} = \frac{4 \times 0.02059}{4 \times 10 \times 10^{-2} \times 30 \times 10^{-2}} = 0.6863 \text{ T}$$

Neglecting reluctance of air gap, we

$$B_{peak} = \frac{3\,\mu_0 F}{2\,g}$$

$$\therefore \quad F = \frac{2\,g\,B_{peak}}{3\,\mu_0} = \frac{2 \times 1 \times 10^{-3} \times 0.6863}{3 \times 4\pi \times 10^{-7}} = 364.09 \text{ At/pole}$$

Let I be the load current per phase,

To calculate I, apply the relation, $F = \frac{4}{\pi} k_w \frac{N_{phs}}{p} \times I$

$$\therefore \quad I = \frac{F \times \pi \times p}{4 \times k_w \times N_{phs}} = \frac{364.09 \times \pi \times 4 \times 3 \times \sin 10^\circ}{4 \times \sin 30^\circ \times \cos 10^\circ \times 96}$$

$$= 12.605 \text{ A}$$

Line current drawn by motor,

$$I_L = \sqrt{3} \times 12.605 = 21.83 \text{ A} \quad \textbf{Ans.}$$

> **Problem 3.9 :** A 3-phase, 4-pole, 50 Hz, star-connected turbogenerator has distributed field winding of 140 turns and winding factor is 0.96. Its other data is as under:
>
> | | |
> |---|---|
> | Armature series turns per phase | = 42 |
> | Armature radius | = 0.4 m |
> | Armature length | = 4 m |
> | Gap length | = 2 cm |
> | Winding factor for armature winding | = 0.955 |
>
> For no-load armature emf of 11 kV, find the value of field current required.

Solution :

Given, $p = 4$, $f = 50$ Hz. $m = 3$,

$N_f = 140$, $k_{wf} = 0.96$, $N_{ph} = 42$,

$l = 4$m, $g = 2 \times 10^{-2}$ m, $r = 0.4$ m,

$k_w = 0.955$ $\quad E_{ph} = \frac{11 \times 1000}{\sqrt{3}}$ V

To calculate flux per pole apply the relation,

$$E_{ph} = \sqrt{2}\, \pi\, f\, N_{ph} k_w \times \phi$$

$$\therefore \quad \phi = \frac{E_{ph}}{\sqrt{2}\, \pi\, f\, N_{ph}\, k_w}$$

Now, Peak value of fundamental flux density wave.

$$B_{1p} = \frac{P\,\phi}{4\,l\,r} = \frac{\mu_0\, F_{1f}}{g}$$

Peak value of fundamental field mmf wave,

$$\therefore \quad F_{1f} = \frac{P\,\phi \times g}{4\,l\,r \times \mu_0}$$

Let I_f be the field current,

$$I_f = \frac{\pi F_{1f} P}{4 k_{wf} \times N_f}$$

$$= \frac{\pi p}{k_{wf} \times N_f \times 4} \times \frac{p \phi \times g}{4lr \times \mu o}$$

$$= \frac{\pi p^2 \times g}{k_{wf} \times N_f \times 4 \times 4 \times l \times r \times \mu o} \times \frac{E_{ph}}{\sqrt{2} \pi f N_{ph} k_w}$$

$$= \frac{\pi \times (4)^2 \times 11000 \times 2 \times 10^{-2}}{\sqrt{3} \times 0.96 \times 140 \times 4 \times 4 \times 4 \times 0.4 \times 4\pi \times 10^{-7} \times \sqrt{2} \times \pi \times 50 \times 42 \times .955}$$

= 165.73 A **Ans.**

Problem 3.10 : The total flux per pole in the air gap of a 50 Hz a.c. generator is 0.069 Wb. If the density distribution is given by

$$B = B_1 \sin\theta + B_3 \sin 3\theta + B_5 \sin 5\theta$$

Where $B_3 = \frac{1}{3} B_1$ and $B_5 = \frac{1}{5} B_1$

and θ is the angle measured from iinterpolar axis, then determine the r.m.s. value of the e.m.f. per turn. The coil span is $\frac{4}{5}$th of a pole pitch.

Solution :

Given, $f = 50$ Hz

Total flux per pole $(\phi) = 0.069$ Wb

$$B = B_1 \sin\theta + B_3 \sin 3\theta + B_5 \sin 5\theta$$

Since, $B_3 = \frac{1}{3} B_1 \quad \therefore \phi_3 = \frac{1}{9}\phi_1, B_5 = \frac{1}{5} B_1 \quad \therefore \phi_5 = \frac{1}{25}\phi_1$

Coil span $= \frac{4}{5}$th of a pole pitch $= \frac{4}{5} \times 180° = 4 \times 36° = 144°$

∴ Chording angle, $\varepsilon = 180° - \text{coil span} = 180° - 144° = 36°$

Now, Fundamental pitch factor $= \cos\frac{\varepsilon}{2} = \cos 18° = 0.951$

Third component pitch factor $= \cos\frac{3\varepsilon}{2} = 0.588$

Fifth component pitch factor $= \cos\frac{5\varepsilon}{2} = 0°$

Total flux = 0.069 or $\phi_1 + \frac{1}{9}\phi_1 + \frac{1}{25}\phi_1 = 0.069$

or $1.151\,\phi_1 = 0.069, \therefore \phi_1 = \frac{0.069}{1.151} = 0.06$ Wb

Fundamental flux per pole = 0.06 Wb

Now applying the relation, Emf per turn $= \sqrt{2}\,\pi\, f\, k_{p1}\phi$

Fundamental emf per turn

$$E_1 = \sqrt{2} \times \pi \times 50 \times 0.951 \times 0.06 = 12.674 \text{ V}.$$

Third harmonic emf per turn,

$$E_3 = \frac{E_1 \times k_{p3}}{k_{p1}} \times \frac{1}{3} = 2.612 \text{ V}$$

Fifth harmonic emf per turn, $= E_{ph5} = 0$

$\therefore$ Total emf per turn, $E_{ph} = \sqrt{E_{ph1}^2 + E_{ph3}^3 + E_{ph5}^2}$

$$\sqrt{(12.674)^2 + (2.612)^2 + 0^2} = 12.94 \text{ V}$$

Problem 3.11 : The alternator has 9 slots per pole and the coil-span is from 1 to 9. If r.m.s. value of the fundamental phase voltage is V_1, then compute:

(*i*) r.m.s. value of the phase voltage, and
(*ii*) r.m.s. value of the line voltage.

The field-flux distribution of a 3-phase star-connected alternator is given by $100\sin\theta + 30\sin 3\theta + 20\sin 5\theta$

Where θ is the angle measured from the interpolar axis.

Solution :

Given, $\phi = 100\sin\theta + 30\sin 3\theta + 20\sin 5\theta$

$\therefore$ $\phi_1 = 100$ wb, $\phi_3 = 30$, $\phi_5 = 20$

Slots per pole (Q) = 9, no. of phases = 3

Slots per pole per phase $(q) = \frac{\text{Slots per pole}}{\text{No. of phases}} = \frac{9}{3} = 3,$

Angular slot pitch, $\gamma = \dfrac{180°}{\text{slots per pole}} = \dfrac{180°}{9} = 20°$

Coil-span $= (9 - 1) = 8$ slots $= 8\gamma = 160°$

Chording angle $(\varepsilon) = 180° - \text{coil-span} = 180° - 160° = 20°$

Fundamental distribution factor,

$$K_{d1} = \frac{\sin\dfrac{q\gamma}{2}}{q\sin\dfrac{\gamma}{2}} = \frac{\sin\dfrac{3\times 20°}{2}}{3\sin\dfrac{20°}{2}} = \frac{\sin 30°}{3\sin 10°}$$

Fundamental pitch-factor, $k_{p1} = \cos\dfrac{\epsilon}{2} = \cos\dfrac{20°}{2} = \cos 10°$

$$\therefore \quad k_{w1} = k_{d1} \times k_{p1} \quad \frac{\sin 30° \times \cos 10°}{3\sin 10°}$$

Third harmonic distribution factor,

$$k_{d3} = \frac{\sin\dfrac{3q\gamma}{2}}{q\sin\dfrac{3\gamma}{2}} = \frac{\sin\dfrac{3\times 3\times 20°}{2}}{3\sin\dfrac{3\times 20°}{2}} = \frac{1}{3\sin 30°}$$

Thrid harmonic pitch-factor,

$$k_{p3} = \cos\frac{3\varepsilon}{2} = \cos\frac{3\times 20°}{2} = \cos 30°$$

$$\therefore \quad kw_3 = kd_3 \times kp_3 = \frac{1}{3\sin 30°} \times \cos 30° = \frac{\cos 30°}{3\sin 30°}$$

Fifth harmonic distribution factor,

$$kd_5 = \frac{\sin\dfrac{5q\gamma}{2}}{q\sin\dfrac{5\gamma}{2}} = \frac{\sin\dfrac{5\times 3\times 20°}{2}}{3\sin\dfrac{5\times 20°}{2}} = \frac{\sin 150°}{3\sin 50°}$$

Fifth harmonic pitch factor,

$$k_{p5} = \cos\frac{5\varepsilon}{2} = \cos\frac{5\times 20°}{2} = \cos 50°$$

$$\therefore \quad k_{w5} = k_{d5} \times k_{p5} = \frac{\sin 150° \times \cos 50°}{3\sin 50°}$$

Now, Applying the relation, $\dfrac{E_{phn}}{E_{ph1}} = n \times \dfrac{k_{wn}}{k_{w1}} \times \dfrac{\phi_n}{\phi_1}$

$$\therefore \quad \frac{V_3}{V_1} = 3 \times \frac{\phi_3}{\phi_1} \times \frac{kw_3}{kw_1}$$

$$\therefore \quad V_3 = 3 \times \frac{30}{100} \times \frac{\cos 30^\circ}{3 \sin 30^\circ} \times \frac{3 \times \sin 10^\circ}{\sin 30^\circ \times \cos 10^\circ} \times V_1 = 0.5497\, V_1$$

Now, $$\frac{V_5}{V_1} = 5 \times \frac{\phi_5}{\phi_1} \times \frac{kw_5}{kw_1}$$

$$\therefore \quad V_5 = 5 \times \frac{20}{100} \times \frac{\sin 150^\circ \times \cos 50^\circ}{3 \sin 50^\circ} \times \frac{3 \times \sin 10^\circ}{\cos 10^\circ \times \sin 30^\circ} \times V_1$$

$$= 0.14796\, V_1$$

(*i*) For star-connection,
R.M.S. value of the phase voltage,

$$= \sqrt{V_1^2 + V_3^2 + V_5^2}$$

$$= \sqrt{V_1^2 + (0.5497\, V_1)^2 + (0.14796\, V_1)^2} = 1.15067\, V_1 \quad \textbf{Ans.}$$

(*ii*) For star-connection, third harmonic emf's do not appear between the lines,

$$\therefore \quad V_{line} = \sqrt{3} \times \sqrt{V_1^2 + V_5^2}$$

$$= \sqrt{3} \times \sqrt{V_1^2 + (0.14796\, V_1)^2} = 1.751\, V_1 \quad \textbf{Ans.}$$

Problem 3.12 : An 8-pole alternator with single layer winding and full pitch coils, has 12 slots per pole. Each coil has 30 turns. The flux is sinusoidally distributed in space and its value *p* per pole is 0.03 Wb. The machine is rotated at a speed of 750 r.p.m. The current per conductor is 50 A. For negligible internal voltage drops, find the kVA output available from the stator if the winding is connected as:
(*a*) single phase, and
(*b*) three phase.

Solution :

(*a*) Given, No. of phase (m) = 1, slots per pole per phase (s) = 12
No. of poles (p) = 8, slots per pole (Q) = 12
No. of turns in each coil = 30

$\phi = 0.03$ wb, $N_r = 750$ r.p.m., $I_c = 50$ A.

$$\text{Frequency} = \frac{Np}{120} = \frac{750 \times 8}{120} = 50 \text{ Hz}$$

For single layer winding,

$$\text{No. of coils} = \frac{1}{2} \times \text{no. of slots}$$

$$= \frac{1}{2} \times (\text{no. of slots per pole}) \times \text{no. of poles}$$

$$= \frac{1}{2} \times 12 \times 8 = 48$$

∴ Total no. of turns = no. of coils × no. of turns in each coil
= 48 × 30 = 1440 turns

(*a*) For single phase, $m = 1$,

$$\text{Slots per pole/phase} = \frac{Q}{m} = \frac{12}{1} = 12$$

$$\text{Angular slot pitch, } \gamma = \frac{180}{12} = 15°$$

$$\text{Winding fator, } k_d = \frac{\sin \frac{12 \times 15°}{2}}{12 \sin \frac{15°}{2}} = \frac{1}{12 \sin 7.5°}$$

For full-pitched coils, $k_p = 1$

$$\therefore \quad k_w = k_p \times k_d = \frac{1}{12 \sin 7.5°}$$

Now, Emf generated

$$\varepsilon = \sqrt{2}\, \pi f\, N_{ph} \times k_w \times \phi$$

$$= \sqrt{2} \times \pi \times 50 \times 1440 \times \frac{1}{12 \sin 7.5°} \times 0.03$$

$$= 6126.885 \text{ V}$$

∴ kVA output, $E \times I_c = 6126.885 \times 50$ VA = 306.344 kVA **Ans.**

(*b*) For three phase, $m = 3$,

$$\text{Slots per pole per phase} \quad (q) = \frac{s}{m \times p} = \frac{s/p}{m} = \frac{12}{3} = 4.$$

$$\text{Angular slot pitch, } \gamma = \frac{180}{12} = 15°$$

$$\text{Distribution factor} \quad k_d = \frac{\sin\frac{q\gamma}{2}}{q\sin\frac{\gamma}{2}} = \frac{\sin\frac{4\times 15^\circ}{2}}{4\sin\frac{15^\circ}{2}} = \frac{1}{8\sin 7.5^\circ}$$

For full-pitched coil, $k_p = 1$

$$\therefore \quad \text{Winding factor } (k_w) = k_p \times k_d = \frac{1}{8\sin 7.5^\circ}$$

Now, Emf generated,

$$\varepsilon = \sqrt{2}\times\pi f\, N\,\phi\, k_w$$

$$= \sqrt{2}\times\pi\times 50\times 1440\times 0.03\times\frac{1}{8\sin 7.5^\circ}$$

$$= 9190.327 \text{ V}$$

$\therefore$ **kVA output available** $= E \times I_c$

$= 9190.327 \times 50 = 459.516$ kVA **Ans.**

> **Problem 3.13 :** A 3-phase, 50 Hz, 10 pole alternator with 90 slots has a star-connected winding to give a generated e.m.f. of 11 kV at no load. The coils are chorded by one slot. If the flux per pole is 0.110 Wb, compute the number of series turns required for each phase of this alternator.

Solution :

Given, $p = 10$, $f = 50$ Hz, $m = 3$

No. of slots $(S) = 90$

$$\text{Generated emf per phase} \quad (E_{ph}) = \frac{11\times 1000}{\sqrt{3}} \text{ V}$$

Flux per pole $(\phi) = 0.110$ Wb

$$\text{Angular slot pitch, } \gamma = \frac{180}{\text{no. of slots}} = \frac{180}{90} = 20^\circ$$

$$\text{Winding factor, } k_d = \frac{\sin\frac{q\gamma}{2}}{q\sin\frac{\gamma}{2}} = \frac{\sin\frac{3\times 20^\circ}{2}}{3\sin\frac{20^\circ}{2}} = \frac{\sin 30^\circ}{3\sin 10^\circ}$$

Chording anlge $(\varepsilon) =$ One slot pitch $= 20^\circ$

Pitch factor, $K_p = \cos\frac{\varepsilon}{2} = \cos\frac{20°}{2} = \cos 10°$

Winding factor $(k_w) = k_p \times k_d = \frac{\cos 10° \times \sin 30°}{3 \sin 10°}$

To calculate number of series turns N_{ph} applying the relation,

$$E_{ph} = \sqrt{2}\,\pi f\, k_w\, N_{ph}\, \phi \quad \therefore N_{ph} = \frac{E_{ph}}{\sqrt{2}\,\pi\, f k_w \times \varphi}$$

$$= \frac{11 \times 1000 \times 3 \sin 10°}{\sqrt{3} \times \sqrt{2} \times \pi \times 50 \times 0.110 \times \cos 10° \times \sin 30°}$$

$$= 274.96 \cong 275$$

For any rotating ac machine, conductors/slot must be a whole number,

Now, conductors per slot $= \frac{\text{no. of series turns} \times m \times 2}{S}$

$$= \frac{275 \times 3 \times 2}{90} = 18.33$$

Which is not a whole number, let us take $N_{ph} = 270$

Conductors per slot $= \frac{270 \times 3 \times 2}{90} = 18$, which is a whole number, Hence, no. of series turns/phase = 270 **Ans.**

Problem 3.14 : A 3-phase, 3kV, 50 Hz, 300 rpm alternator has the following data:

Armature diameter	=	1.90 m,
Core lenght	=	0.39 m
Peak air-gap flux density	=	0.95 T
Number of stator slots	=	180
Air-gap lenght	=	1 cm

This alternator has single-layer winding and two circuits per phase. Find the total armature turns per phase. Assume sinusoidal flux distribution. The winding is short pitched by one slot. Calculate also the peal value of fundamental field mmf.

In case current per conductors is 100 A, estimate kVA rating of the alternator.

Solution :

Gievn, $f = 50$ Hz, $N_r = 300$ rpm, $m = 3$, $r = \frac{d}{2} = \frac{1.9}{2}$ m

$l = 0.39$ m, $B_{peak} = 0.95$ T

No. of stator slots = 180, $g = 1$ cm $= 1 \times 10^{-2}$ cm.

$$\text{No. of poles} \quad (p) = \frac{120 f}{N_r} = \frac{120 \times 50}{300} = 20$$

$$\text{Now, flux per pole} \quad (\phi) = \frac{4}{p} B_{peak} \times l \times r$$

$$= \frac{4}{20} \times 0.95 \times 0.39 \times \frac{1.9}{2} = 0.0704 \text{ Wb}$$

$$\text{Slots per pole per phase} \quad (q) = \frac{s}{m \times p} = \frac{180}{3 \times 20} = 3$$

$$\text{Angular slot pitch,} \quad \gamma = \frac{180 \times p}{s} = \frac{180 \times 20}{180} = 20°$$

$$\text{Chording angle} \quad (\varepsilon) = \text{one slot pitch} = 20°$$

$$\text{Pitch factor,} \quad k_p = \cos\frac{\varepsilon}{2} = \cos\frac{20°}{2} = \cos 10°$$

$$\text{Distribution factor,} \quad k_d = \frac{\sin\frac{q\gamma}{2}}{q \sin\frac{\gamma}{2}} = \frac{\sin\frac{3 \times 20°}{2}}{3 \times \sin\frac{20°}{2}} = \frac{\sin 30°}{3 \sin 10°}$$

$$\therefore \text{Winding factor} \quad (k_w) = k_d \times k_p = \frac{\cos 10° \times \sin 30°}{3 \sin 10°}$$

To calculate armature turns per phase for single circuit.

$$E_{ph} = \sqrt{2}\, \pi\, f\, N_{ph} \times k_w \times \phi$$

$$\therefore \quad N_{ph} = \frac{E_{ph}}{\sqrt{2}\, \pi\, f\, k_w \times \phi}$$

$$= \frac{3000 \times 3 \times \sin 10°}{\sqrt{3} \times \sqrt{2} \times \pi \times 50 \times 0.0704 \times \cos 10° \times \sin 30°}$$

$$= 117.172$$

For double circuit, No. of series turns per phase = 2 × 117.172 = 234.344 ≅ 234. As we know that conductors per slot must be a whole number:

When $N_{phs} = 230$

$$\text{Conductors/slot} = \frac{\text{No. of turns} \times m \times 2}{\text{No. of slots}} = \frac{230 \times 3 \times 2}{180} = 7.667$$

Which is not a whole number.

When $N_{phs} = 234$

$$\text{Conductors/slot} = \frac{234 \times 3 \times 2}{180} = 7.8 \text{ not whole number}$$

When $N_{phs} = 240$ conductors/slot

$$= \frac{240 \times 3 \times 2}{180} = 8 \text{ which is a whole number.}$$

Hence, number of series turns/phase = 240 **Ans.**

To calculate fundamental value of field mmf, applying the relation,

$$f_{1f} = \frac{B_{peak} \times g}{\mu o} = \frac{0.95 \times 1 \times 10^{-2}}{4\,\pi \times 10^{-7}} = 7559.86 \text{ At/pole } \textbf{Ans.}$$

Also, given $I_c = 100$ A, $N_{ph} = 234.344$

$$\text{Slots per pole per phase } (q) = \frac{180}{20 \times 3} = 3$$

$$\text{Angular slot pitch } (\gamma) = \frac{180 \times 20}{180} = 20°$$

$$\text{Distribution factor } (k_d) = \frac{\sin \frac{q\gamma}{2}}{q \sin \frac{\gamma}{2}} = \frac{\sin 30°}{3 \sin 10°}$$

Coil span = one slot pitch = 20

$$\text{Pitch factor } (k_p) = \cos\frac{\varepsilon}{2} = \cos 10°$$

$$\text{Winding factor } (k_w) = k_d \times k_p = \frac{\sin 30° \times \cos 10°}{3 \sin 10°}$$

Now, kVA rating

$$= 3 \times E_{ph} \times I_c$$

$$= 3 \times \sqrt{2} \times \pi \times f \times N_{phs} \times k_w \times \phi \times I_c$$

$$= 3 \times \sqrt{2} \times 50 \times \pi \times 234.344 \times \frac{\sin 30^\circ \times \cos 10^\circ}{3 \sin 10^\circ} \times 100 \times 0.0704$$

$$= 1039.232 \text{ kVA } \textbf{Ans.}$$

> **Problem 3.15 :** The yoke of a 6-pole d.c. machine has its external and internal diameters of 90 and 76 cms respectively and axial length of 30 cms. The armature is wave connected with 72 slots and 6 conductors per slot. The flux density in the yoke is 0.86 T. Calculate the generated e.m.f. at the machine terminals for an armature speed of 900 r.p.m..

Solution :

Given, $P = 6$, $D_i = 76 \text{ cms} = 76 \times 10^{-2}\text{ m}$,

$D_e = 90 \text{ cms} = 90 \times 10^{-2m}$, $l = 30$ cms $= 30 \times 10^{-2}$ m

No. of slots = 72

Conductors per slot = 6, $B_Y = 0.86$ T, N_r 900 r.p.m.

Total no. of conductors Z = (no. of slots) × (conductors per slots)

$Z = 6 \times 72 = 432$

$$\text{Area of the yoke} = \frac{(D_e - D_i)}{2} \times l$$

$$A_Y = \left(\frac{90-76}{2}\right) \times 30 \times 10^{-4} = 0.021 \text{ m}^2$$

Flux though the yoke $= B_Y \times A_Y = 0.021 \times 0.86 = 0.0181$ Wb

Flux per pole = 2 × flux through the yoke

= 0.0362 Wb/pole

To calculate generated emf, applying the relation

$$E_g = \frac{\phi\, NZp}{A \times 60}$$

For wave-winding A = 2

$$E_g = \frac{0.0362 \times 900 \times 432 \times 6}{60 \times 2} = 702.173 \quad \textbf{Ans.}$$

Problem 3.16 : A 3-phase delta-connected induction motor has 6 poles, 36 slots, 2-layer winding and 12-turn coils short-pitched by one slot. Find the flux per pole of the rotating flux density wave, with stator connected to 400 V, 50 Hz.

Solution :

Given, $P = 6$, No. of slots $(s) = 36$

$v_t = 400$ V, $f = 50$ Hz, $m = 3$

For delta-connected motor, $v_{ph} = V_L$

According to Question, emf per phase,

$$E_{ph} = V_t - 400 \times \frac{1}{100} = 400 - 4 = 396\text{V}$$

Slots per pole per phase $(q) = \dfrac{36}{6 \times 3} = 2$

Angular slot pitch $(\gamma) = \dfrac{180 \times 6}{36} = 30°$

Distribution factor, $k_d = \dfrac{\sin\frac{q\gamma}{2}}{q \sin\frac{\gamma}{2}} = \dfrac{\sin\frac{2 \times 30°}{2}}{2 \sin\frac{30°}{2}} = \dfrac{1}{4 \sin 15°}$

Chording angle (ε) = one slot pitch = $30°$

pitch factor, $k_p = \cos\dfrac{\varepsilon}{2} = \cos\dfrac{30°}{2} = \cos 15°$

$\therefore$ Winding factor $(k_w) = k_d \times k_p = \dfrac{1}{4 \sin 15°} \times \cos 15° = \dfrac{\cos 15°}{4 \sin 15°}$

No. of series turns per phase,

$$N_{phs} = \frac{\text{No. of slots} \times \text{no. of coils}}{\text{no. of phases}} = \frac{36 \times 12}{3} = 144.$$

To calculate flux per pole, applying the relation,

$$E_{ph} = \sqrt{2}\, \pi\, f\, k_w\, N_{ph} \times \phi$$

$$\therefore \quad \phi = \frac{E_{ph}}{\sqrt{2} \times \pi\ f k_w\ N_{ph}} = \frac{396 \times 4 \times \sin 15°}{\sqrt{2} \times \pi \times 50 \times \cos 15° \times 144}$$

$= 0.013226$ Wb **Ans.**

Problem 3.17 : A 3-phase, 100 kW, 6-pole, 50 Hz, delta-connected induction motor has 20 conductors per slot arranged in 72 slots. For a line current of 76 A, compute the peak value of the fundamental m.m.f. wave of any phase. Also, calculate the peak value of the resultant m.m.f. wave and its speed. The winding is short-pitched by one slot.

Solution :

Given, $p = 6,\ f = 50$ Hz, $m = 3$

No. of slots $= 72,$

Conductors per slot $= 12,\ I_L = 76$ A

Total No. of conductors $= 20 \times 72$

Total No. of turns $= \frac{20 \times 72}{2} = 720$

No. of turns per phase, $N_{phs} = \frac{720}{3} = 240$

Phase current, $I = \frac{I_L}{\sqrt{3}} = \frac{76}{\sqrt{3}}$

To calculate peak value of fundamental mmf.
Applying the relation,

$$F_{1f} = \frac{4}{\pi} k_w \frac{N_{phs}}{p} \times \sqrt{2}\, I \text{ ATs/pole} \quad (1)$$

Now, Slots per pole per phase $(q) = \frac{\text{no. of slots}}{\text{no. of poles} \times \text{phase}}$

$$= \frac{72}{6 \times 3} = 4$$

Angular slot pitch $\gamma = \frac{180 \times p}{\text{no. of slots}} = \frac{180 \times 6}{7.2} = 15^\circ$

Chording angle (ε) = one slot pitch $= 15^\circ$

Distribution factor, $k_d = \frac{\sin \frac{q\gamma}{2}}{q \sin \frac{\gamma}{2}} = \frac{\sin \frac{4 \times 15}{2}}{4 \sin \frac{15^\circ}{2}} = \frac{1}{8 \sin 7.5}$

Pitch factor, $k_p = \cos \frac{\varepsilon}{2} = \cos \frac{15}{2} = \cos 7.5^\circ$

$\therefore$ Winding factor, $k_w = k_d \times k_p = \frac{1}{8 \sin 7.5} \times \cos 7.5 = \frac{\cos 7.5}{8 \sin 7.5}$

From (*i*), Peak value of fundamental mmf

$$= \frac{4}{\pi} k_w \frac{N_{phs} \times \sqrt{2}\, I}{p}$$

$$= \frac{4}{\pi} \times \frac{\cos 7.5}{8 \sin 7.5} \times \frac{240}{6} \times \sqrt{2} \times \frac{76}{\sqrt{3}} \text{ ATs/pole}$$

$$= 3000.68 \text{ ATs/pole } \textbf{Ans.}$$

Peak value of resultant mmf $= \frac{3}{2} \times$ peak value of fundamental mmf

$$= \frac{3}{2} \times 3000.68 = 4501.0 \text{ ATs/pole } \textbf{Ans.}$$

Speed $(N) = \frac{120 f}{p} = \frac{120 \times 50}{6} = 1000$ r.p.m. **Ans.**

Problem 3.18 : The armature m.m.f. and field flux density vary sinusoidally with respect to angular position in the air gap of a rotating electrical machine having 4 poles and a rated speed of 960 r.p.m. At rated load conditions, the flux per pole is 20 m Wb, the amplitude of the armature m.m.f. wave is 530 ampere turns and the space displacement between the two waveforms is 36.685°. Estimate the power rating of the machine deriving the necessary relations.

Solution :

Given, $P = 4$, $N = 960$ rpm, $\phi = 20 \times 10^{-3}$ Wb

$F_a = 530$ ATS, $S = 36.685$

Speed in rad/sec, $\omega = \frac{960 \times 2\pi}{60} = 32\pi$ rad / sec.

To calculate power rating. Let us calculate the electro-magnetic torque developed.

Torque developed $(T_e) = \frac{\pi}{8} p^2 \phi F_a \sin \delta$ Nm

Power Rating of the machine,

$$p = T_e \times \omega = \frac{\pi}{8} p^2 \varphi F_e \sin \delta \times \omega$$

$$= \frac{\pi}{8} \times (4)^2 \times 0.020 \times 530 \times \sin 36.685 \times 32\,\pi \text{ W}$$

$$= 4000 \text{ W}, \quad = 4 \text{ kW} \textbf{ Ans.}$$

Problem 3.19 : (*a*) The proposed load schedule for a small rolling mill is as under:

Billet in the rolls, 500 kW for 30 sec.

Billet out of the rolls, 80 kW for 40 sec.

Find a continuous kW rating of the motor.

(*b*) The operating cycle for a lift is as under:

Load going up (1min) = 60 kW

Loading period at top (2 min) = 10 kW

Load going down (1 min) = –50 kW

(power pumped back to supply)

Loading period at bottom (2 min) = 10 kW

On the basis of heating, select a suitable motor out of the following available sizes: 25, 30, 35, 40, 50 and 60 kW.

Solution :

(*a*) Given, $p_1 = 500$ kW, $p_2 = 80$ kW

$t_1 = 30$ sec, $t_2 = 40$ sec

Now, Continuous kW rating of the motor $= \left[\dfrac{\Sigma\{(p)^2 \times t\}}{\Sigma t}\right]$

Where p is in kW $= \left[\dfrac{p_1^2 \times t_1 + p_2^2\, t_2}{t_1 + t_2}\right]^{1/2}$

$$= \left[\frac{(500)^2 \times 30 + (80)^2 \times 40}{70}\right]^{1/2} = 332.87 \text{ kW}$$

This method does not give precise results. The error involved in using this method is partically offset by choosing a higher standard rating of the motor.

Hence, Rating of the motor = 350 kW **Ans.**

(*b*) Here $p_1 = 60$ kW, $t_1 = 1$ minute

$p_2 = 10$ kW, $t_2 = 2$ minute

$p_3 = -\ 50$ kW, $t_3 = 1$ min

$p_4 = 10$ kW, $t_4 = 2$ min

Rating of CR motor

$$= \left[\frac{\Sigma \{(p)^2 \times t\}}{\Sigma t}\right]^{1/2}$$

$$= \left[\frac{(60)^2 \times 1 + (10)^2 \times 2 + (-50)^2 \times 1 + (10)^2 \times 2}{1+2+1+2}\right]^{1/2}$$

$$= 32.91 \text{ kW}$$

As we know that this method does not give precise result. The error introduced in using this method is partially offset by choosing a higher value.

From the given option of rating as 25, 30, 35, 40, 50 and 60 kW

Rating of the motor = 35 kW **Ans.**

Problem 3.20 : A motor driving a colliery winding equipment has the following points on its load duty cycle.

Time, sec	0	20	20	60	60	70	90	Repeat
Output, kW	0	1000	1000 to 500	500	500 to − 200	0	0	cycle

The complete load cycle curve can be obtained by joining these points with straight lines.

Specify a continuous kW rating of the motor for this load cycle.

Solution :

The complete load cycle curve from given data can be drawn as

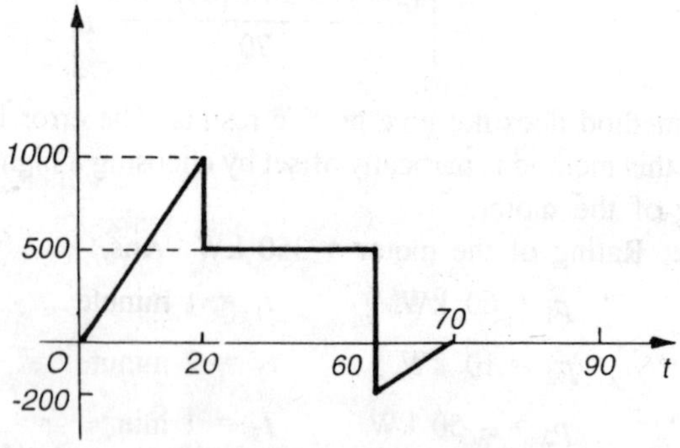

Fig. P. 3.20

Here, $T = 90$ seconds.

For duration 0 to 20 seconds $H_1 = 0$ kW, $H_2 = 1000$ kW,

$\Delta t_1 = 20s$

For duration 20 to 60 seconds $H_1 = H_2 = 500$kW, $\Delta t_2 = 40$ sec

For duration 60 to 70 sec. $H_1 = 200$ kW, $H_2 = 0$ kW,

$\Delta t_3 = 10$ sec and so on.

Applying the formula for continuous kW rating.

kW rating

$$= \left[\frac{1}{T}\left\{\frac{\Delta t_1}{3}\left(H_1^2 + H_2^2 + H_1 \times H_2\right)\right\} + \frac{\Delta t_2}{3}\left(H_1^2 + H_2^2 + H_1 \times H_3\right)\right.$$

$$\left. + \frac{\Delta t_3}{3}\left(H_1^2 + H_2^2 + H_1 \times H_2\right)\ \right]^{1/2}$$

$$= \left[\frac{1}{90}\left\{\frac{20}{3}\left(0^2 + (1000)^2 + 0 \times 1000\right)\right.\right.$$

$$\left.\left. + \frac{40}{3}((500)^2 + (500)^2 - (500 \times 500) + \frac{10}{3}\left[\left\{(200)^2 + (0)^2\right\}\right]\right]$$

$$= 100\left[\frac{1}{90}\left(\frac{2000}{3} + \frac{1000}{0} + \frac{40}{3}\right)\right]^{1/2} = 100\left[\frac{1}{90}\left(\frac{5040}{3}\right)\right]^{1/2}$$

$$= 100\left[\frac{5040}{270}\right]^{1/2} = 432.05 \text{ kW}$$

Since this method does not given the precise result. The error involved in using this method is partially offset by choosing a higher value.

$\therefore$ Rating = 450 kW **Ans.**

Problem 3.21 : A 10 kW, 3-phase induction motor has a heating time constant of 2 hours and attains a final steady temperature rise of 50°C with continuous loading. Find the time for which this motor may be safely loaded at 20 kW. The full-load ohmic loss is twice the core loss and ambient temperature is 30°C.

Solution :

Given, $P_r = 10$ kW, $t_h = 2$ hrs, $\theta_f = 50°$ C

$P_x = 20$ kW, $t_x = ?, W_o = 2 \times W_c$

Now, Here $\alpha = \frac{W_c}{W_o} = \frac{Wc}{2W_c} = \frac{1}{2}$

To calculate t_x, Applying the relation

$$\left(\frac{P_x}{P_r}\right)^2 = \frac{1+\alpha}{1-e^{-\frac{t_x}{t_h}}} - \alpha \text{ or } \left(\frac{20}{10}\right)^2 = \frac{1+\frac{1}{2}}{1-e^{-\frac{t_x}{2}}} - \frac{1}{2}$$

or $$4 + \frac{1}{2} = \frac{3}{2}\left(\frac{1}{1-e^{-\frac{t_x}{2}}}\right) \text{ or } 1-e^{-\frac{t_x}{2}} = \frac{1}{3}$$

or $$e^{-\frac{t_x}{2}} = 1 - \frac{1}{3} = \frac{2}{3}$$

or $$t_x = -2\ln\left(\frac{2}{3}\right) = 0.8109 \text{ hr } \textbf{Ans.}$$

Problem 3.22 : An induction motor has a final steady temperature rise of 40°C when running at its rated output of 20 kW. Determine its half-hour rating for the same temperature rise if the ohmic losses at rated output are 1.25 times its constant losses. The heating time constant is 90 minutes.

Solution :

(a) Given $\theta_f = 40°\text{ C}, P_r = 20\text{ kW}, t_x = \frac{1}{2}\text{ hr} = 30\text{ minutes}$

$$t_h = 90\text{ minutes}, P_x = 2$$

$$W_o = 1.25\,W_c \Rightarrow \alpha = \frac{W_c}{W_o} = \frac{1}{1.25} = \frac{4}{5}$$

Applying the relation,

$$P_x = p_r\left[\frac{1+\alpha}{1-e^{-\frac{t_x}{t_n}}} - \alpha\right]^{1/2} = 20\left[\frac{1+\frac{4}{5}}{1-e^{-\frac{1}{3}}} - \frac{4}{5}\right]^{1/2}$$

$$= 20\,(5.5499)^{1/2} = 47.116 \text{ kW } \textbf{Ans.}$$

(b) With improved cooling by 20%

Total losses, $W_x = W_r + W_r \times \dfrac{20}{100} = 1.2\,W_r$

$\therefore \quad \dfrac{W_x}{W_r} = 1.2$

To calculate power rating. Applying the relation

$$\frac{W_x}{W_r} = \frac{\alpha + \left(\dfrac{P_x}{P_r}\right)^2}{1+\alpha}$$

or $$1.2 \times \left(1 + \frac{4}{5}\right) = \frac{4}{5} + \left(\frac{P_x}{20}\right)^2$$

or $$\left(\frac{P_x}{20}\right)^2 = \frac{10.8}{5} - \frac{4}{5}$$

$\therefore$ $$P_x = 20\sqrt{\frac{6.8}{5}} = 23.234 \text{ kW} \quad \textbf{Ans.}$$

(c) For a temperature rise of 50°C,

Total losses at $\theta = 50°\text{C}, \; W_x = \dfrac{\theta}{\theta_f} \times W_x$

at improved cooling $= \dfrac{50°}{40°} \times 1.2\,W_r = 1.5\,W_r$

Now, Applying the relation, $$\frac{W_x}{W_r} = \frac{\alpha + \left(\dfrac{P_x}{P_r}\right)^2}{1+\alpha}$$

or $$1.5 = \frac{\dfrac{4}{5} + \left(\dfrac{P_x}{20}\right)^2}{1 + \dfrac{4}{5}} \quad \text{or} \quad 1.5 \times \frac{9}{5} = \frac{4}{5} + \left(\frac{P_x}{20}\right)^2$$

or $$\frac{13.5}{5} - \frac{4}{5} + \left(\frac{P_x}{20}\right)^2 \quad \text{or} \quad \left(\frac{P_x}{20}\right)^2 = \frac{9.5}{5}$$

$\therefore$ $$P_x = 20\sqrt{\frac{9.5}{5}} = 20\sqrt{1.9} = 27.568 \text{ kW} \quad \textbf{Ans.}$$

Problem 3.23 : *(a)* Temperature rise of an electric motor at rated load of 30 kW was found to be 20°C after one-hour run and 50°C after three-hour run. In case its full load ohmic loss is 1.5 times the constant loss, calculate its one-hour rating without exceeding the specified temperature rise.

(b) In case temperature rise of 10% is allowed for motor in part *(a)* find its new continuous rating.

Solution :

(a) Given, $P_r = 30\text{ kW},\ \theta_1 = 20°\text{ C},\ t_1 = 1\text{ hr}$

$$\theta_2 = 50°\text{ C},\ t_2 = 3\text{ hr},\ W_o = 1.5 \times W_c$$

Here $$\alpha = \frac{W_c}{W_o} = \frac{W_c}{1.5 \times W_c} = \frac{1}{1.5} = \frac{2}{3}$$

Let θ_f be the final steady state temperature $\therefore\ \theta = \theta_f\left(1 - e^{\frac{-t}{t_h}}\right)$

Where t_h = heating time constant

After $$t_1 = 1\text{ hr},\ 20 = \theta_f\left(1 - e^{\frac{-1}{t_h}}\right) \qquad (1)$$

After $$t_1 = 3\text{ hr},\ 50 = \theta_f\left(1 - e^{\frac{-3}{t_h}}\right) \qquad (2)$$

From (1) and (2)

$$\frac{50}{20} = \frac{1 - e^{\frac{-3}{t_h}}}{1 - e^{\frac{-3}{t_h}}} \text{ or } \frac{5}{2} = \frac{1 - \left(e^{\frac{-1}{t_h}}\right)^3}{1 - e^{\frac{-1}{t_h}}}$$

$$\frac{5}{2} = \frac{1 + e^{\frac{-1}{t_h}} + e^{\frac{-2}{t_h}}}{1}$$

Putting $e^{\frac{-1}{t_h}} = x$ we get $1 + x + x^2 = 2.5$ or $x^2 + x - 1.5 = 0$

This yields $x = \dfrac{-1 \pm \sqrt{1 + 4 \times 1.5}}{2}$

Neglecting negative value, $x = \frac{-1 \pm \sqrt{7}}{2}$ hr $= 0.8229$

$$x = e^{\frac{-1}{t_h}} = 0.8229$$

$$\therefore \quad t_h = -\frac{1}{\ln 0.8229} = 5.1303 \text{ hr}$$

$$\therefore \quad \theta_f = \frac{20}{1 - e^{\frac{-1}{5.1303}}} = 113°\text{ C}$$

Now, power rating of the motor after 1 hour

$$= P_r \left[\frac{1+\alpha}{1 - e^{\frac{-t}{t_h}}} - \alpha \right]^{1/2} = 30 \left[\frac{1 + \frac{2}{3}}{1 - e^{\frac{-1}{5.1303}}} - \frac{2}{3} \right]^{1/2}$$

$$= 30 \left[\frac{1 + \frac{2}{3}}{1 - e^{\frac{-1}{5.1303}}} - \frac{2}{3} \right]^{1/2} = 30\,[8.7442]^{1/2} = 88.71 \text{ kW}$$

(b) After 10% temperature rise, $\theta = 1.1\,\theta_f$

$$\therefore \quad \frac{\theta}{\theta_f} = 1.1$$

Total losses after 10% temperature rise, $W_x = \frac{\theta}{\theta_f} \times W_r = 1.1\,W_r$

$$\therefore \quad \frac{W_x}{W_r} = 1.1$$

Now, Applying the relation $\frac{W_x}{W_r} = \frac{\alpha + \left(\frac{P_x}{P_r}\right)^2}{1+\alpha}$

or $$1.1 \times \left(1 + \frac{2}{3}\right) = \frac{2}{3} + \left(\frac{P_x}{30}\right)^2$$

or $$\left(\frac{P_x}{30}\right)^2 = \frac{5.5}{3} - \frac{2}{3} = \frac{3.5}{3}$$

$$\therefore \quad P_x = 30 \times \sqrt{\frac{3.5}{5}} = 32.40 \text{ kW} \quad \textbf{Ans.}$$

Problem 3.24 : *(a)* The temperature rise of an 80 kW motor is 25° C after 0.5 h and 40°C after 1.0 h on full load. The motor I^2R loss on full-load is twice the core loss. When this motor is fitted with a fan to improve the cooling, the final full-load temperature is found to be 48° C. Estimate the motor rating for final temperature rise to be that obtained without fan. Derive the formula used for obtaining this rating. *(b)* For the motor of part *(a)* determine its one-hour rating (1) without fan, and (2) with fan.

Solution :

(a) Given $P_r = 80 \text{ kW}, \ \theta_1 = 25° \text{C}, \ t_1 = 0.5 \text{ hr}$

$$\theta_2 = 40° \text{C}, \ t_2 = 1.0 \text{ hr} \ W_o = 2 \times W_c \quad \therefore \quad \frac{W_c}{W_o} = \frac{1}{2} = \alpha$$

Let t_h be the heating time constant of motor and θ_f be the final steady temperature rise

Now, Applying the relation, $\theta = \theta_f \left(1 - e^{\frac{-t}{t_h}}\right)$

At temperature 25° C, $$25 = \theta_f \left(1 - e^{\frac{0.5}{t_h}}\right) \quad (1)$$

At temprature 40° C, $$40 = \theta_f \left(1 - e^{\frac{-1}{t_h}}\right) \quad (2)$$

From (1) and (2) $$\frac{40}{25} = \frac{1 - e^{\frac{-1}{t_h}}}{1 - e^{\frac{-1}{2t_h}}} \quad \text{or} \quad \frac{40}{25} = \frac{1 - \left(e^{\frac{-1}{2t_h}}\right)^2}{1 - e^{\frac{-1}{2t_h}}}$$

or $$\frac{8}{5} = 1 + e^{\frac{-1}{2t_h}} \quad \text{or} \quad e^{\frac{-1}{2t_h}} = 0.6$$

$$\therefore \qquad t_h = -\frac{1}{2\ln 0.6} = 0.979 \text{ hr}$$

Form (1), Final steady temprature, $\theta_f = \dfrac{40}{1 + e^{\frac{-1}{0.979}}} = 62.51°C$

When fan is employed, $\theta = 48°$

At $48°C$, $$48 = 62.51\left(1 - e^{\frac{-t}{t_h}}\right)$$

$$\therefore \qquad 1 - e^{\frac{-t}{t_h}} = \frac{\theta}{\theta_f} = \frac{48}{62.51} = 0.7679$$

To calculate motor rating at $48°C$, applying the formula,

$$P_x = P_r\left[\frac{1+\alpha}{1 - e^{\frac{-t}{t_h}}} - \alpha\right]^{1/2} = 80\left[\frac{1+\frac{1}{2}}{0.7679} - \frac{1}{2}\right]^{1/2}$$

$$= 80\sqrt{1.4534} = 96.45 \text{ Kw} \quad \textbf{Ans.}$$

Expression for Rating

The temprature rise at any instant is given by,

$$\theta = \theta_f\left(e^{\frac{-t}{t_n}}\right) \tag{3}$$

Where, θ = Final temprature rise,

θ_f = Final steady temprature rise

t = Time, t_h = heating time constant

To obtain power rating at any instant, we have

$$P_x = P_r\left[\frac{1+\alpha}{1 - e^{\frac{-t}{t_h}}} - \alpha\right]^{1/2}$$

From (3) $$P_x = P_r\left[\frac{(1+\alpha)\times\theta_f}{\theta} - \alpha\right]^{1/2}$$

This is the required expression.

(1) Without fan, we have, $t_h = 0.979$ hr , $t = 1$ hr , $\alpha = \frac{1}{2} = P_r = 80$ kW

Applying the relation,

$$P_x = P_r\left[\frac{1+\propto}{1-e^{\frac{-t}{t_h}}} - \propto\right]^{1/2} = 80\left[\frac{1+\frac{1}{2}}{1-e^{\frac{-1}{0.979}}} - \frac{1}{2}\right]^{1/2}$$

$$= 80 \times \sqrt{1.8440} = 108.64 \text{ kW} \quad \textbf{Ans.}$$

When fan is used, % improve in temperature dissipation,

$$\frac{\theta_f - \theta_f}{\theta_f} = \frac{62.51 - 48}{62.51} = 23.21\%$$

Since losses depends upon temperature

$$\therefore \quad W_e = W_o + 0.2321\, W_o + 0.5\, W_o$$

$$\therefore \quad \frac{W_e}{W_o}(\alpha) = 1.7321$$

We have, $t_h = 0.979$ hr

To calculate power rating, Apply the relation,

$$P_x = P_r\left[\frac{1+\alpha}{1-e^{\frac{-t}{t_h}}} - \alpha\right]^{1/2} = 80\left[\frac{1+1.7321}{1-e^{\frac{-1}{.979}}} - 1.7321\right]^{1/2}$$

$$= 80\sqrt{2.537388} = 127.43 \text{ kW} \quad \textbf{Ans.}$$

Chapter 4

D.C. MACHINES

D.C. MACHINES

DC machines: The principle of operation of dc machines is based upon Faraday's law of electromagnetic induction. It converts dc power into mechanical power or vice versa. The schematic diagram based upon Faraday's law of electromagnetic induction is shown below.

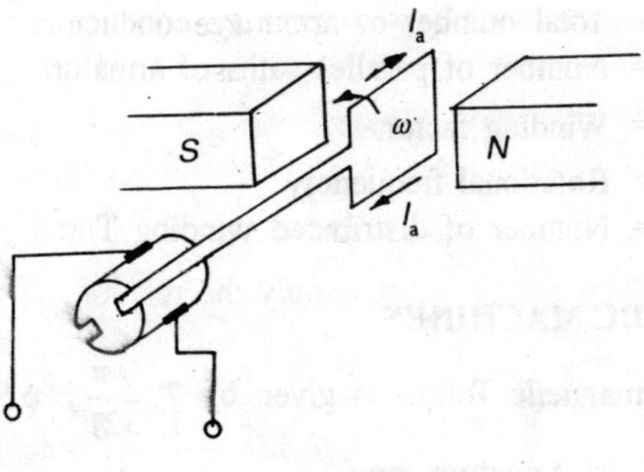

Fig. 4.1

EMF EQUATION OF DC MACHINES

- The average emf E_c generated in one coil of N_c Turns is given by

$$E_c = 2\,N_c\,p.N.\phi$$

- The brush voltage or the Total emf generated in armature is

$$E_a = \frac{c}{a}E_c = \frac{c}{a}(2\,N_c\,p.N.\phi)$$

or $$E_a = \frac{\phi NZ}{60}\frac{P}{A}$$ where N is in rpm.

or $$E_a = 4\,f_r\,N\,\phi$$

- The rms value of the generated emf in a distributed winding is given by $$E_{rms} = \sqrt{2}\,\pi\,K_w f_r\,N\,\phi$$
- The average value of generated emf per conductor is

$$E_{av} = B_{av} \times l \times v$$

$$E_{av} = B_{av} \times l \times 2\pi rN, \quad E_{av} = P\,\phi\,N$$

- The average flux density is given by $B_{av} = \dfrac{P\phi}{2\pi rl}$

Where
P = Number of poles
ϕ = Flux per pole
r = Mean air gap radius
l = Active conductor length
N = Speed
N_c = Number of Turns in one coil
C = Total number of coils on the armature.
Z = Total number of armature conductors
a = Number of parallel paths of armature
K_w = Winding factor
f_r = Rotational frequency
N_w = Number of distributed winding Turns.

TORQUE IN DC MACHINES

- The Electromagnetic Torque is given by $T_e = \dfrac{\pi}{8} p^2 \phi F_a$

Where F_a = Armature mmf

$$F_a = \frac{8}{\pi^2}\left(\frac{2}{2P} \times \frac{I_a}{A}\right) \text{ ATS/pole}$$

Where I_a = Armature current

or $$T_e = \frac{Pz\phi I_a}{2\pi A} \text{ or } T_e = k_a\,\phi\,I_a$$

Where $K_a = \dfrac{ZP}{2\pi a}$ is called armature constants

- From power balance Theorem $T_e = \dfrac{E_b \times I_a}{\omega_n}$

Where ω_n = motor shaft speed in rad/sec

Generating mode: $E_g = V_t + I_a r_a$

Shaft power Input = $E_a \times I_a$ + no-load rotational loss

Where V = Armature terminal voltage

r_a = Armature circuit resistance

Electrical power output = Electromagnetic power – ohmic losses – Brush contact loss

Motoring Mode

$$E_b = V_t - I_a r_a, \quad \text{For shunt motor}$$

$$E_b = V_t - I_a\left(r_a + r_b\right), \quad \text{For series motor}$$

Shaft power input = $E_b \times I_a$ – no-load rotational loss

Electrical power input = Electromagnetic power + ohmic losses + Brush contact loss

Commutation

$$\text{Commutation period } (T_c) = \frac{\text{Brush width}}{\text{Commutator peripheral speed}}$$

Interpoles

Interpole mmf/pole = Armature mmf/pole + air gap mmf

$$I_a N_{cp} = \frac{I_a Z}{ZAp} + \frac{B_{cp}}{\mu_o}.g_{cp}$$

Where B_{cp} = Interpolar flux density

g_{cp} = Interpolar air gap length

N_{cp} = Number of turns on each interpole

Compensating Winding

$$\text{Compensating winding mmf/pole} = \frac{\text{Pole arc}}{\text{Pole pitch}} \times \text{Armature mmf/pole}$$

$$AT_c = \frac{\text{Pole arc}}{\text{Pole pitch}} \times \frac{I_a Z}{ZAp} \text{ or } \frac{Z_{cw} \times I_a}{2} = \frac{\text{Pole arc}}{\text{Pole pitch}} \times \frac{I_a Z}{ZAp}$$

$$Z_{cw} = \frac{Z}{Ap}$$

Where Z_{cw} = Compensating winding mmf/pole

T_c = Number of turns on compensating winding

BASIC EQUATION OF D.C. MACHINES

$$E_a = \frac{\phi n z p}{60A}, \text{ where } n = \text{r.p.m.},$$

$$E_a = \frac{\phi \omega_m z p}{2\pi A} = \left(\frac{zp}{2\pi a}\right) \times \phi \omega_m \Rightarrow E_a = K_a \phi \omega_m$$

$$E_a = k_a \phi \omega_m \text{ where } K_a = \frac{zp}{2\pi a} \text{ is called armature constant}$$

$$T_e = \frac{\phi R Z I_a}{2\pi A} = \frac{PZ}{2\pi A} \times \phi I_a, \quad T_e = K_a \times \phi I_a$$

Now $\frac{E_a}{T_e} = \frac{\omega_m}{I_a}$ for constant flux.

For Compound d.c. Machines

Total d-axis mmf $= N_f I_f \pm I_s N_s$

Where $N_f =$ shunt field turns, $N_s =$ series field turns

Where + ve sign is taken for cumulatively machine

– ve sign is taken for differentially compounded machine

Equivalent current in shunt field winding.

$$= \frac{N_f I_f \pm I_s N_s}{N_f}, \quad I_{sh} = I_f \pm \frac{N_s}{N_f} I_s$$

When demagnetising effect is included,

$$\text{Net mmf} = N_f I_f \pm N_s I_s - AT_d$$

D.C. SHUNT MOTOR

(i) Speed-current characteristics $\omega_m = \frac{V_t - I_a r_a}{K_a \phi}$

At no-load $\omega_{no} = \frac{V_t}{K_a \phi}$

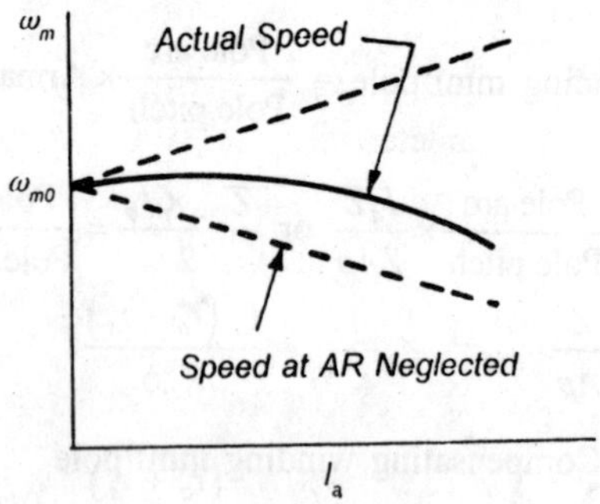

Fig. 4.2

(ii) Torque armature Current Characteristics

$$T_e = K_a \phi I_a, \qquad T_e \propto I_a$$

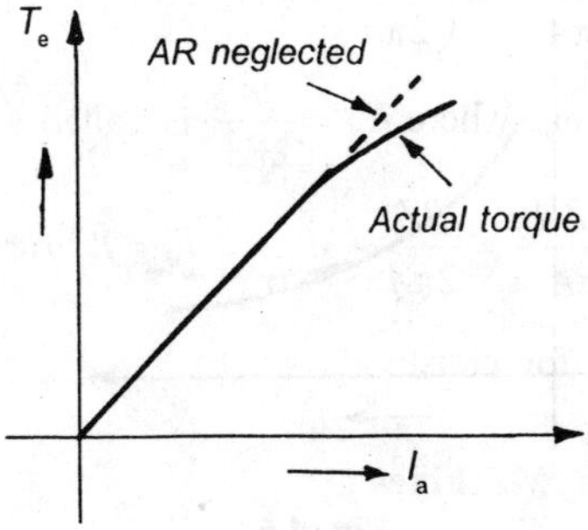

Fig. 4.3

(iii) Speed-torque Characteristics

$$\omega_m = \frac{i}{K_a\phi}\left[V_t - \frac{T_e\, r_a}{K_a\phi}\right] = \omega_{mo} - r_a \frac{T_e}{K_a^{\,2}\phi^2}$$

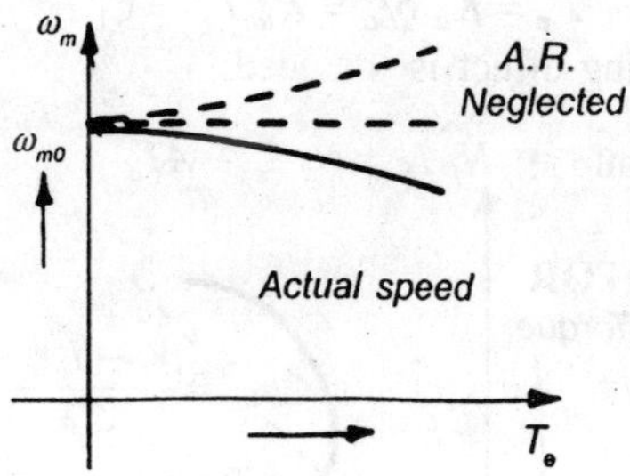

Fig. 4.4

D.C Series Motor

(i) Speed-current characteristics

$$E_a = V_t - I_a(r_a + r_s)$$

or

$$\omega_m = \frac{V_t}{K_a\phi} - \frac{I_a(r_a + r_s)}{K_a\phi}$$

$$= \frac{V_t}{K_{ac}\, I_a} - \frac{(r_a + r_s)}{K_{ac}}$$

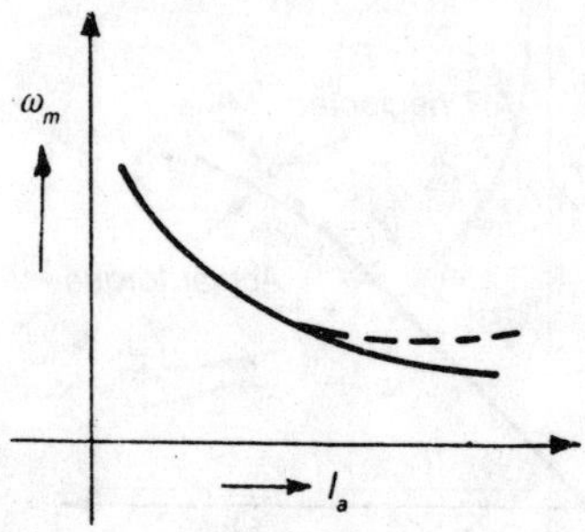

Fig. 4.5

At no-load $$\omega_{mo} = \frac{V_t}{K_a\phi} - \frac{V_t}{K_{ac}\ I_a}$$

(ii) Torque-armature current characteristics

$$T_e = K_a\ \phi I_a = K_{ac} I_a^{\ 2} = C_1\ I_a^{\ 2}$$

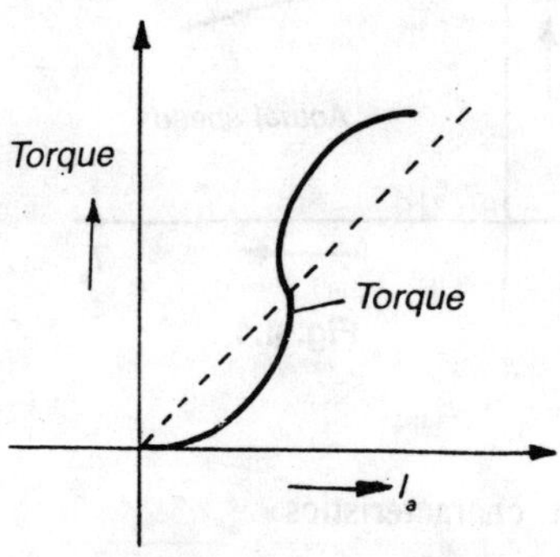

Fig. 4.6

(iii) Speed-torque characteristics

$$\omega_m = \frac{V_t}{\sqrt{K_{ac}\ T_e}} - \frac{r_a + V_s}{K_{ac}} = \frac{V_t}{K_a\phi} - \frac{T_e\left(r_a + V_s\right)}{K_a^{\ 2}\phi}$$

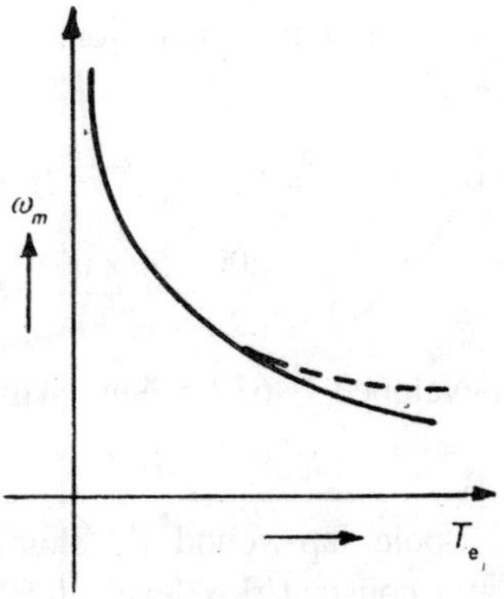

Fig. 4.7

> **Problem 4.1 :** A 10 kW, 6-pole dc generator develops an emf of 200 V at 1500 rpm. The armature has a lap-connected winding. The average flux density over a pole-pitch is 0.9 Tesla. The length and diameter of the armature are 0.25 m and 0.2 m respectively. Calculate *(i)* the flux per pole, *(ii)* the total number of active conductors in the armature, and *(iii)* the torque developed by the machine when the armature supplies a current of 50A. *(GATE, 1991)*

Solution :

Given, $E_b = 200$ V, $N = 1500$ rpm, $P = 6$

$B_{av} = 0.9$ T, $l = 0.25$ m, $d = 0.2$ m

(i) Applying $B_{av} = \dfrac{P\phi}{2\pi rl} \Rightarrow \phi = \dfrac{B_{av} \times 2\pi rl}{P}$

Thus, flux per pole $= \dfrac{0.9 \times 2 \times 3.14 \times 0.1 \times 0.25}{6}$

$= 0.02355$ weber **Ans.**

(ii) Applying $E_b = \dfrac{\phi N Z P}{60 A} = \dfrac{\phi N Z P}{60 A}$

for lap winding, $A = P = 6 \Rightarrow Z = \dfrac{60 E_b A}{\phi NP}$

$$= \frac{60 \times 200 \times 6}{0.02355 \times 1500 \times 6} = 339.70 = 340 \textbf{ Ans.}$$

Thus, Total number of active conductors in the Armature = 340

(iii) Given, $I_a = 50$ A.

Applying equation, $T \times \omega = E_b \times I_a$, $T \times 2\pi N = E_b \times I_a$

$$T = \frac{200 \times 50 \times 60}{2\pi \times 1500} = 63.66 \text{ Nm}$$

Thus, Torque developed = 63.66 Nm **Ans.**

Problem 4.2 : A 2-pole lap-wound dc shunt motor with 360 conductors operates at a constant flux level of 50 mWb. The motor armature has a resistance of 0.12 Ω and is designed to operate at 240 V, taking a current of 60 A at full load.

(i) Determine the value of external resistance to be inserted in the armature circuit so that armature current does not exceed twice its full-load value at starting.

(ii) The external resistance is completely cut out when the motor reaches its final speed, with the armature current at the full-load value. Calculate the motor speed under this condition.

(GATE, 1992)

Solution :

Given, $Z = 360, \phi = 50 \times 10^{-3}$ Wb, $P = 2, r_a = 0.12\ \Omega$

for lap winding $A = P = 2$, $V_t = 240$ V, $I_{fl} = 60$ A

(i) Let R_{ext} be the external resistance to be inserted in armature circuit.

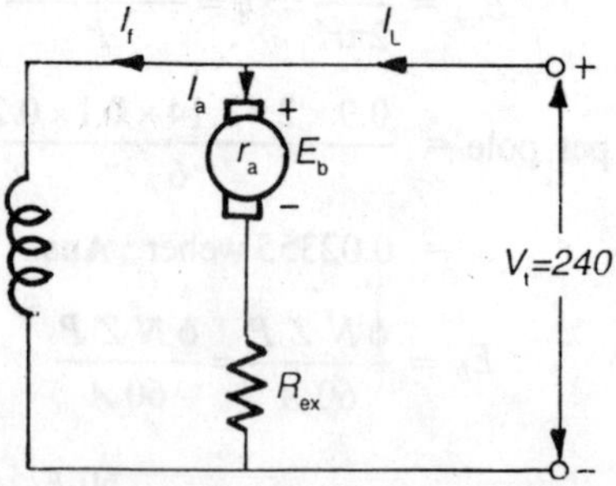

Fig. P. 4.2

Assuming field current to be negligible in comparison to armature current at starting. $I_a = 2 \times 60 = 120$ A

From motor circuit $V_t = I_a\left(R_{ext} + r_a\right) + E_b$

At starting $E_b = 0$, From above

$$R_{ext} + r_a = \frac{V_t}{I_a} \Rightarrow R_{ext} = \frac{V_t}{I_a} - r_a = \frac{240}{120} - 0.12 = 1.88\ \Omega$$

Thus, External resistance to be inserted in Armature Circuit

$= 1.88\ \Omega$

(ii) When external resistance is completely cut out

Armature resistance $= 0.12\ \Omega$

Neglecting field current

Armature current $I_L = I_a = 60$ A

To calculate speed, Applying $E_b = \dfrac{\phi N Z P}{60\, A}$

$$\therefore N = \frac{E_b \times 60\, A}{\phi Z P} = \frac{\left(V_t - I_a r_a\right) \times 60 \times A}{\phi Z P} \qquad \left[\because V_t = E_b + I_a r_a\right]$$

$$= \frac{(240 - 60 \times 0.12) \times 60 \times 2}{50 \times 10^{-3} \times 360 \times 2} = 776 \text{ rpm } \textbf{Ans.}$$

Thus, final speed = 776 rpm **Ans.**

Problem 4.3 : A dc shunt generator gives an open-circuit voltage of 240 V. Determine the load current in case armature-circuit and field-winding resistances are $0.1\ \Omega$ and $50\ \Omega$ respectively.

Solution :

For given situation the circuit can be drawn as below.

From fig.

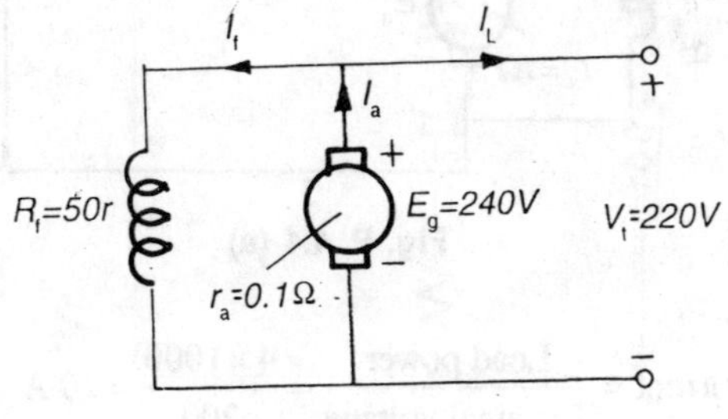

Fig. P. 4.3

Field current $(I_f) = \dfrac{220}{50} = 4.4$ A

For generator, $E_g = V_t + I_a r_a$, or $240 = 220 + 0.1 \times I_a$

$\therefore \quad I_a = \dfrac{240 - 220}{0.1} = 200$ A

and $\quad I_L = I_a - I_f = 200 - 4.4 = 195.6$ A

Thus Load current = 195.6 A **Ans.**

Problem 4.4 : A 4-pole compound generator has armature, series-field and shunt-field resistance of 1 Ω, 0.5 Ω and 100 Ω respectively. This generator delivers 4 kW at a terminal voltage of 200 V. Allowing 1 V per brush for contact drop, calculate for both short-shunt and long-shunt connections.

(i) The generated emf, and

(ii) The flux per pole if the armature has 200 lap-connected conductors and is driven at 750 rpm.

Solution :

Calculation for short shunt

The equivalent circuit for short shunt can be drawn as

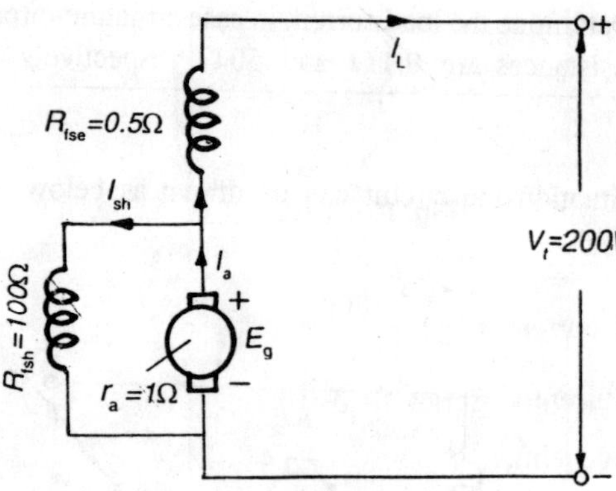

Fig. P. 4.4 (a)

$$\text{Load current} = \frac{\text{Load power}}{\text{Terminal voltage}} = \frac{4 \times 1000}{200} = 20 \text{ A}$$

(i) Voltage drop across series field resistance

$$= I_{se} \times R_{fse} = 0.5 \times 20 = 10 \text{ V}$$

Voltage drop across shunt field resistance

$$= V_t - V_{se} = 200 - 10 = 190 \text{ V}$$

Shunt field current $(I_{sh}) = \dfrac{V_f}{R_{fsh}} = \dfrac{190}{100} = 1.9$ A

Series field current = Load current = 20 A

Armature current $(I_a) = 20 + 1.9 = 21.9$ A

Total voltage drop across brushes = 1×2 V = 2 V

For generator, $E_g = V_b + I_a r_a + I_{se} \times R_{fse} + V_t$

$$= 2 + 2.19 \times 1 + 10 + 200 = 233.9 \text{ V}$$

Thus, genereated emf = 233.9 V **Ans.**

Calculation for Long Shunt:

The equivalent circuit can be drawn as below.

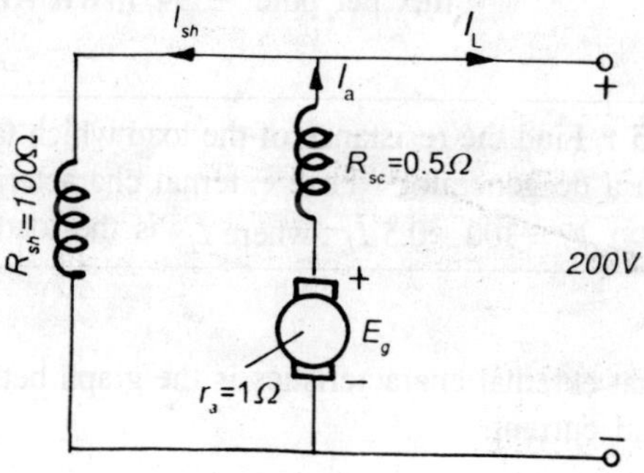

Fig. P. 4.4 (b)

Shunt field current $(I_{sh}) = \dfrac{200}{100} = 2$ A

Armature current = Series field current = 2 + 20 = 22 A

For generated emf, $E_g = V_t + V_b + I_a(R_{se} + r_a)$

$$= 200 + 2 + 22(1 + 0.5) = 235 \text{ V}$$

Thus, generated emf = 235 V **Ans.**

(ii) To calculate flux per pole

Applying the equation $E_g = \dfrac{\phi N Z P}{60 A}$

For lap winding, $A = P = 4, \phi = \dfrac{E_g \times 60\, A}{Z \times P \times N}$

$$= \frac{233.9 \times 60 \times 4}{200 \times 4 \times 750} = 93.56 \text{ mWb}$$

Thus, flux per pole = 93.56 mWb

Again, To calculate flux per pole,

Applying the equation, $E_g = \dfrac{\phi NZP}{60\, A}$

For lap winding, $A = P = 4 \Rightarrow \phi = \dfrac{E_g \times 60\, A}{Z \times P \times N}$

$$= \frac{235 \times 60 \times 4}{750 \times 4 \times 200} = 94 \text{ mWb}$$

Thus, flux per pole = 94 mWb **Ans.**

> **Problem 4.5 :** Find the resistance of the load which takes a power of 6 kW from a dc generator whose external characteristic is given by the equation $V_t = 300 - 0.5\, I_L$, where I_L is the load current.

Solution :

As we know that external characteristics is the graph between terminal voltage and load current.

Given, $V_t = 300 - 0.5\, I_L$ (1)

Power delivered to load = 6 kW

$$\Rightarrow V_t \times I_L = 6000 \Rightarrow (300 - 0.5\, I_L) \times I_L = 6000$$

$$\Rightarrow I_L^2 - 600\, I_L + 12000 = 0$$

On solving, we get $I_L = 21$ A

Now, Terminal voltage = $300 - 0.5 \times 21 = 289.5$ V

Load Resistance = $\dfrac{V_t}{I_L} = \dfrac{289.5}{21} = 13.8\ \Omega$

Thus, $R_L = 13.8\ \Omega$

Problem 4.6 : A compensated d.c. machine has 19,000 armature ampere-turns per pole. The ratio of pole-are to pole-pitch is 0.7. Interpolar air-gap length and flux density are respectively 1 cm and 0.3 T. For rated armature current of 1000 A, calculate the compensating winding conductors per pole and the number of turns on each interpole.

Solution :

To calculate compensating winding conductors per pole,
Applying the relation

$$\frac{AT_{cw}}{\text{Pole}} = \frac{AT_a}{\text{Pole}} \times \frac{\text{Pole Are}}{\text{Pole pitch}} = 19000 \times 0.7 = 13300$$

Number of turns of compensating winding per pole

$$\frac{AT_{cw}/\text{pole}}{\text{Armature current}} = \frac{13300}{1000} = 13.3 \cong 13$$

$\therefore$ Number of compensating conductor per pole $= 13 \times 2 = 26$

Number of compensating conductor per pole $= 26$

Now, Ampere Turn required for the Air gap under the interpole

$$\frac{B_g}{\mu o} \times l_g = \frac{0.3}{4\pi \times 10^{-7}} \times 1 \times 10^{-2} = 2387.324 \text{ ATs}$$

Without compensating winding, the interpole ampere-turns must be equal to (19000 + 2387.324) ATs. But compensating winding has already neutralised 13,300 ATS.

Now, the net interpole amp-turn/pole = 19000 + 2387.324 – 13300

$\therefore$ Number of turns on each interpole

$$= \frac{19000 + 2387.324 - 13300}{1000} \cong 9 \text{ turns}$$

Thus number of turns on each interpole = 9

Problem 4.7 : In a 220 V compound generator, the armature, series and shunt windings have resistance of 0.25 Ω, 0.15 Ω and 50 Ω respectively. The load consists of 100 lamps, each rated at 60 W, 220 V. Find the total emf and armature current when the machine is connected for

(i) long shunt, and *(ii)* short shunt.

Solution :

Given $V_t = 220$ V, $r_0 = 0.25\,\Omega$,

$R_{se} = 0.15\,\Omega$, $R_{sh} = 50\,\Omega$

(i) Calculation for long shunt

The equivalent circuit can be drawn as below

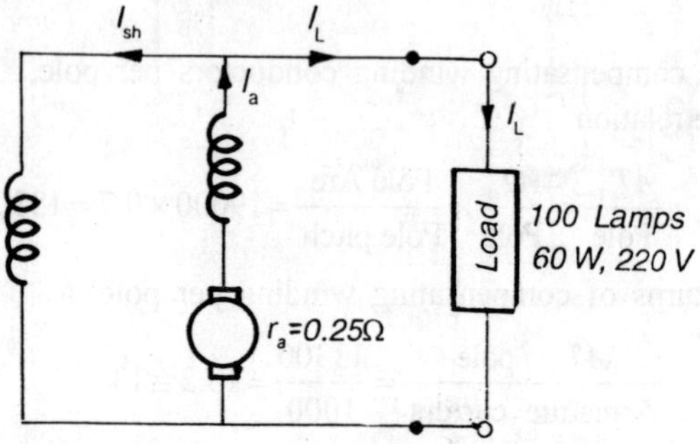

Fig. P. 4.7 (a)

$$\text{Load current } (I_L) = \frac{\text{Load Power}}{\text{Terminal Voltage}}$$

$$= \frac{100 \times 60}{220} = 27.27 \text{ A}$$

$$\text{Shunt field current } (I_{sh}) = \frac{V_t}{R_{sh}} = \frac{220}{50} = 4.4 \text{ A}$$

$$\text{Armature current} = I_{sh} + I_L = 27.27 + 4.4$$

$$= 31.67 \text{ A}$$

For generator, (Neglecting brushes drop)

$$E_g = V_t + I_a (r_a + R_{se})$$

$$= 220 + 31.67 (0.15 + 0.25)$$

$$= 232.67 \text{ V}$$

Thus, Total emf = 232.67 V, Armature current = 31.67 A

Calculation for Short Shunt

The equivalent circuit can be drawn as:

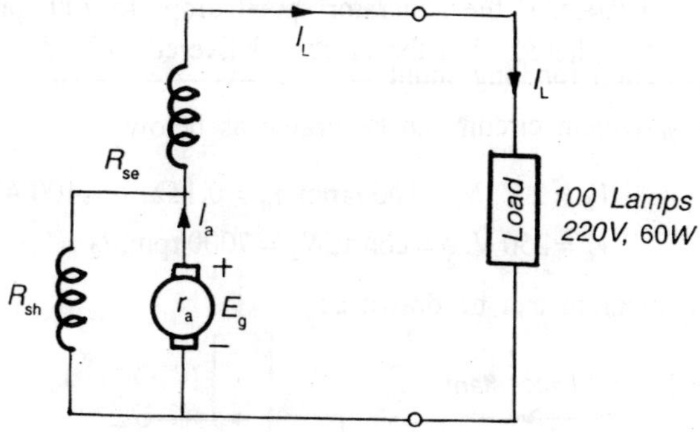

Fig. P. 4.7 (b)

Voltage drop across series field resistance = $0.15 \times 27.27 = 4.09$ V

Voltage across shunt resistance = $220 - 4.09 = 215.91$

Current through shunt resistance = $\frac{215.91}{50} = 4.32$ A

Armature current = 4.32 + 27.27 = 31.59 A **Ans.**

Total emf $= V_t + I_a r_a + I_L R_{se}$

$= 220 + 31.59 \times 0.25 + 27.27 \times 0.15 = 231.99$ V **Ans.**

Problem 4.8 : A 220 V, 1.5 kW, 859 rpm, separately excited dc motor has armature resistance of 2.5 ohms and it draws a current of 8 A at rated load condition. If the field current and the armature voltage are fixed at the value of rated speed at rated load, what will be the no-load speed of the motor? Assume losses remain constant between no-load and full-load operation. *(GATE, 1995)*

Solution :

Given, $V_t = 220$ V, $N_{rated} = 859$ rpm.

At rated Load condition $E_{b1} = V_t - I_a r_a = 220 - 8 \times 2.5 = 200$ V

At No-Load condition, $I_a = 0$, $E_{b2} = V_t - 0 \times r_a = 220$ V

Since, field current is constant $\therefore E_b \propto N$

Now, $\frac{E_{b1}}{E_{b2}} = \frac{N_1}{N_2} \Rightarrow N_2 = N_1 \times \frac{E_{b2}}{E_{b1}} = \frac{220 \times 859}{200} = 944.9$ rpm

Thus, No-Load speed of the motor = 944.9 rpm

Problem 4.9 : A separately excited dc generator has armature circuit resistance of 0.1 Ω and total drop at brushes is 2 V. When running at 1000 rpm, it delivers a current of 100 A at 250 V to a load of constant resistance. If the generator speed drops to 700 rpm, with field current unaltered, find the current delivered to load.

Solution :

Given, $V_b = 2\text{ V}, N_1 = 1000\text{ rpm}, r_a = 0.1\,\Omega, I_1 = 100\text{ A}$

$V_t = 250\text{ V}, \phi = \text{const}, N_2 = 7000\text{ rpm}, I_2 = ?$

The circuit diagram can be drawn as

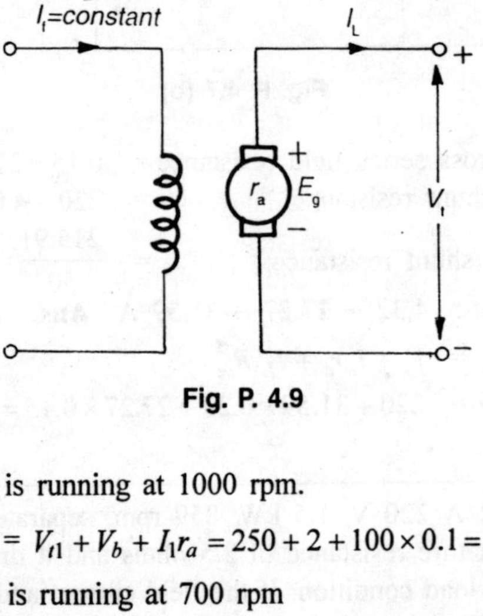

Fig. P. 4.9

When machine is running at 1000 rpm.

$$E_{g1} = V_{t1} + V_b + I_1 r_a = 250 + 2 + 100 \times 0.1 = 262\text{ V} \quad (1)$$

When Machine is running at 700 rpm

$$E_{g2} = V_{t2} + V_b + I_2 r_a = V_{t2} + 0.1 \times I_2 + 2 \quad (2)$$

Since, flux is constant $E \propto N$

$$\frac{E_{g1}}{E_{g2}} = \frac{N_1}{N_2} \Rightarrow E_{g2} = \frac{N_2}{N_1} \times E_{g1} = \frac{700}{1000} \times 262 = 183.4\text{ V}$$

From (2) $V_{t2} = 183.4 - 2 - 0.1 \times I_2 = 181.4 - 0.1 \times I_2$

Now, Load current $(I_2) = \dfrac{V_{t2}}{R}$

$$\Rightarrow \quad I_2 = \frac{181.4 - 0.1\,I_2}{R}$$

$$\Rightarrow \quad I_2 \times R = 181.4 - 0.1\,I_2$$

$$\Rightarrow \quad I_2 \times 2.5 = 181.4 - 0.1\,I_2$$

$$\Rightarrow \qquad 2.6\, I_2 = 181.4 \quad \therefore\ I_2 = \frac{181.4}{2.6} = 69.77\ \text{A}$$

Thus, Load current delivered to the Load = 69.77 A **Ans.**

Problem 4.10 : The open circuit characteristic for a d.c. shunt generator at 800 r.p.m. is given by following data:

I_f amp.	0	0.2	0.40	0.65	1.02	1.75	3.15	5.00
E_a volts	10	40	80	120	160	200	240	260

Determine the critical field resistance at

(a) 800 r.p.m. and,

(b) 900 r.p.m.

(c) If the field winding resistance is 55 Ω, find the range of field rheostat to vary the voltage from 200 to 250 V, on open circuit at a speed of 800 r.p.m.

(d) Determine the terminal voltage and field current of the generator for a total armature resistance of 0.6 Ω, armature current of 50 A and a speed of 800 r.p.m. Neglect armature reaction.

Solution :

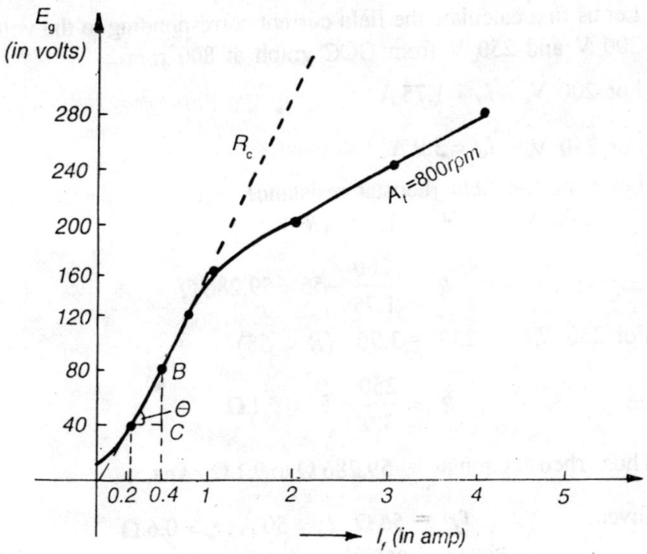

Fig. P. 4.10 (a)

(a) From graph, critical resistance can be obtained by knowing the slope of line AB.

Hence, Critical resistance

$$\tan\theta = \frac{BC}{AC} = \frac{80-40}{0.4-0.2} = \frac{40}{0.2} = 200\ \Omega$$

Thus, $R_c = 200\ \Omega$

(b) To find critical resistance at 900 rpm.

Let us calculate the generated emf at 900 rpm corresponding to the field currents 0.2 A and 0.4 A.

Now, Applying formula $\dfrac{E_{g1}}{E_{g2}} = \dfrac{N_1}{N_2}$

Hence for 40 V $\quad E_g = \dfrac{900}{800} \times 40 = 45\ \text{V}$

For 80 V $\quad E_g = \dfrac{900}{800} \times 80 = 90\ \text{V}$

Now, critical resistance $= \dfrac{90-45}{0.4-0.2} = \dfrac{45}{0.2} = 225\ \Omega$

Thus, $R_c = 225\ \Omega$

(c) To calculate the range of field rheostat,

Let us first calculate the field current corresponding to the voltage 200 V and 250 V from OCC graph at 800 rpm

For 200 V, $I_f = 1.75\ \text{A}$

For 250 V, $I_f = 3.9\ \text{A}$

Let R be the field rheostat resistance.

For 200 V, $\quad 200 = 1.75\ (R + 55)$

$$\Rightarrow \quad R = \frac{200}{1.75} - 55 = 59.286\ \Omega$$

For 250 V, $\quad 250 = 3.90\ (R + 55)$

$$\Rightarrow \quad R = \frac{250}{3.9} - 55 = 9.1\ \Omega$$

Thus, rheostat range $= 59.286\ \Omega$ to $9.1\ \Omega$ **Ans.**

(d) Given, $\quad r_f = 55\ \Omega,\ I_a = 50\ \text{A},\ r_a = 0.6\ \Omega$

Speed = 800 rpm

When the armature current is 50 A.

Let E_g be the generated emf and I_f be the corresponding field current.

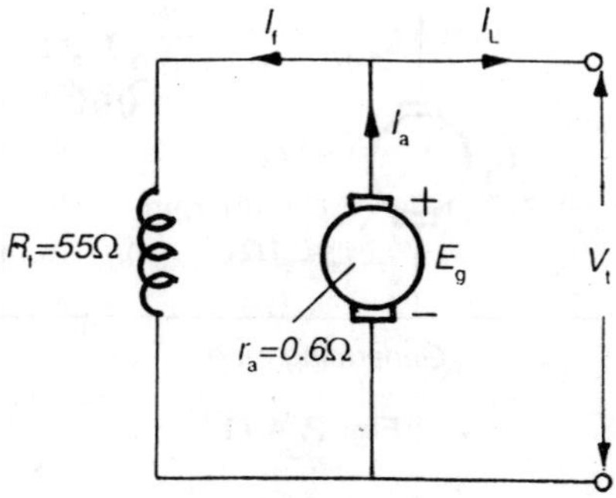

Fig. P. 4.10 (b)

Now, for generator $E_g - I_a r_a = V_t$

or $E_g - I_a r_a = I_f \times R_f$

or $E_g - 50 \times 0.6 = I_f \times 55$

or $E_g - 30 = 55\, I_f$ (1)

We choose the values of I_f and E_g from the graph such that these values satisfy the equation *(i)*,

From graph, we get $E_g = 250\text{ V}, I_f = 4\text{ A}$

Now, Terminal voltage $I_f \times R_f = 4 \times 55 = 220\text{ V}$ **Ans.**

Field current $(I_f) = 4\text{ A}$ **Ans.**

Problem 4.11 : A belt-driven 60 kW shunt-wound generator running at 500 r.p.m. is supplying full load to a bus bar at 200 V. At what speed will it run if the belt breaks and the machine continues to run taking 5 kW from the bus bar? The armature and field resistances are 0.1 Ω and 100 Ω respectively. Brush contact drop may be taken as 2 V. Neglect armature reaction.

Solution :

Given, $V_t = 200\text{ V}$, Power = 60 kW

For generating mode

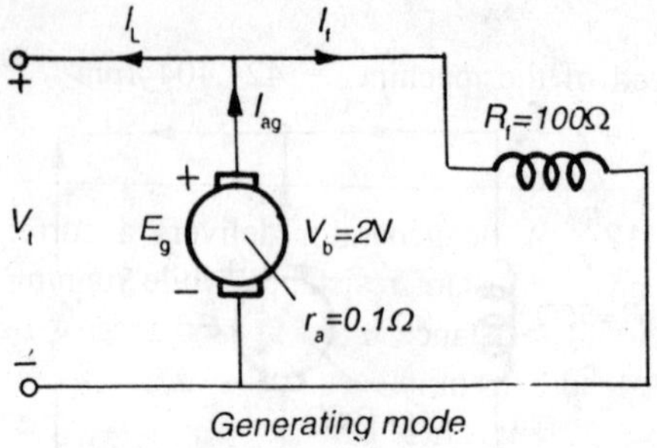

Fig. P. 4.11

Load current $$(I_L) = \frac{\text{Power}}{\text{Terminal Voltage}}$$

$$I_L = \frac{60 \times 1000}{200} = 300 \text{ A}$$

Field current $$\left(I_f\right) = \frac{V_t}{R_f},\ I_f = \frac{200}{100} = 2 \text{ A}$$

Armature current $$I_{ag} = I_2 + I_f = 302 \text{ A}$$

Generated emf $$E_g = V_t + V_b + I_{ag}\, r_a \propto \phi\, N_g \qquad (1)$$

When the belt breaks, it operates as d.c. motor

In this case, Line current $$I_L = \frac{5 \times 1000}{200} = 25 \text{ A}$$

Armature current $$I_{an} = I_L - I_f = 25 - 2 = 23 \text{ A}$$

Back emf of the motor $$E_{bm} = V_t - V_b - I_{am}\, r_a \propto \phi\, N_m \qquad (2)$$

From (1) and (2) $$\frac{E_{bm}}{E_g} = \frac{\phi\, N_m}{\phi\, N_g}$$

Since, Terminal voltage is constant. The flux will be also constant in both motoring and generating case.

$$\therefore \quad \frac{V_t - V_b - I_{am}\, r_a}{V_t + V_b + I_{ag}\, r_a} = \frac{N_m}{N_g}$$

or $$\frac{N_m}{N_g} = \frac{200 - 2 - 23 \times 0.1}{200 + 2 + 302 \times 0.1} = \frac{195.7}{232.2}$$

$$\therefore \qquad N_m = \frac{195.7}{232.2} \times 500 = 421.404 \text{ rpm } \textbf{Ans.}$$

Thus, Speed of the machine = 421.404 rpm

Problem 4.12 : A dc generator delivers a current of 50 A at 200 V to a load of constant resistance while running at 1000 rpm. Its armature circuit resistance is 0.5 Ω and that of field winding is 50 Ω. Find the load current in case speed of dc generator falls to 800 rpm but the field current is held constant at its previours value. Neglect the effect of armature reaction.

Solution :

When the generator is running at 100 rpm.

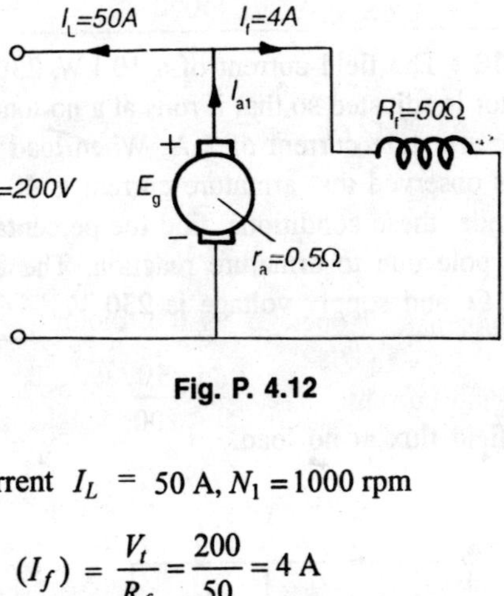

Fig. P. 4.12

Given, Load current $I_L = 50 \text{ A}, N_1 = 1000 \text{ rpm}$

Field current $(I_f) = \frac{V_t}{R_f} = \frac{200}{50} = 4 \text{ A}$

Armature current $I_{a1} = I_L + I_f = 50 + 4 = 54 \text{ A}$

Generated emf $E_{g1} = V_t + I_{a1} r_a + V_b \propto \phi_1 N_1$

(i) When the generator is running at $N_2 = 800$ rpm

$$E_{g2} = V_t + I_{a2} r_a + V_b \propto \phi_2 N_2$$

(2) From (1) and (2) $\frac{E_{g2}}{E_{g1}} = \frac{\phi_2 N_2}{\phi_1 N_1}$

Since, the terminal voltage is constant. Flux is also constant.

$$\therefore \qquad \frac{E_{g2}}{E_{g1}} = \frac{N_2}{N_1}$$

or $$\frac{V_t - I_{a2}\, r_a + V_b}{V_t + I_{a1}\, r_a + V_b} = \frac{N_2}{N_1}$$

or $$200 + I_{a2} \times 0.5 = \frac{800}{1000}(200 + 54 \times 0.5)$$

or $$I_{a2} = -36.8 \text{ A}$$

This negative sign shows that when speed drops the generated emf drops causing drawing current from the load.

Now, Load current = 36.8 + 4 = 40.8 A

Thus, Load current at speed 800 rpm = 40.8 A

Problem 4.13 : The field current of a 10 kW, 230 V, 1500 r.p.m. d.c. shunt-motor is adjusted so that it runs at a no-load speed of 1600 r.p.m. with ìan armature current of 2 A. When load on the motor is increased, it is observed that armature current is 40 A and speed is 1400 r.p.m. Under these conditions, find the percentage reduction in field flux per pole due to armature reaction. The armature circuit resistance is 1 Ω and supply voltage is 230 V.

Solution :

Let ϕ, be the field flux at no load.

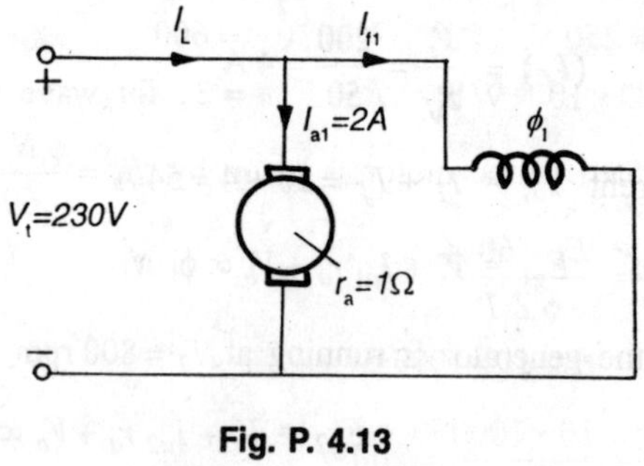

Fig. P. 4.13

At No load, $\qquad N_1 = 1600$ rpm

Armature current $\qquad I_{a1} = 2$ A

$$E_{b1} = V_t - I_{a1} \times r_a \; 230 - 2 \times 1 = 228 \text{ V} \; \alpha \; \phi_1 \, N_1 \qquad (1)$$

Let ϕ_2 be the reduced field flux at load $N_2 = 1400$ rpm

Armature current $\quad I_{a2} = 40$ A, $E_{b2} = V_t - I_{a2} \times r_a$

$$= 230 - 40 \times 1 = 190 \text{ V}$$

For dc shunt Motor $E_b \propto \phi N$

$$\frac{E_{b1}}{E_{b2}} = \frac{\phi_1 N_2}{\phi_2 N_2}$$

or

$$\frac{\phi_1}{\phi_2} = \frac{E_{b1} \times N_2}{E_{b2} \times N_1} = \frac{228 \times 1400}{190 \times 1600} = \frac{228 \times 14}{190 \times 16}$$

Now, % reduction in flux $= \dfrac{\phi_1 - \phi_2}{\phi_1} \times 100\%$

$$= \frac{(228 \times 14 - 190 \times 16)}{228 \times 14} \times 100\%$$

$$= 4.762\% \textbf{ Ans.}$$

Problem 4.14 : A 4-pole dc series motor has wave-connected winding with 600 conductors. Total resistance of motor is 0.8 Ω. When fed from 250 V dc source, the motor supplies a load of 10 kW and takes 50 A with a flux per pole of 3 mWb. For these operating conditions, calculate the developed torque and the shaft torque

Solution :

Given, $\quad V_t = 250$ V, $\quad N = ?, Z = 600$

$\phi = 3 \times 10^{-3}$ Wb, $\quad P = 4, A = 2$, for wave winding

Back emf generated in the motor is given by $E_b = \dfrac{\phi N Z P}{60 A}$

or

$$N = \frac{E_b \times 60 A}{\phi Z P}$$

Since, $\quad E_b \times I_a = P_i = 10 \times 1000$

$$E_b = \frac{10 \times 1000}{50} = 200 \text{ V}$$

$$N = \frac{200 \times 60 \times 2}{3 \times 10^{-3} \times 600 \times 4} = \frac{10 \times 1000}{3} \text{ rpm} = \frac{10 \times 1000}{3 \times 60} \text{ rps}$$

As we know that, Developed Torque × speed = power

$$\text{or, Developed Torque} = \frac{10 \times 1000 \times 3 \times 60}{2 \times \pi \times 10 \times 1000} = 28.65 \text{ Nm}$$

Developed Torque = 28.65 Nm **Ans.**

$$\text{As we know that, Shaft Torque} = \frac{\text{Gross Power} - \text{Rotational Loss}}{\text{Speed}}$$

When 250 V dc supply is given to the motor.

$$E_b = V_t - I_a r_a, = 250 - 0.8 \times 50 = 210 \text{ V}$$

$$\therefore \text{ Rotational Loss} = (210 - 200) \times 50 = 500 \text{ watt}$$

$$\text{Now, Shaft Torque} = \frac{\text{Gross Power} - \text{Rotational Loss}}{\text{Speed}}$$

$$\text{Shaft Torque} = \frac{10 \times 1000 - 500}{2\pi N} = \frac{9500 \times 3 \times 60}{2\pi \times 10 \times 1000} = 27.22 \text{ Nm}$$

Thus, Shaft Torque = 27.22 Nm **Ans.**

Problem 4.15 : A d.c. shunt motor runs at 750 r.p.m. from 250 V supply and is taking a full-load line current of 60 amperes. Its armature and field resistance are 0.4 Ω and 125 Ω respectively.

Assuming 2 V brush drop and negligible armature reaction effect, calculate

(i) The no-load speed for a no-load line current of 6 amperes.

(ii) The resistance to be added in series with the armature circuit to reduce the full-load speed to 600 r.p.m.

(iii) The percentage reduction in flux per pole needed, in order that the speed may be 900 r.p.m. when the armature current is 30 amperes, with no-added resistance in the armature circuit.

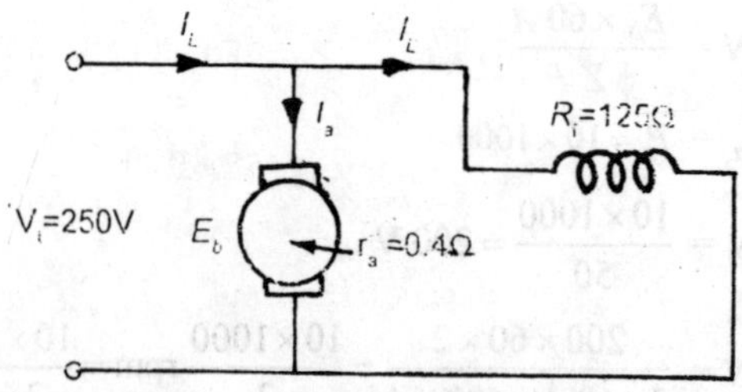

Fig. 4.15

Solution :

At full-load $N_1 = 750$ rpm

Line current $I_{L1} = 60$ A

Field current $I_f = \frac{250}{125} = 2$ A

Armature current $I_{a1} = 60 - 2 = 58$ A

Back emf developed in the motor

$$E_{b1} = V_t - V_b - I_{a1}r_a$$

or $$E_{b1} = 250 - 2 - 58 \times 0.4 \propto \phi_1 N_1 \quad (1)$$

At No-load $N_2 = ?$

Line current $I_{L2} = 6$ A

Field current $I_f = 2$ A

Armature current $I_{a2} = 6 - 2 = 4$ A

No-load back emf $E_{b2} = V_t - V_b - I_{a2}\, r_a \propto \phi_2 N_2$

$$= 250 - 2 - 4 \times 0.4 \propto \phi_2 N_2 \quad (2)$$

For shunt Motor, Field remains constant.

$$\therefore \quad \frac{E_{b2}}{E_{b1}} = \frac{\phi_2 N_2}{\phi_1 N_1} \qquad [\because \phi_1 = \phi_2]$$

$$\Rightarrow \quad N_2 = \frac{E_{b2}}{E_{b1}} \times N_1 = \left(\frac{248 - 1.6}{248 - 23.2}\right) \times 750 = 822.1 \text{ rpm}$$

No-load speed = 822.1 rpm **Ans.**

(ii) Let R_a be the resistance added in series with the armature current to reduce to speed 600 rpm.

$$\frac{E_{b3}}{E_{b1}} = \frac{N_3}{N_1} \Rightarrow E_{b3} = \frac{600}{750} \times E_{b1}$$

or, $$E_{b3} = \frac{4}{5}(250 - 2 - 58 \times 0.4)$$

or, $$= 250 - 2 - 58\,(0.4 + R_a)$$

or, $$= \frac{4}{5}(250 - 2 - 58 \times 0.4) = 179.84$$

or, $$R_a = 0.7759\ \Omega \quad \textbf{Ans.}$$

(iii) Let ϕ_2 be the flux per pole when armature current is 30 A.

Back emf in this case, $N_4 = 900$ rpm

$$E_{b4} = 250 - 2 - 30 \times .4 = 236 \text{ V}$$

For dc shunt machines, $E_b \propto N \phi$

$$\frac{E_{b1}}{E_{b4}} = \frac{\phi_1}{\phi_2} \times \frac{N_1}{N_4}$$

or, $$\frac{\phi_1}{\phi_2} = \frac{E_{b1}}{E_{b4}} \times \frac{N_4}{N_1} = \frac{224.8 \times 900}{236 \times 750}$$

Now, % Reduction in flux $= \dfrac{\phi_1 - \phi_2}{\phi_1} \times 100\% = 12.51\,\%$

Thus, % reduction in flux $= 12.51\%$ **Ans.**

> **Problem 4.16 :** A 220-V, 1500 rpm dc motor has full-load armature current of 30 A. It is proposed to design a starter which restricts the maximum armature current during starting to 60 A. For design purposes, the minimum current during starting is to be restricted to 30 A, the series resistance in the armature circuit being cut in steps whenever the current falls to 30 A. Assuming that the armature resistance of the dc motor is 0.5 Ω, calculate the maximum series resistance used in the starter and the amount of resistance cut during each of the first two steps. *(I.A.S., 1987)*

Solution :

Given, $V_t = 220 \text{ V}, N = 1500 \text{ rpm}$

Maximum starting current $I_1 = 60$ A

Minimum starting current $I_2 = 30$ A

Armature resistance $r_2 = 0.5$ A

Total series resistance at the starting $R_1 = \dfrac{220}{60} = 3.667$ A

∴ Maximum series resistance used in the starter

$$= 3.667 - 0.5 = 3.167 \text{ A} \quad \textbf{Ans.}$$

Using relation $$\frac{\text{Minimum armature current}}{\text{Maximum armature current}} = \frac{R_2}{R_1} = \frac{R_3}{R_2} = \frac{R_4}{R_5} = ---$$

Now, $$\frac{30}{60} = \frac{R_2}{220} \times 60$$

or, $$R_2 = \frac{30 \times 220}{60 \times 60} = 1.833\ \Omega$$

During first step amount of resistance cut = 1.833 Ω **Ans.**

Now, $$\frac{R_3}{R_2} = \frac{30}{60}$$

$$R_3 = \frac{30}{60} \times R_2 = \frac{1}{2} \times 1.833 = 0.9165\ \Omega$$

During second step, Amount of resistance cut = 0.9165 Ω **Ans.**

Problem 4.17 : A 5 kW, 250 V d.c. shunt motor takes a no-load armature current of 4 A at rated voltage and runs at 1200 r.p.m. The armature circuit resistance is 0.4 Ω and the field resistance is 250 Ω. At rated load and rated voltage, the motor takes 26 A and the armature reaction weakens the field flux by 2%.

Calculate the full-load speed and the corresponding electromagnetic torque of the motor.

Solution :

At no load $N_1 = 1200$ rpm.

Armature current $I_{ao} = 4$ A

No Load Back-emf $E_{bo} = V_t - I_{ao} \times r_a \propto N_1 \phi_1$

$$= 250 - 4 \times 0.4 \propto N_1 \phi_1 = 248.4 \propto N_1 \phi_1 \quad (1)$$

At Load, According to the Question

$$\phi_2 = (1 - 0.02)\phi_1 = 0.98\phi_1$$

Line current, $I_L = 26$ A

Field current $$I_f = \frac{V_t}{R_f} = \frac{250}{250} = 1\text{ A}$$

Armature Current $I_{a1} = 26 - 1 = 25$ A, $N_2 = ?$

Load back-emf $E_{b1} = V_t - I_{a1} \times r_a \propto N_2 \phi_2$

$$= 250 - 25 \times 0.4 \propto N_2 \phi_2 = 240 \propto N_2 \phi_2 \quad (2)$$

From (1) and (2) $$\frac{E_{bo}}{E_{b1}} = \frac{N_1 \phi_1}{N_2 \phi_2}$$

$$\therefore \quad N_2 = \frac{\phi_1}{0.08\,\phi_1} \times \frac{E_{b1}}{E_{b0}} \times N_1$$

$$= \frac{1}{0.08} \times \frac{(250 - 25 \times 0.4)}{(250 - 4 \times 0.4)} \times 1200 = 1183.08 \text{ rpm}$$

Full-Load speed $= 1183.08$ **Ans.**

Now, Electromagnetic Torque

$$(T_e) = \frac{E_{b1} \times I_{a1}}{2\pi N} = \frac{240 \times 25 \times 60}{2 \times \pi \times 1183.1} = 48.429 \text{ Nm}$$

Thus, $T_e = 48.429$ Nm **Ans.**

> **Problem 4.18 :** A dc shunt motor takes an armature current of 50 A at its rated voltage of 240 V. Its armature-circuit resistance is 0.2 Ω. If an external resistance of 1 Ω is inserted in series with the armature and the field flux remains unchanged, then calculate:
> (*a*) Percentage decrease of increase in speed for the same load torque,
> (*b*) Percentage decrease or increase in speed for half of the load torque.

Solution :

Given, $V_t = 240$ V

Armature resistance $= r_a = 0.2\ \Omega$

Additional armature resistance $= 1\ \Omega$

When, no resistance is added to the armature speed $= N$

Back emf, $E_{b1} = 240 - 50 \times 0.2 = 230$ V (1)

$N_1 = ?$

Let I_{a2} be the current when resistance is added to armature speed $= N_2$

Back emf in this case $E_{b2} = 240 - I_{a2}\,(0.2 + 1) = 240 - 1.2\, I_{a2}$ (2)

In DC shunt Machines $= E \propto \phi N$

From (1) and (2) $\dfrac{E_{b1}}{E_{b2}} = \dfrac{\phi N_1}{\phi N_2} = \dfrac{N_1}{N_2}$

Since, field current is constant.

(*i*) For same load Torque, the armature curmet remains same Hence, flux is constant

$$I_{a2} = I_{a1} = 50 \text{ A}$$

Now, $$\frac{N_1}{N_2} = \frac{E_{b1}}{E_{b2}} = \frac{230}{240 - 50 \times 1.2} = \frac{230}{180}$$

$$\frac{N_1 - N_2}{N_1} = \frac{230 - 180}{230}$$

$$\text{\% Reduction in speed} = \left(\frac{230 - 180}{230}\right) \times 100\% = 21.739\%$$

Thus, % reduction in speed = 21.739 % **Ans.**

(ii) For half of the load Torque the armature current is reduced to half as field flux remains constant

$$I_{a2} = \frac{I_{a1}}{2} = \frac{50}{2} = 25\text{ A}$$

Now, $$\frac{N_1}{N_2} = \frac{230}{240 - 25 \times 1.2} = \frac{230}{210}$$

$$\frac{N_1 - N_2}{N_1} = \frac{230 - 210}{230}$$

$$\text{\% Reduction in speed} = \left(\frac{N_1 - N_2}{N_1}\right) \times 100\% = \frac{20}{230} \times 100\%$$

$$= 8.696\ \%$$

Thus, % reduction in speed = 8.696 % **Ans.**

Problem 4.19 : The e.m.f. developed in the armature of a shunt generator at 1155 r.p.m. is 240 volts for a field current of 4.5 amperes and 255 volts for a field current of 5 amperes. The generator is now used as a motor on a 260 V supply and takes an armature current of 75 amperes. Find the motor speed, when the field current is adjusted to 4.8 A. Armature resistance (including brushes) is 0.12 Ω.

(Ranchi Univ.)

Solution :

Given, At speed (N_1) = 1155 rpm.

At $I_{f1} = 4.5$ A, $E_{g1} = 240$ V

At $I_{f2} = 5$ A, $E_{g2} = 255$ V

At $I'_{f1} = 4.8$ A

$$E'g_1 = E_{g1} + \frac{E_{g2} - E_{g1}}{I_{f2} - I_{f1}} \times \left(I'_{f1} - I_{f1}\right)$$

$$E'g_1 = 240 + \frac{255 - 240}{5 - 4.5} \times 0.3 = 249 \text{ V} \propto \phi_1 N_1, \text{ or } N_1 = 1155 \text{ rpm}$$

When the generator is operated as motor.

Back emf $E_{b1} = V_t - I_{a1} \times r_a = 260 - 75 \times 0.12 = 251 \text{ V} \propto \phi_2 N_2$

Now, $\dfrac{E_{b1}}{E'_{g1}} = \dfrac{\phi_2 N_2}{\phi_1 N_1}$

Since, field is constant $\dfrac{E_{b1}}{E'_{g1}} = \dfrac{N_2}{N_1} \Rightarrow N_2 = N_1 \times \dfrac{E_{b1}}{E'_{g1}}$

$$N_2 = 1155 \times \frac{251}{249} = 1164.3 \text{ rpm} \quad \textbf{Ans.}$$

Problem 4.20 : A d.c. series motor has the following data for its magnetization curve:

Current, p.u.	0.40	0.60	0.80	1.00
Flux, p.u.	0.62	0.80	0.92	1.00

The per unit values are expressed in terms of their respective full-load values.

A diverter is used to raise its speed to 1.2 p.u. from full-load speed of 1.00 p.u., at constant full-load torque. Neglecting losses, calculate diverter resistance in terms of series field resistance.

Solution :

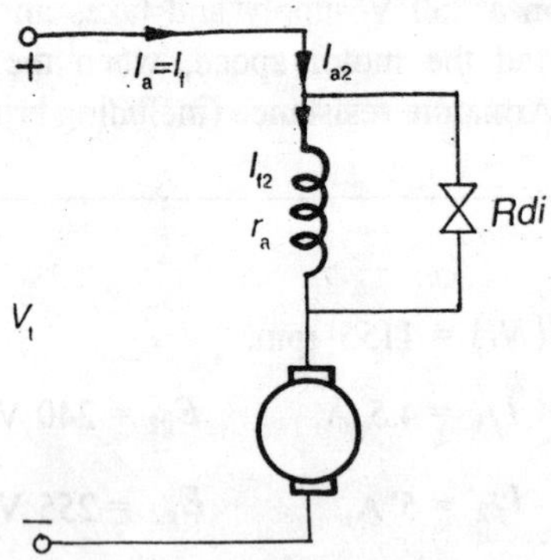

Fig. P. 4.20

Here, At full load $N_1 = 1.00$ pu, $\phi_1 = 1.00$ pu,

$I_1 = 1.00$ pu, $V_t = E_{b1} \propto N_1 \phi_1$

At given load $N_2 = 1.2$ pu, $\phi_2 = ?$

$V_t = E_{b2} \propto N_2 \phi_2$

For dc Machines $E_b \propto N\phi$ $\quad \dfrac{E_{b1}}{E_{b2}} = \dfrac{N_1 \phi_1}{N_2 \phi_2}$

or,
$$\frac{V_{t1}}{V_t} = \frac{N_1 \phi_1}{N_2 \phi_2}$$

or,
$$\phi_2 = \frac{1}{1.2}\phi_1 = \frac{1}{1.2} \times 1,$$

$$\phi_2 = 0.833 \text{ pu}$$

Now, Let us draw magnetisation curve for given data and the field current corresponding to flux 0.833 pu is about $I_{f2} = 0.65$ pu.

For constant full load Torque

$$T_e = \phi_1 I_1 = \phi_2 I_2$$

or,
$$I_2 = \frac{\phi_1}{\phi_2} I_1 = 1.2 \text{ pu}$$

Now, from circuit $\quad I_{f2} = I_{a2} \times \dfrac{R_{di}}{R_{di} + r_a} = 1.2 \times \dfrac{R_{di}}{R_{di} + r_a}$

On solving $\quad 0.652 = 1.2 \times \dfrac{R_{di}}{R_{di} + r_a}$

$\Rightarrow \quad R_{di} = 1.182\, r_a$ **Ans.**

> **Problem 4.21:** A 4-kW d.c. series motor has four field coils. The motor runs at 900 r.p.m. and takes 20 A from a 230 V d.c. source when field coils are in series under normol operation. Estimate the speed and current taken by the motor if field coils are reconnected in two parallel groups of two in series. Assume that the flux is directly proportional to the current and all losses are neglected. Take the load torque as proportional to (*a*) square of speed, (*b*) speed, (*c*) speed raised to power zero.

Solution :

Let r_c be the resistance of each coil. Under normal operation

$$N_1 = 900 \text{ rpm}, \ I_{f1} = I_{a1} = 20 \text{ A}$$

Back emf developed $\quad E_{b1} = V_t - I_{a1}(r_a + 4\,r_c) \propto N_1\,\phi_1 \qquad (1)$

When the armature resistance are connected in parallel.

$$E_{b2} = V_t - I_{a2}(r_a + r_c) \propto N_2\,\phi_2 \qquad (2)$$

Since, r and r_c are not given $\quad \dfrac{E_{b1}}{E_{b2}} = \dfrac{N_1\,\phi_1}{N_2\,\phi_2}$

Since flux is proportional to field curent i.e. armature current in dc series motor.

$$\frac{E_{b1}}{E_{b2}} = \frac{N_1}{N_2} \times \frac{I_{a1}}{I_{a2}} \times 2, \text{ or } \frac{230}{230} = \frac{N_1}{N_2} \times \frac{I_{a1}}{I_{a2}} \times 2$$

or,
$$= \frac{I_{a1}}{I_{a2}} \times \frac{N_1}{N_2} = \frac{1}{2},$$

$\Rightarrow$
$$N_2 = \frac{2 \times 20 \times 900}{I_{a2}} \qquad (3)$$

(*a*) According to Question $\quad T_L \propto N^2$

Electromagnetic Torque $T_e \propto \phi\, I$

under steady state condition

$$T_e = T_L, \text{ or, } N^2 \propto \phi\, I, \ \frac{N_1^2}{N_2^2} = \frac{\phi_1\, I_{a1}}{\phi_2\, I_{a2}}$$

For Series Motor $\quad \left(\dfrac{N_1}{N_2}\right)^2 = \dfrac{I_{a1} \times I_{a1}}{\dfrac{I_{a2}}{2} \times I_{a2}}$

or,
$$\frac{N_1}{N_2} = \sqrt{2}\ \frac{I_{a1}}{I_{a2}} \text{ or, } \frac{900 \times I_{a2}}{\sqrt{2} \times 20} = N_2$$

From eqn (3)

or,
$$\frac{900 \times I_{a2}}{\sqrt{2} \times 20} = \frac{40 \times 900}{I_{a2}} \text{ or, } I_{a2}^2 = 40 \times \sqrt{2} \times 20$$

$$I_{a2} = 33.64 \text{ A},$$

Putting $\quad I_{a2} = 33.64$ in (3) we get

$$N_2 = \frac{40 \times 900}{33.64} = 1070.15 \text{ rpm}$$

Thus, $N_2 = 1070.15$ rpm, $I_{a2} = 33.64$ A **Ans.**

(*b*) When $T_L \propto N$, Electromagnetic Torque $T_e \propto \phi I$

Hence, $N \propto \phi I \quad \frac{N_1}{N_2} = \frac{\phi_1 I_{a1}}{\phi_2 I_{a2}}$ or, $\frac{N_1}{N_2} = \frac{I_{a1} \times I_{a1}}{I_{a2} \times \frac{I_{a2}}{2}}$

or, $$2\frac{(I_{a1})^2}{(I_{a2})^2} = \frac{N_1}{N_2} \text{ or, } \frac{900}{N_2} = 2 \times \frac{400}{I_{a2}^2}$$

or, $$N_2 = \frac{900 \times I_{a2}^2}{2 \times 400}$$

From (3) $$\frac{40 \times 900}{I_{a2}} = \frac{900 \times I_{a2}^2}{2 \times 400},$$

or, $$I_{a2}^3 = 2 \times 40 \times 400,$$

$$I_{a2} = 31.75 \text{ A}$$

Putting $I_{a2} = 31.75$ A in (3)

$$N_2 = \frac{40 \times 900}{31.75} = 1134 \text{ rpm}$$

Thus, speed = 1134 rpm, armature current = 31.75 A **Ans.**

(*c*) When $T_L \propto N°$

Electromagnetic Torque $T_e \propto \phi I$

Under steady state condition $T_L = T_e$ or, $N° = \phi I$

$$\phi I = \text{constant or, } \phi_1 I_{a1} = \phi_2 I_{a2}$$

or, $$I_{a1} \times I_{a1} = \left(\frac{I_{a2}}{2}\right) I_{a2} \text{ or, } (I_{a2})^2 = 2 \times (I_{a1})^2$$

$$I_{a2} = \sqrt{2} \times 20 = 28.28 \text{ A}$$

Putting $I_{a2} = 28.28$ A in eqn (3)

$$N_2 = \frac{40 \times 900}{28.28} = 1273 \text{ rpm}$$

Thus, Speed = 273 rpm,

Armature current = 28.28 A **Ans.**

Problem 4.22 : A 4-pole dc series motor runs at a speed of N rpm on a fixed dc supply with all field coils in series. Estimate the speed at which the motor would run if fed from the same supply with the field coils re-connected in two parallel groups of two coils each. Assume that the load torque varies as the square of the speed, that all losses are negligible and that the magnetization characteristic is liner. (*I.A.S.*, 1989)

Solution :

Let r_c be the resistance of each coil. When coils are connected in Series.

$$\text{Speed} = N$$

Back emf
$$E_{b1} = V_t - I_1 (r_a + 4\,r_c) \propto N_1\, \phi_1 \qquad (1)$$

When coils are connected in two parallel groups of two coils each

$$N_2 = ?$$

$$E_{b2} = V_t - I_2 (r_a + r_c) \propto N_2\, \phi_2 \qquad (2)$$

From (1) and (2)
$$\frac{E_{b1}}{E_{b2}} = \frac{\phi_1\, N_1}{\phi_2\, N_2}$$

Since, r_a and r_c are not given and Terminal voltage is constant.

$$\therefore \qquad \frac{V_t}{V_t} = \frac{I_1\, N_1}{\frac{I_2}{2} \times N_2}, \quad I_2\, N_2 = 2\, I_1\, N_1$$

$$\frac{I_1}{I_2} = \frac{N_2}{2\, N_1} \qquad (3)$$

Accoeding to Question, Load Torque $T_L \propto N^2$

Electromagnetic Torque $T_e \propto \phi\, I$

Under steady state conditions $N^2 \propto \phi\, I$

$$\text{or,} \qquad \left(\frac{N_1}{N_2}\right)^2 = \frac{\phi_1\, I_1}{\phi_2\, I_2}, \text{ or, } \left(\frac{N_1}{N_2}\right)^2 = \frac{I_1 \times I_1}{\frac{I_1}{2} \times I_2}$$

$$\text{or,} \qquad \left(\frac{N_1}{N_2}\right)^2 = 2\left(\frac{I_1}{I_2}\right)^2$$

$$\text{From (3)} \qquad \left(\frac{N_1}{N_2}\right)^2 = 2\left(\frac{N_2}{2N_1}\right)^2 \text{ or, } (N_2)^4 = 2\,(N_1)^4$$

or, $N_2 = 2^{1/4}\,(N_1)$ putting $N_1 = N, N_2 = (2)^{1/4}\,N$

The motor speed is 1.189 times the initial speed.

> **Problem 4.23 :** A 460-V series motor has its field coils split into two equal halves and takes 40 A when running at 600 rpm with the field coils in series. Determine the speed when the two halves of the field winding are connected in parallel, assuming the load torque to remain constant and the magnetic circuit to be unsaturated. The resistance of the armature circuit is 0.5 Ω and that of the field coils in series is 0.2 Ω. *(I.A.S 1980)*

Solution :

Given $V_t = 460$ V

For dc series machine $I_a = I_f$

When coils are connected in series

$$N_1 = 600 \text{ rpm}, \quad I_1 = 40 \text{ A}$$

$$E_{b1} = V_t - I_1\left(r_a + r_c\right) = 460 - 40\left(0.5 + 0.2\right)$$

$$E_{b1} = 432 \propto I_1\, N_1 \qquad (1)$$

When coil are connected in parallel $N_2 = ?$

$$r_{eq} = \frac{0.1 \times 0.1}{0.2} = 0.05, \; E_{b2} = V_t - I_2\left(r_a + r_{eq}\right)$$

$$= 460 - I_2\left(0.5 + 0.05\right) \propto \frac{I_2}{2}\, N_2$$

$$= 460 - 0.55\, I_2 \propto \frac{I_2}{2}\, N_2 \qquad (2)$$

According to Question, Load Torque T_L = constant

Electromagnetic Torque $T_e \propto \phi\, I$

Under steady state conditions $T_L = T_e$

or, $\phi\, I$ = Const., $\phi_1\, I_1 = \phi_2\, I_2$

For dc series motor $I_1 \times I_1 = \frac{I_2}{2} \times I_2, I_2 = \sqrt{2} \times I_1$

Putting $I_1 = 40 \text{ A}, I_2 = \sqrt{2} \times 40 = 56.569 \text{ A}$

From (1) and (2) $\frac{460 - 0.55\, I_2}{432} = \frac{I_2\, N_2}{40 \times N_1 \times 2}$

$$\Rightarrow \quad N_2 = \frac{(460 - 0.55\, I_2) \times 40 \times 600 \times 2}{432 \times 56.569}$$

$$= \frac{(460 - 0.55 \times 56.569) \times 40 \times 600 \times 2}{432 \times 56.569} = 842.4 \text{ rpm} \quad \textbf{Ans.}$$

Problem 4.24 : A 200 V d.c. shunt motor takes 27 A at rated voltage and runs at 800 r.p.m. Its field resistance is 100 Ω. If an additional resistance of 20 Ω is inserted in the armature circuit, compute the motor speed and the line current in case load torque varies as the square of the speed.

Solution :

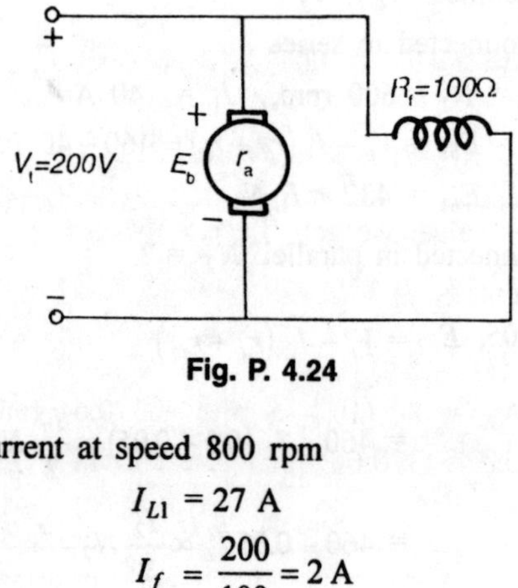

Fig. P. 4.24

Given, Line current at speed 800 rpm

$$I_{L1} = 27 \text{ A}$$

Field current $\quad I_f = \dfrac{200}{100} = 2 \text{ A}$

Armature current $\quad I_2 = 27 - 2 = 25 \text{ A}$

Back emf developed at $\quad N_1 = 800$ rpm

$$E_{b1} = 200 - I_2\, r_a \propto \phi_1\, N_1 \quad (1)$$

When resistance is added to armature circuit $\quad N_2 = ?$

$$E_{b2} = 200 - I_2\,(r_a + 20) \propto \phi_2\, N_2 \quad (2)$$

From (1) and (2) $\quad \dfrac{E_{b1}}{E_{b2}} = \dfrac{200 - I_1\, r_a}{200 - I_2\,(r_a + 20)} = \dfrac{N_1}{N_2}$

$\therefore$ Field current is constant $\quad \phi_1 = \phi_2$

Since, armature resistance is not given, neglecting it.

$$\frac{200}{200-20\,I_2} = \frac{800}{N_2}, \text{ or, } N_2 = 80\left(10 - I_2\right) \quad (3)$$

Now, According to Question $T_L \propto (\text{speed})^2 \propto N^2$

Electromagnetic Torque $T_e \, \alpha \, I$ for constant field

Under steady state condition.

$$T_e = T_L \text{ or, } I \propto N_2 \text{ or, } \frac{I_1}{I_2} = \frac{N_1^2}{N_2^2}$$

or,

$$N_2 = N_1 \times \frac{\sqrt{I_2}}{\sqrt{I_1}} = \frac{800 \times \sqrt{I_2}}{\sqrt{25}} = 160\sqrt{I_2} \quad (4)$$

From (3) and (4)

$$160\sqrt{I_2} = 80\left(10 - I_2\right)$$

Squaring both side $4\,I_2 = 100 - 20\,I_2 + I_2^2$ or, $I_2^2 - 24\,I_2 + 100 = 0$

$$I_2 = \frac{24 \pm \sqrt{24^2 - 400}}{2}$$

Neglecting higher value, we get, $I_2 = 5.367$ A

Now, Line current $= I_2 + I_f = 5.367 + 2 = 7.367$ A

Putting $I_2 = 5.367$ A in (3)

$N_2 = 80\ (10 - 5.367) = 370.64$ rpm.

Thus, Speed = 370.64 rpm, Line current = 7.367 A

> **Problem 4.25 :** A d.c. shunt motor with an armature resistance of 0.2 Ω drives a load at 1245 r.p.m., drawing an armature current of 125 A from a 440 V supply. If the excitation is reduced to 75% of the initial value and the total torque developed by the armature remains unaltered, calculate new speed.

Solution :

At normal operation $N_1 = 1245$ rpm, $I_{a1} = 125$ A

$V_t = 440$ V, $r_a = 0.2$, $\phi_1 = 0$

Applying the relation, $E_b = V_t - I_a\,r_a$

Back emf $E_{b1} = 440 - 125 \times 0.2 \propto \phi_1\,N_1 \quad (1)$

When the excitation is reduced to 75 %

$$\phi_2 = 0.75\,\phi,\ N_2 = ?$$

$$E_{b2} = 440 - I_2 \times 0.2 \propto \phi_2\ N_2 \qquad (2)$$

Now, from (1) and (2)

$$\frac{E_{b1}}{E_{b2}} = \frac{\phi_1\ N_1}{\phi_2\ N_2} \Rightarrow \frac{440-125\times 0.2}{440-0.2\ I_2} = \frac{\phi\times 1245}{0.75\,\phi\times N}$$

$$N = \frac{1245\,(440-0.2\ I_2)}{0.75\times(440-125\times 0.2)} \qquad (3)$$

Since, total torque is constant.

$$\phi\ I = \text{Constant.}$$

or, $\phi_1\ I_1 = \phi_2\ I_2$ or, $I_2 = \dfrac{\phi_1}{\phi_2}\ I_1 = \dfrac{\phi}{0.75\,\phi}\times 125 = \dfrac{500}{3} = 166.667$ A

Putting $I_2 = 166.7$ A in eqn (3)

$$N = \frac{1245\times(440-0.2\times 166.667)}{0.75\times(440-0.2\times 125)}$$

$$N = 1626.67 \text{ rpm}$$

Speed = 1626.67 rpm **Ans.**

> **Problem 4.26 :** A dc series motor is driving a fan load where the load torque is proportional to the cube of speed. The resistance of the armature and field in series is 1 Ω and the motor takes 10 A and runs at 1000 rpm when operating from a 200 V supply.
>
> Calculate the value of resistance to be inserted in series with the armature to reduce the operating speed to 800 rpm. *(I.A.S., 1988)*

Solution :

At speed $N_1 = 1000$ rpm, $I_a = 10$ A, $r_a + r_f = 1\ \Omega$

In dc series machines $\phi \propto I_f = I_a$

Back emf $E_{b1} = 200 - 10\times 1 \propto \phi_1\ N_1 \propto I_1\ N_1 \qquad (1)$

Let r_s be the resistance added to the armature circuit to reduced speed

$$N_2 = 800 \text{ rpm}$$

Now, Back emf $E_{b2} = 200 - I_2\,(1+r_s) \propto I_2\ N_2 \qquad (2)$

$$\frac{E_{b1}}{E_{b2}} = \frac{I_1\ N_1}{I_2\ N_2} \Rightarrow \frac{190}{200 - I_2\,(1+r_s)} = \frac{I_1\ N_1}{I_2\ N_2} = \frac{10\times 1000}{I_2\times 800}$$

$$\text{or,} \qquad \frac{190}{200 - I_2(1 + r_s)} = \frac{25}{2\, I_2} \qquad (3)$$

Now, According to Question, Load Torque $T_L \propto N^3$

Electromagnetic Torque $T_e \propto \phi\, I$

Under steady state condition $T_e = T_L$

$$\text{or,} \qquad N^3 \propto \phi\, I, \left(\frac{N_1}{N_2}\right)^3 = \frac{I_1^2}{I_2^2}$$

$$\text{or,} \qquad \left(\frac{10000}{800}\right)^3 = \frac{100}{I_2^2} \text{ or, } I_2^2 = \frac{256}{5}$$

$$I_2 = \frac{16}{\sqrt{5}} = 7.1554 \text{ A}$$

$$\text{Putting } I_2 = \frac{16}{\sqrt{5}} \text{ in (3) and solving}$$

We get, $r_s = 11.751\ \Omega$ **Ans.**

> **Problem 4.27 :** A dc shunt motor, when fed from 200 V dc source, delivers full-load torque to a load at 1000 rpm. If supply voltage is reduced to 100 V, the motor speed becomes 800 rpm with the load torque remaining constant. Calculate the armature voltage drop at rated torque. Assume no magnetic saturation.

Solution :

When $\quad V_t = 200$ V, $\quad N_1 = 1000$ rpm

Let r_a be the armature resistance.

$$\text{Back emf} \quad E_{b1} = 200 - I_1 \times r_a \propto \phi_1 N_1 \propto I_{f1} \times N_1 \qquad (1)$$

When $V_t = 100$ V, $N_2 = 800$ rpm

$$\text{Back emf} \quad E_{b2} = 100 - I_2 \times r_a \propto \phi_2 N_2 \propto \frac{I_{f1}}{2} \times N_2 \qquad (2)$$

Source voltage is reduced to half.

$$\text{Now,} \quad \frac{E_{b1}}{E_{b2}} = \frac{I_{f1} \times N_1}{I_{f2} \times N_2}$$

or, $$\frac{200 - I_1 r_a}{100 - I_2 r_a} = \frac{I_{f1} \times 1000}{\frac{I_{f1}}{2} \times 800}$$

or, $$\frac{200 - I_1 r_a}{100 - I_2 r_a} = 2.5 \qquad (3)$$

Now, Since Torque is constant

$$\phi_1 I_1 = \phi_2 I_2, \; I_{f1} \times I_1 = I_{f2} \times I_2, \; I_{f1} \times I_1 = \frac{I_{f1}}{2} \times I_2$$

$$I_2 = 2I_1 \qquad (4)$$

Putting $I_2 = 2I_1$ in (3)

$$\frac{200 - I_1 r_a}{100 - 2 I_1 r_a} = 2.5$$

or, $200 - I_1 r_a = 250 - 5 I_1 r_a$ or $I_1 r_a = 12.5$ V

Thus, Armature drop at rated Torque = 12.5 V **Ans.**

> **Problem 4.28 :** A 250V, DC shunt motor has an armature current of 20 A when running at 1000 rpm against some load torque. The armature resistance is 0.5 Ω and brush contact drop is one volt per brush. By how much must the main flux be reduced to raise the speed by 50 % if the developed torque is constant? Ignore effects of armature reaction and magnetic saturation.

Solution :

Given, $r_a = 0.5\ \Omega$, $V_t = 250$ V

At speed $N_1 = 1000$ rpm, $I_1 = 20$ A, $\phi = \phi_1$

Back emf $$E_{b1} = 250 - 20 \times 0.5 \propto \phi_1 N_1 \qquad (1)$$

When the speed is increased by 50%

$$N_2 = 1000 + 500 = 1500 \text{ rpm, and } \phi = \phi_2$$

Back emf $$E_{b2} = 250 - I_2 \times 0.5 \propto \phi_2 N_2 \qquad (2)$$

Now, $$\frac{E_{b1}}{E_{b2}} = \frac{\phi_1 N_1}{\phi_2 N_2} \text{ or } \frac{240}{250 - 0.5 I_2} = \frac{\phi_1}{\phi_2} \times \frac{1000}{1500}$$

or, $$\frac{240}{250 - 0.5 I_2} = \frac{2\phi_1}{3\phi_2} \text{ or } \frac{\phi_1}{\phi_2} = \frac{240 \times 3}{500 - I_2} \qquad (3)$$

Now, Since developed Torque is constant

$$T_e \propto \phi I = \text{constant}$$

Hence, $\phi_1 I_1 = \phi_2 I_2, \quad \dfrac{\phi_1}{\phi_2} = \dfrac{I_2}{20}$ (4)

From (4) and (3) $\dfrac{I_2}{20} = \dfrac{240 \times 3}{500 - I_2}$ or, $I_2^2 - 500\, I_2 + 20 \times 240 \times 3 = 0$

$$I_2 = \frac{500 \pm \sqrt{500^2 - 4 \times 20 \times 240 \times 3}}{2}$$

Neglecting higher value, $I_2 = 30.683$ A

From (4) $\dfrac{\phi_1}{\phi_2} = \dfrac{30.683}{20}$

Now, % reduction in flux $= \dfrac{\phi_1 - \phi_2}{\phi_1} \times 100$

$$= \frac{30.683 - 20}{30.683} = 34.817\ \%$$ **Ans.**

Problem 4.29 : Two adjustable-speed dc shunt motors have speed-control range of 1500 to 500 rpm. In both the motors, speed adjustment is obtained by field-flux control. Motor *A* drives a load requiring constant power over the entire speed range; motor *B* drive a load requiring constant torque over the entire speed range. Ignore all losses and armature reaction in both the motors.

(*a*) If power outputs are equal at 1500 rpm and armature currents are each 90 A, find their armature currents at 500 rpm.

(*b*) If power outputs are equal at 500 rpm and armature currents are each 90 A, Find their armature currents at 1500 rpm.

Solution :

In field flux control method, $\phi \times N$ = constant. Let ϕ be the initial flux,

(*a*) At speed $N_1 = 1500$ rpm, $I_{A1} = I_{B1} = 90$ A

At speed $N_2 = 500$ rpm, $I_{A2} = ?, I_{B2} = ?, \phi_2 = ?$

Applying Relation $\phi_1 N_1 = \phi_2 N_2,\ \phi \times 1500 \subset \phi_2 \times 500$

or, $\phi_2 = 3\,\phi$

When speed is reduced to 500 rpm, the flux is increased by three times. Since, Motor A drives a load requiring constant power, This may be possible when either $E_g\ I_a$ remains constant or $T x \omega$ remains constant. For Motor A, when speed is reduced, the flux is increased by three times, Hence, the back emf E_g remains constant.

$$I_A = 90 \text{ A}$$

Since, Motor B drives a load at constant Torque.

$$T_e \propto \phi\, I = \text{Constant}$$

Now, Applying $\phi_1\, I_{B1} = \phi_2\, I_{B2},\ \phi \times 90 = 3\,\phi \times I_{B2}$

$\Rightarrow$ $I_{B2} = 30 \text{ A}$ **Ans.**

Here, $N_1 = 500 \text{ rpm},\ I_{A1} = I_{B1} = 90\text{A},\ \phi_1 = \phi$

$$N_2 = 1500 \text{ rpm},\ I_{A2} = ?,\ I_{B2} = ?,\ \phi_2 = ?$$

field-flux control Method

$$\phi_1\, N_1 = \phi_2\, N_2,\ \phi \times 500 = \phi_2 \times 1500$$

$\Rightarrow$ $\phi_2 = \dfrac{\phi}{3}$

Since, the motor A drives at const. power $E_g\ I_a$ or *Tw* remains constant. on varying the flux of motor *A*, the generated emf remains unaltered.

Hence, $I_{A2} = 90 \text{ A}$

Since the motor *B* drives at constant Torque i.e.

$$T \propto \phi\, I = \text{constant},\quad \phi_1\, I_{B1} = \phi_2\, I_{B2}$$

or, $\phi \times 90 = \dfrac{\phi}{3} \times I_{B2}$

$\therefore$ $I_{B2} = 270 \text{ A}$ **Ans.**

Problem 4.30 : A 250-V, 1500 r.p.m. d.c. machine has an armature resistance (including brushes) of 0.2 Ω and a brush drop under load of 2.0 volts. The rotational losses for a speed of 1500 r.p.m. and for an armature generated voltage of 250 V are 600 watts. Neglect the field circuit losses.

(*a*) For an armature current of 30 A and an armature generated voltage of 250 V, find the electromagnetic torque of the motor at rated speed.

(*b*) If the machine in part (*a*) is working as a generator, determine the shaft power input and the electric power output.

(*c*) If machine in part (*a*) is working as a motor, determine electric power input and the shaft power output.

(*d*) The motor of part (*c*) drives a load requiring constant torque at all the motor speeds. Calculate the motor speed it the armature terminal voltage is halved and the field flux is kept constant.

(*e*) The motor of part (*c*) drives a load requiring a constant power. Calculate the motor speed if the field flux is reduced to 0.6 of its previous value and the armature terminal voltage is kept constant.

Solution :

(*a*) $I_a = 30 \text{ A}, \quad E_b = 250 \text{ V}, \quad N_{\text{rated}} = 1500 \text{ rpm}$

Applying, $T_e\,\omega = E_b\,I_a$ or, $T_e = \dfrac{250 \times 30 \times 60}{1500 \times 2 \times \pi}$

$= 47.75$ Nm **Ans.**

(*b*) For generatorshaft power input = armature generated power + rotational losses = 250 × 30 + 600 = 7500 + 600

= 8100 watts **Ans.**

Electric power output

$$P_o = \text{armature generated power} - I_a^2\,r_a - V_b\,I_a$$

$= 7500 - 900 \times 0.2 - 2 \times 30 = 7260$ W **Ans.**

(*c*) As a motor, Electric power input

$$P_{in} = \left(E_b + I_a\,r_a + V_b\right) \times I_a = (250 + 30 \times 0.2 + 2) \times 30$$

$= 7740$ W **Ans.**

Shaft power output

$$P_{sho} = \text{power input} - I_a^2\,r_a - V_b\,I_a - \text{rotational losses}$$

$= 7740 - 900 \times 0.2 - 2 \times 30 - 600 = 6900$ W **Ans.**

(*d*) At speed $N_1 = 1500$ rpm, $E_{b1} = 250$ V (Given)

When armature Terminal voltage is halfed

$$E_{b2} = \frac{V_t}{2} - I_a\,r_a - V_b$$

$$= \frac{(250 \times 30 \times 0.2 + 2)}{2} - 30 \times 0.2 - 2 = 121 \text{ V}, N_2 = ?$$

Applying $E_b \propto \phi\,N$, Since ϕ is constant

$$\frac{E_{b1}}{E_{b2}} = \frac{N_1}{N_2} \text{ or, } \frac{250}{121} = \frac{1500}{N_2}$$

or, $N_2 = \dfrac{1500 \times 121}{250} = 726$ rpm **Ans.**

(*e*) Since, power is constant

$$E_{g2} \times I_{a2} = 7500 \qquad (1)$$

Let I_{a2} be the armature current when flux is reduced.

$$E_{g2} = V_t - V_b - I_a r_a, \quad E_{g2} = 258 - 2 - 0.2\ I_a$$

or, $$E_{g2} = 258 - 2 - 0.2 \times \frac{7500}{E_{g2}}$$

or, $$E_{g2}^2 = 256\ E_{g2} + 1500 = 0$$

$$E_{g2} = \frac{256 \pm \sqrt{(256)^2 - 4 \times 1500}}{2}$$

Neglecting lower value.

$$E_{g2} = 250.0 \text{ V}$$

Now, $$\frac{E_{g1}}{E_{g2}} = \frac{\phi_1 N_1}{\phi_2 N_2} \text{ or, } N_2 = \frac{\phi_1 N_1 E_{g2}}{E_{g1} \times \phi_2}$$

$$= \frac{\phi_1 \times 1500 \times 250}{0.6\ \phi_1 \times 250} = 2500 \text{ rpm} \quad \textbf{Ans.}$$

Problem 4.31 : For a 400-V long shunt compound generator, the constant loss, comprising of excitation and no-load rotational losses, is 10 kW. The resistances of the armature, series and shunt field windings are respectively 0.1 Ω, 0.05 Ω and 50 Ω. Find the maximum efficiency and the output at which it occurs.

Solution :

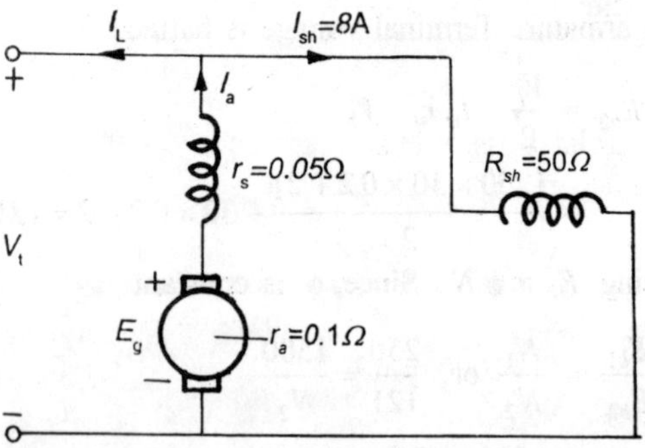

Fig. P. 4.31

Given $V_{sh} = 50\ \Omega$, $V_t = 400$ V,

$r_s = 0.05\,\Omega$, $r_a = 0.1\,\Omega$

Constant loss = 10 kW.

For Maximum efficiency.

Variable armature loss = const. loss

or, $I_a^2\left(r_a + r_s\right) = 10 \times 1000$, or, $I_a^2 = \dfrac{10 \times 1000}{0.15}$

$$\Rightarrow \quad I_a = \sqrt{\frac{10 \times 1000}{0.15}} = 258.199 \text{ A} = 258.2 \text{ Amp}.$$

Now, shunt field current

$$I_{sh} = \frac{400}{50} = 8 \text{ A}$$

Line current $= I_a - I_{sh} = 258.199 - 8 = 250.199$ A

Output power $P_o = V_t \times I_L = 400 \times 250.199 = 100.08$ kW.

Now, Maximum efficiency $= \dfrac{P_o}{P_o + 2 \times \text{const. loss}}$

$$= \frac{100.08 \times 1000}{100.08 \times 1000 + 20 \times 1000}$$

$= 83.34\%$ **Ans.**

Power outpout = 100.08 kW

> **Problem 4.32 :** In a d.c. motor, hysteresis and eddy current losses at 1000 r.p.m. are 200 W and 100 W respectively. Find the speed at which the total losses would be reduced to half on the assumption of constant field flux.

Solution :

We know that Total Loss = Hysteresis Loss + eddy current loss

$$P = P_h + P_e,\ P = A_f + B_f^2$$

Where $P_h = A_f$, $P_e = B_f^{\,2}$ where f is frequency

Given, $f = 1000 \text{ rpm} = \dfrac{1000}{60} \text{ Hz} = \dfrac{50}{3} \text{ Hz}$

Now, $P_h = 200$ or $A \times \dfrac{50}{3} = 200 \Rightarrow \text{A} = 12$

$$P_e = 100, B \times \left(\frac{50}{3}\right)^2 = 100$$

$$B = \frac{100 \times 9}{2500} = \frac{9}{25}$$

Now, where loss is reduced to half

$$P = \frac{300}{2} = 150 \text{ watt}, 150 = 12\,f + \frac{9}{25} f^2$$

or $$150 \times 25 = 12 \times 25\,f + 9\,f^2$$

or $$3\,f^2 + 100\,f - 1250 = 0$$

$$f = -100 \pm \frac{\sqrt{(100)^2 + 4 \times 3 \times 1250}}{2 \times 3}$$

Neglecting negative value

We get $f = 9.686$ rev/sec

$\Rightarrow$ $f = 9.686 \times 60 \text{ rpm} = 581.14 \text{ rpm}$ **Ans.**

Problem 4.33 : In a test on a d.c. shunt generator, whose full load output is 200 kW at 250 V, the following figure were obtained.

(a) With the machine at rest, a p.d. of 8 volts produced an armature current of 400 A.

(b) With the motor running at no-load and at rated speed, the line current was 36 A, the field current 12 amperes and the supply voltage 250 V.

Obtain the generator efficiency at full load and half full load.

Solution :

Given, $V_t = 250 \text{ V}, \quad P_o = 200 \text{ kW}$

Full Load Line Current $I_{fL} = \dfrac{P_o}{V_t} = \dfrac{200 \times 1000}{250} = 800 \text{ A}$

Given at No-load

Armature voltage $= 8$ V

Armature current $= 400$ A, $r_a = \dfrac{V_a}{I_a} = \dfrac{8}{400} = 0.02\ \Omega$

Full Load copper loss $= (I_{fL} + I_{sh})^2 \times r_a$

$$= (800 + 12)^2 \times 0.02 = 13186.88 \text{ W}$$

No load armature current $= I_L - I_{sh} I_{ao} = 36 - 12 = 24$ A

No load armature loss $= (I_{ao})^2 \times r_a = (24)^2 \times 0.02 = 11.52$ kW

Generated Power at No load $= V_t \times I_L = 250 \times 36$ W

Constant Losses $= 36 \times 250 - 11.52 = 8988.48$ W

Total Losses $= 8988.48 + 13186.88 = 22175.36$ W

$$\text{Now, \% full Load efficiency} = \left(1 - \frac{\text{Total Losses}}{\text{Total Losses} + P_{\text{out}}}\right) \times 100\%$$

$$= \left(1 - \frac{22175.36}{22175.36 + 200{,}000}\right) \times 100\%$$

$$= 90.019\%$$

% full load efficiency = 90.019% **Ans.**

At half of full load,

$$\text{Load current} = \frac{800}{2} = 400 \text{ A}$$

Armature current $= I_L + I_{sh} = 400 + 12 = 412$ A

Now, Armature loss $= I_a{}^2 \times r_a = (412)^2 \times 0.02$

$= 3394.88$ W

Total Losses $= 3394.88 + 8988.48$

$= 12383.36$ W

Power output $= 250 \times 400 = 100{,}000$ W

Percentage efficiency of machine

$$= \left(\frac{\text{Power output}}{\text{Total Loss} + \text{power output}}\right) \times 100$$

$$= \left(\frac{100{,}000}{12383.36 + 100{,}000}\right) \times 100$$

$= 88.981$ % **Ans.**

Problem 4.34 : A 230-V d.c., shunt motor is taking 5 A when running light (i.e., at no load). The armature resistance (including brushes) is 0.2 Ω and field circuit resistance is 115 Ω. For an input current of 72 A, calculate the shaft output and efficiency. Also calculate the armature current at which the efficiency is maximum.

Solution :

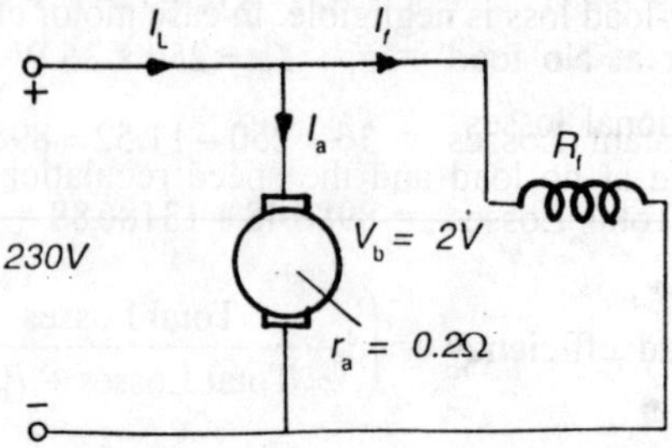

Fig. P. 4.34

Given, $r_a = 0.2\,\Omega$, $R_f = 115\,\Omega$

No load armature current $= 5 - \dfrac{230}{115} = 3$ A

No-Load power input $P_h = 230 \times 5 = 1150$ watts

No-load Armature Loss $= (3)^2 \times 0.2 = 1.8$ watts

Constant Losses At Load $= 1150 - 1.8 = 1148.2$ watts

Load Armature current $I_a = 72 - 2 = 70$ A

Armature Loss $= (70)^2 \times 0.2 = 980$ watts

Total Losses = 1148.2 + 980 = 2128.2 W

Power Input = 230 × 72 = 16560 watt

$$\% \text{ Efficiency of the motor} = \left(1 - \frac{\text{Total Losses}}{\text{Power Input}}\right) \times 100\%$$

$$= \left(1 - \frac{2128.2}{16560}\right) \times 100\% = 87.149\% \quad \textbf{Ans.}$$

Shaft power output

$P_{sho} = P_{in}$ – armature loss – const. loss – brush drop loss

$= P_{in}$ – Total Losses

= 16560 – 2128.2 = 14431.8 W = 14.432 kW

For Max^m efficiency

Armature copper Loss = constant loss

$$I_a^2 r_a = \omega_o + V_f I_f, \; 0.2 I_a^2 = 1148.2$$

$$I_a = \sqrt{\frac{1148.2}{0.2}} = 75.769 \text{ Amp} \quad \textbf{Ans.}$$

Problem 4.35 : A 600 V dc shunt motor drives a 60 kW load at 1000 rpm. The field resistance is 100 Ω and armature resistance is 0.15 Ω. Stray-load loss is negligible. In case motor efficiency is 85%, determine

(a) the rotational losses

(b) the speed at no load and the speed regulation.

Solution :

Given, $\eta_m = 85\% = 0.85$

Total Losses $= \left(\frac{1}{\eta_m} - 1\right) \times 60 \times 1000 = 10588.24$ W

Total power Input $= \left(\frac{1}{0.85} - 1\right) \times 60 \times 1000 + 60 \times 1000$

$= 60{,}000 + 10588.24 = 70588.24$ W

Line current $I_L = \frac{\text{Total power Input}}{\text{Line voltage}} = \frac{70588.24}{600} = 117.647$ A

Armature current $I_a = I_L - I_{sh} = 117.647 - \frac{600}{100} \left[\because I_{sh} = \frac{V_t}{R_{sh}}\right]$

$= 111.647$ A

Armature copper Loss $= I_a^2 r_a = (111.647)^2 \times 0.15 = 1869.758$ W

Shunt Loss $= V_f I_f = 600 \times 6 = 3600$ W

Brush drop Loss $= V_b \times I_a = 223.28$ W

Now, Rotational Loss = Total Losses – Armature Copper Loss – Shunt Loss = 10588.24 – 1869.758 – 3600

= 5118.49 W **Ans.**

(b) At Load $N_1 = 1000$ rpm

Applying $E_b = V_t - I_a r_a$

$$E_{b1} = 600 - I_a \times 0.15 \propto \phi_1 N_1 \quad (1)$$

At No-load $I_{a2} = 0,\ N_2 = ?$

$$E_{b2} = 600 - 0 \times 0.15 \propto \phi_2 N_2 \quad (2)$$

Now, $\frac{E_{b1}}{E_{b2}} = \frac{\phi_1 N_1}{\phi_2 N_2}$

Since, field current is constant

$$\phi_1 = \phi_2 = \phi$$

$$\Rightarrow \quad \frac{600 - I_a \times 0.15}{600} = \frac{1000}{N_2}$$

$$\Rightarrow \quad N_2 = \frac{1000 \times 600}{600 - 0.15 \times 111.647} = 1028.713 \text{ rpm}$$

$$N_2 = 1028.713 \text{ rpm} \quad \textbf{Ans.}$$

Now, % speed regulation $= \left(\frac{N_2 - N_1}{N_1}\right) \times 100\%$

$$= \left(\frac{1028.713 - 1000}{1000}\right) \times 100\%$$

$$= 2.871\% \quad \textbf{Ans.}$$

Problem 4.36 : Hopkinson's test on two similar dc shunt machines gave the following data:
Line voltage 230 V; Line current, excluding both the field currents. 40 A; motor armature current 350 A; field currents 5 A and 4.2 A. Calculate the efficiency of each machine. Armature resistance of each machine is 0.02 Ω.

Solution :

Given, $V_L = 230$ V $\quad I_L = 40$ A

$I_{fg} = 5$ V $\quad I_{fm} = 4.2$ A

$r_{ga} = r_{ma} = 0.02\ \Omega$

Motor armature current $(I_m) = 350$ A

Generator armature current $(I_g) = I_M - I_L = 350 - 40 = 310$ A

Motor armature circuit Loss $= (350)^2 \times 0.02\ (350)^2 \times 0.02$

$= 2450$ W

Generator armature circuit Loss

$P_{ag} = (310)^2 \times 0.02 = 1922$ W

Power Input from supply $\quad P_{in} = 230 \times 40 = 9200$ W

Now, No-load rotational Losses in both Machine

$W_{no} = 9200 - (1922 + 2450) = 4828$ W

Generator output $= 230 \times 310 = 71300$ W

$$\text{Total Losses in generator} = \frac{W_{no}}{2} + V_L \times I_{fg} + P_{ag}$$

$$= \frac{4828}{2} + 230 \times 5 + 1922 = 5486 \text{ W}$$

$$\eta_g = \left(1 - \frac{\text{Total Losses}}{\text{Total Losses + output}}\right) \times 100\%$$

$$= \left(1 - \frac{5486}{5486 + 71300}\right) \times 100\%$$

$$= 92.855\% \quad \textbf{Ans.}$$

$$\text{For Motor, Motor Input} = 230 \times (350 + 4.2) = 81466 \text{ W}$$

$$\text{Total Losses} = \frac{W_{no}}{2} + V_L \times I_{fm} + P_{ag}$$

$$= \frac{4828}{2} + 230 \times 4.2 + 2450 = 5830$$

$$\therefore \text{Efficiency of motor} = \left(1 - \frac{\text{Total Losses}}{\text{Motor output}}\right) \times 100\%$$

$$= \left(1 - \frac{5830}{81466}\right) \times 100\% = 92.844\% \quad \textbf{Ans.}$$

Problem 4.37 : *(a)* In cross-field machines, one pole consists of two polar projections. Explain the reason.

(b) A 4 kW, 250 V, 3000 r.p.m. amplidyne has the following constants:

$$r_f = 60, \ \Omega \ r_a = 5\,\Omega, \ r_c = 1\,\Omega$$

Voltage constants $K_{qf} = 300$ V/field amp.

$K_{dq} = 120$ V/amp.

$K_{qd} = 50$ V/amp.

Calculate the field current and power gain at rated output. Also obtain these values when the compensation is zero.

Solution :

Given, P = 4 kW $\qquad K_{qf} = 300$ V/field amp.

$V_t = 250$ V $\qquad K_{dq} = 120$ V/amp.

$$N_{rated} = 3000 \text{ rpm}, \quad K_{qd} = 50 \text{ V/amp.},$$

$$r_f = 60\,\Omega, \quad r_d = 5\,\Omega,$$

$$r_c = 1\Omega, \quad I_f = ? \quad P_{gain} = ?$$

At rated voltage 250 V rated current, $(I_d) = \dfrac{4 \times 1000}{250} = 16\text{ A}$

Voltage amplification, $A = \dfrac{K_{dq} \times K_{qf}}{r_d \times V_f} = \dfrac{120 \times 300}{5 \times 60} = 120$

For an amplidyane, voltage equation is

$$V_d = A V_f - I_d (r_d + r_c)$$

$\Rightarrow$ $$250 = 120 V_f - 16(5+1)$$

or $$250 = 120 V_f - 96$$

$\Rightarrow$ $$V_f = \frac{346}{120} = 2.883\text{ V}$$

Field current $$I_f = \frac{V_f}{r_f} = \frac{2.883}{60} = 48.050\text{ mA}$$

$$I_f = 48.050\text{ mA} \quad \textbf{Ans.}$$

Power gain with compensation $= \dfrac{V_d\, I_q}{V_f\, I_f} = \dfrac{250 \times 16}{2.883 \times 48.050 \times 10^{-3}}$

$$= 28874.998 \quad \textbf{Ans.}$$

For Amplidyne with zero compensation

$$K_1 = \frac{K_{dq} \times K_{qd}}{r_d} = \frac{120 \times 50}{5} = 1200$$

Now, Voltage equation is $V_d = A V_f - I_d (K_1 + r_d)$

or $$250 = 120 V_f - 16(1200 + 5)$$

or $$V_f = \frac{19530}{120} = 162.75\text{ V}$$

Field current $$I_f = \frac{162.75}{60} = 2.7125\text{ V} \quad \textbf{Ans.}$$

Power gain with zero compensation =

$$\frac{V_d\, I_d}{V_f\, I_f} = \frac{250 \times 16}{162.75 \times 2.7125} = 9.061 \quad \textbf{Ans.}$$

Field current $I_f = \dfrac{V_f}{r_f} = \dfrac{162.75}{60} = 2.7125\text{ A}$ **Ans.**

> **Problem 4.38 :** A small permanent magnet dc motor runs at a no-load speed of 10,600 rpm and takes 10 mA when fed from 6 V dc source. Its armature circuit resistance is 6 Ω. Calculate its
> *(a)* No-load rotational losses
> *(b)* Stall torque
> *(c)* Speed to achieve a shaft power of 1 W
> *(d)* Efficiency under the conditions of part *(c)*

Solution :

Given, $I_a = 10 \times 10^{-3}\text{ A}, \quad V_t = 6\text{ V}$

$N = 10{,}600\text{ rpm}, \quad r_a = 6\,\Omega$

(a) Back emf $E_b = V_t - I_a r_a$

$= 6 - 10 \times 10^{-3} \times 6 = 5.94\text{ V}$

At No-load, all the electromagnetic power developed is used to supply rotational losses.

∴ No-Load rotational Losses =

$E_b \times I_a = 5.94 \times 10 \times 10^{-3} = 0.0594\text{ W}$ **Ans.**

(b) We know that $K_m = \dfrac{E_b}{2\pi N} = \dfrac{5.94 \times 60}{2\pi \times 10{,}600}$

When motor stalls $E_b = 0,\ V_t = I_a r_a$

or $I_a = \dfrac{6}{6} = 1\text{ A}$

(c) Given, Shaft power $(P_{sh}) = 1\text{ W}$

As we know that

Shaft power output = Electromaguetic power – no-load rotational losses

∴ Electromagnetic power = Shaft power output + W_{no}

$= 1 + 0.0594 = 1.0594\text{ W}$

Let E_b' be the back emf and I_2' be the armature current.

∴ $E_b' \times I_2' = 1.0594$ (1)

Since supply voltage $= 6\text{ V}$

$$\therefore \qquad E_b' = V_t - I_2' r_a = 6 - 6 I_2'$$

or
$$\frac{E_b'}{6} = 1 - I_2'$$

$$\therefore \qquad I_2' = 1 - \frac{E_b'}{6}$$

Putting $I_2' = 1 - \dfrac{E_b'}{6}$ in equation (1)

We get
$$E_b'\left(1 - \frac{E_b'}{6}\right) = 1.0594$$

or
$$6\,E_b' - E_b'^2 = 6 \times 1.0594$$

or
$$E_b'^2 - 6\,E_b' + 6 \times 1.0594 = 0$$

This yield
$$E_b' = \frac{6 \pm \sqrt{36 - 4 \times 6 \times 1.0594}}{2}$$

Neglecting lower value $E_b' = 4.626$ V

Now, Applying the relation
$$\frac{E_b'}{E_b} = \frac{N_2'}{N_1}$$

or
$$N_2' = \frac{E_b'}{E_b} \times N_1$$

We have $E_b = 5.94$ V,

$N_1 = 10{,}600$ rpm,

$E_b' = 4.646$ V

$$N_2' = \frac{4.626}{5.94} \times 10600 = 8255.2 \text{ rpm} \quad \textbf{Ans.}$$

(d) We have, power output = 1 W

$\therefore$ Total losses = No-load rotational losses $+ I_2'^2\, r_a$

$\therefore$ Power input = Power output + No-load rotational losses $+ I_2'^2\, r_a$

$$= 1 + 0.0594 + \left(\frac{1.0594}{E_b'}\right)^2 \times r_a$$

$$= 1 + 0.0594 + \left(\frac{1.0594}{4.626}\right)^2 \times 6$$

$$= 1.374073 \text{ W}$$

Now, $$\text{Efficiency} = \frac{\text{Power Output}}{\text{Power Input}} = \frac{1}{1.374073}$$

$$= 0.72776 = 72.776\% \quad \textbf{Ans.}$$

Problem 4.39 : A 4-pole d.c. series fan motor takes an armature current of 50 A, when running steadily at 1500 r.p.m. on a 240 V supply. The four field coils are now connected in two parallel groups of two in series. Assuming that the flux produced is directly proportional to the exciting current and the load torque increases as the square of the speed, find the new speed and the armature current. Neglect losses and assume constant supply voltage.

Solution :

The electromagnetic torque is given by $T_e \propto \phi\, I_a$

As we know that in d.c. series motor $\phi \propto I_a, \quad \therefore\ T_e \propto I_a^2$ (1)

Also, Given load toraue $\alpha\,(\text{speed})^2, \quad \therefore\ T_L \propto N^2$ (2)

Under steady state conditions, $T_e = T_L$

From relations (1) and (2) $I_a^2 \propto N^2$

or $$I_f\, I_a \propto N^2, \text{ since } \phi \propto I_f = I_a$$

Let I_{a1} be the armature current with four field coils in series then

$$I_{a1}^2 \propto N_1^2 \quad (3)$$

Again Let I_{a2} be the armatune current when four field coils are connected in two parallel groups of two in series. In this case

$$\left(\frac{I_{a2}}{2}\right) \times I_{a2} \propto N_2^2$$

or $$I_a^2 \propto 2\, N_2^2 \quad (4)$$

Now, Dividing (3) by (4), we get

$$\left(\frac{I_{a2}}{I_{a1}}\right)^2 = 2\left(\frac{N_2}{N_1}\right)^2$$

or $$\frac{I_{a2}}{I_{a1}} = \sqrt{2} \times \frac{N_2}{N_1} \text{ or } I_{a2} = \sqrt{2} \times N_2 \times \frac{50}{1500}$$

or $$I_{a2} = \frac{\sqrt{2}\, N_2}{30} \quad (5)$$

For motor $E_b = V_t - I_a\,(r_a + r_f) \propto N\,\phi$

When four coils are connected is series

$$E_{b1} = V_t - I_{a1}\,(r_a + r_f) \propto N_1\,\phi_1, \propto N_1\, I_{a1}$$

When coils are connected in two parallel groups

$$E_{b2} = V_t - I_{a2}\,(r_a + \frac{r_f}{4})\; \alpha\; N_2\,\phi_2 \times \propto N_2 \times \frac{I_{a2}}{2}$$

Since r_a and r_f are not given

$$\therefore \quad \frac{E_{b1}}{E_{b2}} = \frac{2\,N_1\,I_{a1}}{N_2\,I_{a2}} \text{ or } \frac{240}{240} = \frac{2 \times 1500 \times 50 \times 30}{N_2\,\sqrt{2} \times N_2}$$

$$\therefore \quad N_2 = \frac{\sqrt{2 \times 1500 \times 50 \times 30}}{\sqrt{2}} = 1783.81 \text{ rpm}$$

Putting N_2 in relation (5) $I_{a2} = \frac{\sqrt{2} \times 1783.81}{30} = 84.09$ A

Then Speed = 1783.81 rpm, armature current = 84.09 A **Ans.**

Problem 4.40 : A 220 V d.c. shunt machine has armature circuit resistance of 0.5 Ω and field circuit resistance is 110 Ω. If the machine is connected to 220 V. Supply mains, find the ratio of speed as a generator to the speed as a motor. The line current in each case is 50 A.

Solution :

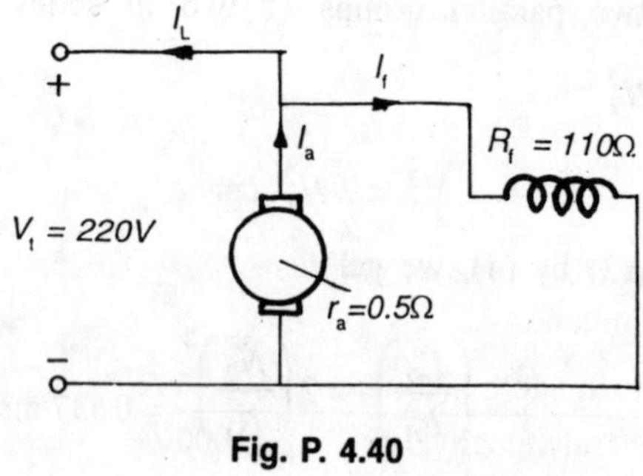

Fig. P. 4.40

Given, Line current, $I_L = 50\,\text{A}$

Now, Field current, $I_f = \dfrac{220}{110} = 2\text{ A}$

For Generator, Armature current, $I_{ag} = 50 + 2 = 52\text{ A}$

Generated voltage $\quad E_g = V_t + I_{ag}\, r_a \propto N_g\, I_f \qquad (1)$

For motoring mode, Armature current,

$$I_{am} = 50 - 2 = 48\text{ A}$$

Back emf $\quad E_b = V_t + I_{am}\, r_a \propto N_M I_f \qquad (2)$

From (1) and (2) $\quad \dfrac{E_g}{E_b} = \dfrac{N_g\, I_f}{N_M\, I_f}$

or $$\frac{V_t + I_{ag}\, r_a}{V_t - I_{am}\, r_a} = \frac{N_g}{N_M}$$

or $$\frac{N_g}{N_M} = \frac{220 + 52 \times 0.5}{220 - 48 \times 0.5}$$

or $$\frac{N_g}{N_M} = 1.255 \quad \textbf{Ans.}$$

Problem 4.41 : A commutator with a diameter of 30 cm rotates at 1000 rpm. For a brush width of 1 cm, find the time of commutation.

Solution :

Here $$f = \frac{1000}{60} \text{ rev/sec.,}$$

$$D = 30 \text{ cm}$$

Perepheral velocity of the commutator

$$V_p = \pi \times D \times f$$

$$= \pi \times 30 \times \frac{1000}{60} \text{ cm/s}$$

As we know that V_p × time of commutation = Brush width

∴ Time of commutation

$$t_c = \frac{\text{Brush width}}{\text{Peripheral velocity}} = \frac{1 \times 60}{\pi \times 30 \times 1000} = 0.637 \text{ ms} \quad \textbf{Ans.}$$

Problem 4.42 : A 2-pole dc generator charges a 200 V battery of negligible internal resistance. The armature of the machine is made up of 1,000 conductors, each of two milli-ohm resistance. The charging current is found to be 10 A and 20 A for generator speeds of 1055 and 1105 r.p.m. respectively. Find the field circuit resistance and flux per pole of the generator. Neglect armature reaction effects.

Solution :

Given, Total No. of conductors $= 1000$

Resistance of each conductor $= 2 \times 10^{-3}$ ohm

$$\therefore \text{Armature circuit resistance } (r_a) = \frac{1000}{2} \times 2 \times 10^{-3} \times \frac{1}{2} = 0.5\,\Omega$$

For d.c. generator, $E_g = V_t + I_a r_a = V_t + (I_L + I_f) r_a$

Let I_f be the shunt field current.

Now, Generated emf at speed

$$N_1 = 1055 \text{ r.p.m.}$$

$$E_{g1} = 200 + (10 + I_f) \times 0.5 \propto I_f \times N_1 \quad (1)$$

Generated emf at speed $N_2 = 1105$ r.p.m.

$$E_{g2} = 200 + (20 + I_f) \times 0.5 \propto I_f \times N_2 \quad (2)$$

From (1) and (2)

$$\frac{200 + (10 + I_f) \times 0.5}{200 + (20 + I_f) \times 0.5} = \frac{N_1}{N_2}$$

or $$\frac{400 + 10 + I_f}{400 + 20 + I_f} = \frac{1055}{1105} \text{ or } \frac{410 + I_f}{420 + I_f} = \frac{211}{221}$$

or $$410 \times 221 + 221\, I_f = 420 \times 211 + 211\, I_f$$

or $$10\, I_f = 420 \times 211 - 410 \times 221$$

$$\therefore \quad I_f = 8862 - 9061, \; I_f = -199 \text{ A}$$

The Negative sign shows that the field direction is reversed.

Now From equ (1)

$$E_{g1} = 200 + (10 - 199) \times 0.5 = 200 - 94.5 = 105.5 \text{ V}$$

Again, Applying $E_g = \dfrac{\phi\, nzp}{60\, A}$

$$\therefore \quad \phi = \frac{60 \times 2 \times 105.5}{1055 \times 1000 \times 2} = 6 \times 10^{-3} = 6 \text{ mWb} \quad \textbf{Ans.}$$

Thus, Flux per pole = 6 mWb **Ans.**

Field circuit resistance $(V_f) = \frac{200}{199} = 1.01\,\Omega$ **Ans.**

Problem 4.43 : A 200 V shunt motor has an armature resistance of 0.2 Ω and field resistance of 100 Ω. At no load the motor takes 10 A and runs at 1500 r.p.m. If the motor draws 42 A at rated voltage and rated load. Calculate the motor speed and rated shaft torque in Nm. The rotational losses at no load and full load are same. Neglect armature resistance.

Solution :

Given, $V_t = 200$ V, $R_f = 100\,\Omega$

Shunt field current $I_f = \frac{200}{100} = 2$ A

At no-load, $N_0 = 1500$ r.p.m.

Armature current $I_{a0} = 10 - 2 = 8$ A

Back emf developed at No-load

$$E_{bo} = 200 - 0.2 \times 8 = 200 - 1.6 = 198.4 \text{ V}$$

Rotational lossess $= E_{bo} \times I_{ao} = 8 \times 198.4 = 1587.2$ watts

At Full load, Armature current

$$I_a = 42 - 2 = 40 \text{ A}$$

Back emf developed $E_b = 200 - 40 \times 0.2 = 200 - 8 = 192$ V

For motor, $E \propto \phi N$

Here, the field current is constant, hence no-load flux is equal to full load flux.

$$\therefore \quad \frac{E_b}{E_{bo}} = \frac{N}{N_o} \text{ or } N = \frac{E_b}{E_{bo}} \times N_o$$

$$N = \frac{192}{198.4} \times 1500 \cong 1452 \text{ r.p.m.} \quad \textbf{Ans.}$$

Now, Shaft power $P_{sh} = P_e - \text{rotational losses}$

$$= 40 \times 192 - 1587.2 = 6092.8 \text{ Watts}$$

Rated shaft torque $T_{sh} = \frac{6092.8 \times 60}{2 \times \pi \times 1452}$ Nm = 40.1 Nm **Ans.**

Problem 4.44 : A 200 V, 4-pole, shunt motor has two circuit armature winding with 500 conductors. The armature circuit resistance is 0.2 Ω, field resistance is 100 Ω and flux per pole is 0.002 weber. Armature reaction is neglected. If the motor draws 12 A from the mains then compute

(*a*) Speed and internal torque developed

(*b*) The shaft power, shaft torque and efficiency with rotational losses equal to 200 watts.

Solution :

Given, $V_t = 200$ V, $P = 4$, $I_L = 12$ A

$R_f = 100\ \Omega$, $r_a = 0.2\ \Omega$, $\phi = 0.002$ Wb

Field current, $I_f = \dfrac{200}{100} = 2$ A

(*a*) Armature current $I_a = I_L - I_f = 12 - 2 = 10$ A

Back emf developed $E_b = V_t - I_a r_a$

$= 200 - 10 \times 0.2 = 198$ V

Applying, $E_b = \dfrac{\phi NZP}{60A}$

$$\therefore \quad N = \frac{E_b \times 60 \times A}{\phi ZP} = \frac{198 \times 60 \times 2}{0.002 \times 500 \times 4} \text{ r.p.m.}$$

$= 5940$ r.p.m.

Electromagnetic power,

$$P_e = E_b \times I_a = 198 \times 10 = 1980 \text{ W}$$

Internal torque developed,

$$T_e = \frac{P_e}{\omega_n} = \frac{1980 \times 60}{2 \times \pi \times 5940} = 3.18 \text{ Nm} \quad \textbf{Ans.}$$

(*b*) Shaft power, P_{sh} = Electromagnetic power − Rotational losses

= 1980 − 200 = 1780 W

$$\therefore \text{ Shaft tarque, } T_{sh} = \frac{1780 \times 60}{2 \times \pi \times 5940} \text{ Nm} = 2.86 \text{ Nm} \quad \textbf{Ans.}$$

$$\text{Efficiency} \quad \eta = \frac{\text{Output power}}{\text{Input power}} = \frac{\phi NZP}{60A}$$

= 74.17% **Ans.**

Problem 4.45 : A 12 kW, 200 V, 70 A, 1000 r.p.m. d.c. series motor has following full load losses expressed in percentage of motor input:

Armature circuit ohmic loss (including brush loss) = 2.6%

Field ohmic loss = 2.4%

Rotational loss = 2.0%

Neglect the armature reaction and magnetic saturation and assume the rotational loss to remain constant.

If the motor draws half the rated current at rated voltage, determine

(i) Speed in r.p.m., and

(ii) Shaft power output.

Solution :

Given $V_t = 200$ V, $I_{rated} = 70$ A,

$N_{rated} = 1000$ r.p.m.

At full load, $N_1 = 1000$ r.p.m.

Full load input $= 70 \times 200 = 14000$ W

Total ohmic loss $= I_L^2 (r_a + r_s)$

or $$I_L^2 (r_a + r_s) = 14000 \times \frac{5}{100} = 700$$

or $$r_a + r_s = \frac{700}{(70)^2} = 0.142\ \Omega$$

Back emf developed $$E_{b1} = 200 - 0.142 \times 70 = 190.1 \propto \phi_1 N_1 \quad (1)$$

At half of the full load, Armature current $I_{a2} = 70 \times \frac{1}{2} = 35$ A

Back emf developed $$E_{b2} = 200 - 35 \times 0.142 = 195\ \alpha\ \phi_2 N_2 \quad (2)$$

From (1) and (2) $$\frac{E_{b2}}{E_{b1}} = \frac{\phi_2 N_2}{\phi_1 N_1}$$

Since $I_f\ \alpha\ \phi$ $$\frac{E_{b2}}{E_{b1}} = \frac{35 \times N_2}{70 \times 1000}$$

or $$\frac{195}{190.1} = \frac{35 \times N_2}{70 \times 1000}$$

$\therefore$ $$N_2 = \frac{70 \times 1000 \times 195}{35 \times 190.1} = 2052 \text{ r.p.m}$$ **Ans.**

(b) Shaft pover output P_{sh} = Electromagnetic power–rotational loss

$$= 195 \times 35 - 14000 \times \frac{2}{100}$$

$$= 6545 \text{ watts} \quad \textbf{Ans.}$$

Problem 4.46 : In a 240 V, 10 kW d.c. shunt motor, it is required that the starting armature current should not exceed twice its rated armature current. During the starting of the motor, the starting resistance is cut out in steps, as soon as the armature current drops to its rated value. The field resistance is 120 Ω, and the total armature circuit resistance is 0.34 Ω. Neglect armature inductance.

(i) Find the external resistance required at the time of starting the motor.

(ii) Determine the value of the first resistance element that must be cut out, when the armature current drops to rated value.

(iii) Find the external resistance to be cut out in the second step.

(iv) Find the total no. of steps required and their resistance value.

Solution :

Given, $P_{rated} = 10 \text{ kW}, \quad V_t = 240 \text{ V}$

$R_f = 120\,\Omega, \quad r_a = 0.34\,\Omega$

∴ Rated line current $(I_L) = \dfrac{10 \times 1000}{240} = 41.67 \text{ A}$

Shunt field current $(I_f) = \dfrac{240}{120} = 2 \text{ A}$

∴ Rated armature current $(I_a) = 41.67 - 2 = 39.67 \text{ A}$

(i) Starting armature current $I_{st} = 39.67 \times 2 = 79.34 \text{ A}$

External resistance required at the time of starting

$$= \frac{V_t}{I_{st}} - r_a = \frac{240}{79.34} - 0.34 = 2.68\,\Omega \quad \textbf{Ans.}$$

(ii) When the armature current drops to rated current at stud 1, then

Back emf, $E_{b1} = V_t - I_a \times$ (external resistance at stud 1 + armature resistance) $= 240 - 39.67 \times (2.68 + 0.34)$

$$= 240 - 39.67 \times 3.02 = 120.2 \text{ V}$$

When the handle is moved to stud 2, armature current shoots up to starting current, but the counter emf remains same.

Total resistance at the 2nd stud.

$$r_2 + r_3 + \ldots\ldots + r_n + r_a = \frac{V_t - E_{b1}}{I_{st}} = \frac{240 - 120.2}{79.34} = 1.51\,\Omega$$

$\therefore$ External resistance = 1.51 – 0.34 = 1.17 Ω **Ans.**

(iii) When armature current drops to rated current at stud 2,

Back emf, $E_{b2} = V_t - I_a \times$ (Total resistance at 2nd stud)

$= 240 - 39.67 \times 1.51 = 180.1$ V **Ans.**

When the handle is moved to stud 3, armature current shoots upto starting current but the counter emf remains same.

$\therefore$ total resistance at the 3rd stud

$$r_3 + r_4 + \ldots\ldots + r_n + r_a = \frac{V_t - E_{b2}}{I_{st}}$$

$$= \frac{240 - 180.1}{79.34} = 0.75\,\Omega$$

Resistance that must be cut out in the second step

= 1.51 – 0.75 = 0.76 Ω **Ans.**

(iv) At when armature current drops to rated current at stud 3.

Back emf. $E_{m3} = V_t - I_a\,(r_3 + r_4 + \_\_\_\_ + r_n \times r_a)$

$= 240 - 39.67 \times 0.75 = 210.2$ V

When the handle is moved to fourth stud

$$r_4 + r_5 + r_6 + \_\_\_\_ + r_n + r_a = \frac{V_t - E_{n3}}{I_{st}} = \frac{240 - 210.2}{73.34} = 0.38 \cong r_a$$

Therefore, when the handle is moved to fourth stud, the armature currentbeing very nearly equal 0.34 is sufficient to limit the starting current.

Total No. of steps = 4 **Ans.**

Since there are four studs, three resistance elements are required, the resistance of third element is = $0.75 - 0.34 = 0.41\,\Omega$

Thus, $r_1 = 1.17\,\Omega$, $r_2 = 0.76\,\Omega$, $r_3 = 0.41\,\Omega$ **Ans.**

Problem 4.47 : A 240 V d.c. shunt motor has an armature resistance of 0.5 Ω and a field resistance of 120 Ω. When driving a constant load torque at 800 r.p.m. the motor draws 22 A. What will be the new speed of the motor if an addition resistance 120 Ω is inserted in the field circuit.

Solution :

Given, $V_t = 240$ V, $r_a = 0.5\,\Omega$, $r_f = 120\,\Omega$, $I_L = 22$ A

Field current $I_{f1} = \dfrac{240}{120} = 2$ A

Armature current, $I_{a1} = (22 - 2) = 20$ A

Back emf in this case

$$E_{b1} = V_t - I_{a1} \times r_a$$

$$= 240 - 20 \times 0.5 = 240 - 10 = 230 \text{ V} \propto \phi_1 N_1 \qquad (1)$$

$$N_1 = 800 \text{ r.p.m.}$$

When resistance is added in the field circuit

$$I_{f2} = \frac{240}{120 + 120} = 1 \text{ A}$$

Armature current $I_{a2} = 22 - 1 = 21$ A

Back emf in this case $E_{n2} = V_t - I_{a2} \times r_a$

$$= 240 - 21 \times 0.5 = 229.5 \text{ v} \,\alpha\, \phi N_2 \qquad (2)$$

Now, $\dfrac{E_{n2}}{E_{n1}} = \dfrac{\phi_2 N_2}{\phi_1 N_1}$

Neglecting magnetic saturation $\phi \,\alpha\, I_f$

$$\frac{E_{n2}}{E_{n1}} = \frac{I_{f2}}{I_{f1}} \times \frac{N_2}{N_1}$$

$$\therefore \quad N_2 = \frac{E_{n2}}{E_{n1}} \times \frac{I_{f1}}{I_{f2}} \times N_1 = \frac{229.5}{230} \times \frac{2}{1} \times 800 = 1597 \text{ r.p.m.} \quad \textbf{Ans.}$$

Problem 4.48 : A 240 V d.c. shunt motor runs at 1000 r.p.m. at no load, while taking an armature current of 2 A. The armature resistance including brushes is 0.5 Ω. At rated load and rated voltage the armature current is 20 A. Find the speed regulation if

(a) 1 Ω placed in series with armature and 2 Ω in parallel with it,

(b) Only 2 Ω is inserted in series with the armature. Neglect rotational losses and armature reaction.

Solution :

Given, $V_t = 240$ V, $N_o = 1000$ r.p.m. $I_{ao} = 2$ A,

$r_a = 0.5\,\Omega$, $I_a = 20$ A,

At No load $E_{bo} = V_t - I_{ao} \times r_a = 240 - 2 \times 0.5 = 239$ V

Also $E_{bo} = k_a \phi \omega_o$

$$\therefore \quad k_a \phi = \frac{E_{bo}}{\omega_o} = \frac{239 \times 60}{2 \times \pi \times 1000} = 2.28$$

(a) Here $R_1 = 1\,\Omega$, $R_2 = 2\,\Omega$

$$\therefore \quad X = \frac{R_2}{R_1 + R_2} = \frac{2}{3} = 0.67$$

Applying the relation

$$\omega_m = \frac{1}{K_a \phi}\left[X V_t - I_a (X R_1 + r_a)\right]$$

$\therefore$ No load speed

$$\omega_{mo} = \frac{1}{2.28}[0.67 \times 240 - 2\,(0.67 \times 1 + 0.5)]$$

$$= \frac{1}{2.28}[160.8 - 2.34] = 69.5 \text{ rad / sec}$$

Full load speed

$$\omega_{fl} = \frac{1}{2.28}[0.67 \times 240 - 20 \times \ (0.67 \times 1 + 0.5))]$$

$$= \frac{1}{2.28}[160.8 - 20 \times 1.17] = 60.26 \text{ rad / s}$$

$$\% \text{ Speed regulation} = \left(\frac{69.5 - 60.26}{60.26}\right) \times 100\% = 15.33\% \quad \textbf{Ans.}$$

(b) Here, No-load speed

$$\omega_o = \frac{V_t - I_a\,(r_a + R))}{K_a \phi}$$

$$= \frac{240 - 2\,(2 + 0.5)}{2.28} = \frac{240 - 5}{2.28} = \frac{235}{2.28} = 103 \text{ rad / s}$$

Full load speed, $\omega_{fl} = \dfrac{240 - 20 \times 2.5}{2.28} = \dfrac{190}{2.28} = 83.33 \text{ rad / s}$

$$\% \text{ Speed regulation} = \left(\frac{103 - 83.33}{83.33}\right) \times 100\% = 23.60\% \quad \textbf{. Ans.}$$

Problem 4.49 : A 250 V d.c. shunt motor with an armature resistance of 0.2 Ω is running at 1000 r.p.m. and takes an armature current of 50 A. If the field flux is suddenly reduced by 20%, obtain

(a) The maximum value of current at this instant and corresponding torque, and

(b) Ultimate speed and armature current after the transients are over. Assume negligible armature inductance and constant load torque.

Solution :

Given, $V_t = 250$ V, $r_a = 0.2\,\Omega$

$N_1 = 1000$ r.p.m., $I_a = 50$ A

(a) Back emf at 1000 r.p.m., $E_{b1} = 250 - 50 \times 0.2 = 225$ V

When the field is reduced by 20%, the speed remains at 1000 r.p.m. due to inertia, the back emf is reduced as

$$E_{b1} = 0.8 \times 225 = 180 \text{ V}$$

$$\text{Maximum value of armature current} = \frac{250 - 180}{0.2} = 350 \text{ A}$$

In terms of initial current.

Maximum value of armature current= $\frac{350}{50} \times$ (initial value of current) = 7 × (initial value of current)

As we know that $T \propto \phi I \Rightarrow \frac{T_i}{T_N} = \frac{\phi_i\, I_i}{\phi_N\, T_N}$

$$T_N = \frac{\phi_N}{\phi_i} \times \frac{I_N}{I_i} \times T_i = \frac{80}{100} \times 7 \times T_i$$

$$= 0.8 \times 7 \times T_i = 5.6 \times \text{initial torque}$$

New value of torque = 5.6 × initial value of torque

Thus, with a sudden reducation of flux by 20%, the armature current and torque shoots up to 7 and 5.6 times of their respective initial values. For a constant load torque the increased dectromagnetic torque accelerates the motor, till the armature current decreases to I_{a2} given by

$$N_2 = I_{a1} \times \frac{\phi_1}{\phi_2} = 50 \times \frac{\phi_1}{0.8\,\phi_1} = 62.5 \text{ A}$$

Ultimate current = 62.5 A **Ans.**

Now, Ultimate speed after the transients are over, is given by

$$\frac{E_{b2}}{E_{b1}} = \frac{V_t - I_{a2}\, r_a}{V_t - I_{a1}\, r_a} = \frac{N_1\, \phi_1}{N_2\, \phi_2}$$

$$\therefore \quad N_2 = \frac{\phi_1}{\phi_2} \times \left(\frac{V_t - I_{a1}\, r_a}{V_t - I_{a2}\, r_a} \right) \times N_1$$

$$= \frac{\phi_1}{0.8\,\phi_1} \times \frac{(240 - 50 \times 0.2)}{(240 - 62.5 \times 0.2)} \times 1000$$

$$= \frac{1}{0.8} \times \frac{230}{227.5} \times 1000 = 1264 \text{ r.p.m.} \quad \textbf{Ans.}$$

Problem 4.50 : A d.c. shunt motor takes 40 A on full load from 240 V mains. Its speed is to be raised by 20% by weakening of the field flux. If the torque at the increased speed is 20% more than at initial speed. Find the percentage change in field flux. The armature resistance is 0.5

Solution :

At Initial speed N_1, $\quad T_{e1} \propto \phi_1\, I_1 \qquad (1)$

At increased speed N_2, $\quad T_{e2} \propto \phi_2\, I_2 \qquad (2)$

From (1) and (2) $\quad \dfrac{T_{e1}}{T_{e2}} = \dfrac{\phi_1\, I_1}{\phi_2\, I_2}$ or $\dfrac{T_{e1}}{1.2\, T_{e1}} = \dfrac{40 \times \phi_1}{I_{a2} \times \phi_2}$

or $\quad I_{a2} = 1.2 \times 40 \times \dfrac{\phi_1}{\phi_2} = 48\dfrac{\phi_1}{\phi_2}$ putting $\dfrac{\phi_1}{\phi_2} = A$

$\therefore \quad I_{a2} = 48$ A

At Initial Speed N_1

Back emf $\quad E_{b1} = 240 - 40 \times 0.5 = 220 \text{ V} \propto \phi_1\, N_1 \qquad (3)$

At increased speed N_2, Back emf

$$E_{b2} = 240 - I_{a2} \times 0.5 = 240 - 48 \times 0.5\, A$$

$$= 240 - 24A \propto \phi_2\, N_2 \qquad (4)$$

Now from (3) and (4)

$$\frac{E_{b2}}{E_{b1}} = \frac{\phi_2\, N_2}{\phi_1\, N_1}$$

or

$$\frac{240 - 24\, A}{220} = \frac{\phi_2 \times 1.2\, N_1}{\phi_1 \times N_1}$$

$$\frac{240-24A}{220} = 1.2\frac{\phi_2}{\phi_1}$$

Since $\quad \frac{\phi_2}{\phi_1} = \frac{1}{A}, \quad \frac{240-24A}{220} = \frac{1.2}{A}$

$$(240 - 24A)A = 1.2 \times 220$$

$$A^2 - 10A + 11 = 0, \; A = \frac{10 \pm 7.48}{2}$$

Now Neglecting higher value, $\quad A = \frac{\phi_1}{\phi_2} = 1.26$

$\therefore \quad \phi_2 = \frac{\phi_1}{1.26} = 0.7937\,\phi_1$

Now % reducation in field flux

$$= \left(\frac{\phi_1 - \phi_2}{\phi_1}\right) \times 100\%$$

$$= \left(\frac{1-0.7937}{1}\right) \times 100\% = 20.63\% \quad \textbf{Ans.}$$

Problem 4.51 : A 6 pole, 240 V d.c. series motor has a flux per pile of 2 mWb/amp over the working range of the Magnetisation curve which is assumed to be linear. The load torque is proportional to the speed squared and its value is 26 Nm at 1000 r.p.m. There are 440 wave connected conductors and total resistance of motor is $1\,\Omega$. Determine the motor speed and current when this motor is connected to rated supply voltage.

Solution :

Given $\quad V_t = 240$ V, $\quad P = 6$

$Z = 440, \quad \phi = 2 \times 10^{-3} \times I_a$ Wb

$r_a = 1\,\Omega$

For armature current I_a, Back emf $\quad E_b = V_t - I_a r_a = 240 - I_a \times 1$

$$E_b = 240 - I_a \qquad (1)$$

Now, $\quad E_b = \frac{\phi n z p}{60A}$ or $240 - I_a = \frac{2 \times 10^{-3} I_a \times n \times 440 \times 6}{60 \times 2}$

or $$240 - I_a = 0.044\ I_a N \quad \text{or} \quad I_a\ (1 + 0.044n) = 240$$

$$\therefore \qquad I_a = \frac{240}{1 + 0.044N} \tag{2}$$

Now, Electromagnetic torgue

$$T_e = \frac{E_b\ I_a}{\omega m}$$

$$= \frac{I_a}{2\pi N} \times \frac{2 \times 10^{-3} \times N \times 440 \times 6 \times I_a}{60 \times 2} = \frac{0.022\ I_a}{3.14}$$

$$T_e = 0.007\ I_a^{\ 2} \tag{3}$$

It is given that $T_i \propto N^2$, $T_i = kN^2$

or $$20 = k \times (1000)^2$$

$$\therefore \qquad k = 2.6 \times 10^{-5}\ \text{Nm/r.p.m.}$$

Now, Under steady state condition $T_e = T_L$

or $$0.007\ I_a^2 = 2.6 \times 10^{-5} \times N^2$$

or $$0.007 \times \left(\frac{240}{1 + 0.044N}\right)^2 = 2.6 \times 10^{-5} \times N^2$$

or $$\left(\frac{240}{1 + 0.044N}\right)^2 = \frac{2.6 \times 10^{-5}}{0.007} \times N^2 = 0.004 \times N^2$$

or $$\frac{240}{1 + 0.044N} = 0.063 \times N \quad \text{or} \quad 240 = 0.063\ N\ (1 + 0.044\ N)$$

or $$240 = 0.063N + 0.003N^2 \quad \text{or} \quad N^2 + 21N - 80000 = 0$$

This yields $$N = \frac{-21 \pm \sqrt{(21)^2 + 4 \times 80000}}{2} = \frac{-21 \pm 566}{2}$$

Neglecting negative value

$$N = \frac{545}{2} = 272.5\ \text{r.p.m.}$$

speed = 272.5 r.p.m. **Ans.**

Putting $N = 272.5$ in (2)

$$I_a = \frac{240}{1 + .004 \times 272.5} = 18.48\ \text{A} \quad \textbf{Ans.}$$

Problem 4.52 : A 240 V d.c series develops its rated output at 1000 r.p.m. while taking 25 A. Armature and series field resistances are 0.3 Ω and 0.2 respectively. Neglecting saturation, determine the resistance that must be added to obtain rated torque
(a) at starting, and *(b)* at 500 r.p.m.

Solution :

Give, $V_t = 240$ V, $I_{a1} = 25$ A $r_a = 0.3\Omega$,
$r_s = 0.2\Omega$, $N_1 = 1000$ r.p.m

In dc motors $T_e = k_a \phi I_a$

As magnetic saturation is neglected, $\phi \propto I_a$

$$\therefore \quad T_e = k_a I_a^2$$

(a) At Stating $E_b = 0$

$$\therefore \quad V_t = I_{a1}(r_a + r_s + r_{ext})$$

$$\text{or} \quad r_a + r_s + r_{ext} = \frac{V_t}{I_{a1}} = \frac{240}{25} = 9.6$$

$$\therefore \quad r_{ext} = 9.6 - 0.5 = 9.1\,\Omega \quad \textbf{Ans.}$$

(b) For developing rated torque at 500 rpm, current remains same at 500 r.p.m. $r_a = 0.3\Omega$, $\therefore I_{a2} = I_{a1} = 25$ A

For saturated magnetic circuit, flux is constant $\therefore T_e \propto I_a$

Under steady state $T_e = T_L$, $I_a \propto N^2$

$$\text{Hence,} \quad \therefore \frac{I_{a1}}{I_{a2}} = \frac{N_1^2}{N_2^2} \text{ or } \therefore I_{a2} = \frac{N_2^2}{N_1^2} \times I_{a1} = \left(\frac{N_2}{N_1}\right)^2 \times I_{a1}$$

$$= \left(\frac{1000}{500}\right)^2 \times I_{a1} = 4 \times 20 = 80 \text{ A.}$$

Back emf at 1000 rpm, $E_{b2} = V_{t2} - I_{a2} \times r$ (1)

For dc machines $E_b \propto \phi N$

For saturated magnetic circuit, flux remains constant $E_b \propto N$

$$\therefore \quad \frac{E_{b2}}{E_{b1}} = \frac{N_2}{N_1}$$

$$\therefore \quad E_{b2} = \frac{1000}{500} \times 230 = 460 \text{ V}$$

Putting this in (1) $V_{t2} = 460 + 80 \times 0.5 = 500$ V

$$V_{t2} = 500 \text{ V}, I_{a2} = 80 \text{ A} \quad \textbf{Ans.}$$

(b) When magnetic circuit is not saturated $T_b \propto \phi I_a$

Under steady state $T_L = T_e$

As we know that $E_b \propto \phi N$

$$\frac{E_{a2}}{E_{a1}} = \frac{\phi_2 N_2}{\phi_1 N_1}$$

or Since $I_f \alpha \phi$ $\quad \frac{E_{a2}}{E_{a1}} = \frac{N_2}{N_1}$

$$\therefore \quad E_{a2} = \frac{500}{100} \times E_{a1}, \quad E_{a2} = \frac{1}{2} E_{a1}$$

Now, $\quad E_{a1} = 240 - 25 \times 0.5 = 227.5$

$$\therefore \quad E_{a2} = \frac{1}{2} \times 227.5 = 113.75 \text{ V}$$

Also $\quad E_{a2} = V_t - I_{a2}(r_s + r_a + r_{ext})$

or $\quad 113.75 = 240 - 25(r_s + v_a + r_{ext})$

or, $\quad (r_s + r_a + r_{ext}) = 5.05$

$\therefore \quad r_{ext} = 5.05 - 0.5 = 4.55\,\Omega$ **Ans.**

Problem 4.53 : A d.c. series motor drives a fan at 500 rpm and takes 20 A when fed from rated voltage of 240 V. The motor resistance is $0.5\,\Omega$. The motor speed is to be raised to 1000 rpm by voltage control. Find the voltage and current in case magnetic circuit is unsaturated.

Solution :

Given $\quad V_t = 240$ V, $\quad N_1 = 500$ rpm, $\quad I_a = 20$ A

$r_a + r_s = 0.5\,\Omega = r, \quad N_2 = 1000$ rpm

At $\quad N_1 = 500$ rpm,

$$E_{b1} = 240 - 20 \times 0.5 = 230 \text{ V}$$

For fan $T_L \propto N^2, T_e \propto \phi\, I_a, \phi\, I \propto N^2$

or $$\frac{\phi_1\, I_{a1}}{\phi_2\, I_a} = \left(\frac{N_1}{N_2}\right)^2$$

For series motor

or $$\left(\frac{I_{a1}}{I_{a2}}\right)^2 = \left(\frac{N_1}{N_2}\right)^2 \text{ or } \frac{I_{a1}}{I_{a2}} = \frac{N_1}{N_2}$$

$\therefore$ $$I_{a2} = \frac{N_2}{N_1} \times I_{a1}, = \frac{1000}{500} \times 20 = 40 \text{ A} \quad \textbf{Ans.}$$

For dc motors, $E_b \,\alpha\, \phi\, N$

$\therefore$ $$\frac{E_{b2}}{E_{b1}} = \frac{\phi_2\, N_2}{\phi_1\, N_1} \text{ or } \frac{E_{b2}}{E_{b1}} = \frac{I_{a2}}{I_{a1}} \times \frac{N_2}{N_1}$$

$\therefore$ $$E_{b2} = \frac{40}{20} \times \frac{21000}{500} \times 230 = 920 \text{ V}$$

and also $$E_{b2} = V_t - I_{a2}\, r$$

$\therefore$ $$V_t = E_{b2} + I_a\, r = 920 + 40 \times 0.5 = 940 \text{ V} \quad \textbf{Ans.}$$

Problem 4.54 : A 5 kW, 240 V, 1500 r.p.m. separately excited d.c. motor is fed from 260 V ac supply through a single-phase full converter. At no-load and with zero firing angle delay, the motor draws 2 A and runs at 2000 rpm. The armature circuit rersistance is 0.5 Ω. The voltage drop in conducting thyristor is 2 V. For a firing angle delay of 30° and rated armature current of 25 A, Compute

(a) The motor torque, and

(b) Motor speed

Solution :

Given, $V_t = 240$ V, $N_1 = 1500$ rpm $I_{a1} = 2$ A,

$N_2 = 2000$ rpm $r_a = 0.5\ \Omega$, $\alpha = 30°$,

$I_{rated} = 25$ A, $V_T = 2$ V

For d.c. motor $V_t = E_b + I_a\, r_a + V_T$

or $$\frac{2\, V_m \cos\alpha}{\pi} = E_b + I_a\, r_a + V_T$$

For zero firing angle delay, $\alpha = 0$

or $$\frac{2V_m}{\pi} = E_b + I_a r_a + 2$$

or $$E_b = \frac{2V_m}{\pi} - I_a r_a - 2 = \frac{2\times\sqrt{2}\,V_{rms}}{\pi} - 2\times 0.5 - 2$$

$$= \frac{2\times\sqrt{2}\times 260}{\pi} - 1 - 2 = 231.2 \text{ V}$$

Since, $$E_b = k_a\,\phi\,\omega_{m2}$$

$\therefore$ $$k_a\,\phi = \frac{E_b}{\omega_{m2}} = \frac{231.2}{2\times\pi\times N_2} = \frac{231.2\times 60}{2\times\pi\times 2000}$$

$$= 1.104 \text{ V-sec/rad}$$

(a) For d.c. motor, Electromagnetic torque

$$T_e = k_a\,\phi\,I_a = 1.104\times 25 = 27.6 \text{ Nm} \quad \textbf{Ans.}$$

(b) To calculate motor speed, we have

$$\omega_m = \frac{\frac{2V_m}{\pi}\cos\alpha - V_T}{k_a\,\phi} - \frac{r_a}{(k_a\,\phi)^2}\,T_e$$

$$\omega_m = \frac{\frac{2\times\sqrt{2}\times 260}{\pi} - 2}{1.104} - \frac{0.5}{(1.104)^2}\times 27.6 = 119 \text{ rad/sec}$$

We know $$\omega_m = \frac{2\pi N}{60}$$

$\therefore$ $$N = \frac{60\times 199}{2\pi} = 1901 \text{ r.p.m.} \quad \textbf{Ans.}$$

Problem 4.55 : A d.c. series motor hoists a load at 5m/sec while taking 50 A at 500 V from dc mains. The total resistance of the motor is $1\,\Omega$. Calculate the resistance to be added in the motor circuit in order to slow the hoisting speed from 5 m/s to 4 m/s. Assume no. magnetic saturation

Solution :

Given, $V_t = 500$ V, $I_{a1} = 50$ A $v_1 = 5$ m/sec

$v_2 = 4$ m/s $r_a = 1\,\Omega$

At speed $v_1 = 5$ m/sec

Back emf $E_{b1} = 500 - 50 \times 1 = 450$ V

Let M be the load in kgs to be lifted up.
Under steady state,

Electromagnetic power = power required lifting up the load

or $E_{b1} \times I_{a1} = FV = Mg\,V$

or $E_{b1} \times I_{a1} = M \times 9.81 \times 5$ (1)

Again, Let R_{ex} be the resistance to be added in motor circuit. In this case $E_{b2} \times I_{a1} = M \times 9.81 \times 4$ (2)

From (1) and (2)
$$\frac{E_{b2} \times I_{a1}}{E_{b1} \times I_{a1}} = \frac{M \times 9.81 \times 4}{M \times 9.81 \times 5}$$

or
$$\frac{500 - 50\,(r_a + R_{ext})}{450} = \frac{4}{5}$$

or
$$500 - 50 - 50\,R_{ext} = \frac{4}{5} \times 450 = 360$$

or
$$R_{ext} = \frac{450 - 360}{50} = \frac{90}{50} = 1.8\,\Omega$$

Thus, $R_{ext} = 1.8\,\Omega$ **Ans.**

Problem 4.56 : A 6 pole d.c. motor runs at 500 rpm on full load and takes 16 A at 500 V. The armature is lap wound with 500 conductors and flux per pole is given by 16 $I^{0.5}$ mWb, where I is the motor current. If the supply voltage and torque both are halved, calculate the speed at which the motor will run. Neglect stray losses.

Solution :

Given, $P = 6$, $N_1 = 500$ rpm, $I_a = 16$ A,
$V_t = 500$ V, $z = 500$,
$\phi = 16 \times I^{0.5} \times 10^{-3}$ Wb $= 0.016 \times \sqrt{I}$ Wb

For $I_a = 16$ A

Flux per pole $\phi = 0.016 \times \sqrt{16} = 0.064$ Wb

Back emf in this case

$$E_{b1} = \frac{\phi\, n\, z\, p}{60 \times A} = \frac{0.064 \times 500 \times 500 \times 6}{60 \times 6} = 266.67\text{ V}$$

For d.c. motor $\quad V_t = E_b + I_a r_a = 266.67 + 16 r_a$

or $\quad 500 = 266.67 + 16 r_a$

$\therefore \quad r_a = 14.6\,\Omega$

Now, electromagnetic torque

$$T_{e1} = k_a \phi I_{a1} = k_a \times 0.016 \times \sqrt{I_{a1}} \times I_{a1} \qquad (1)$$

When torque is halfed $\quad \frac{T_{e1}}{2} = k_a \times 0.016 \times \sqrt{I_{a2}} \times I_{a2} \qquad (2)$

From (1) and (2) $\quad \frac{T_{e1}}{\frac{T_{e1}}{2}} = \left(\frac{I_{a1}}{I_{a2}}\right)^{\frac{1}{2}} \times \frac{I_{a1}}{I_{a2}}$

or $\quad 2 = \frac{4 \times 16}{(I_{a2})^{\frac{3}{2}}}$ or $(I_{a2})^{\frac{3}{2}} = 2 \times 16$

$\therefore \quad I_{a2} = (32)^{\frac{2}{3}} = 10.08$ A

$\therefore$ New armature current $\quad I_{a2} = 10.08$ A

When terminal voltage is halfed

$$E_{b2} = 250 - 10.08 \times 14.6 = 102.8 \text{ V}, N_2 = ?$$

$$E_{b1} = 266.67 \text{ V}, \; N_1 = 500 \text{ rpm}, \; \phi_1 = 0.064 \text{ Wb}$$

$$\phi_2 = 0.016\sqrt{10.08} = 0.051 \text{ Wb}$$

For dc motors, $E_b \propto \phi N$

$\therefore \quad \frac{E_{b2}}{E_{b1}} = \frac{\phi_2 N_2}{\phi_1 N_2}$ or $N_2 = \frac{E_{b2}}{E_{b1}} \times \frac{\phi_1}{\phi_2} \times N_1$

$$= \frac{102.8}{266.67} \times \frac{0.064}{0.051} \times 500 = 242 \text{ rpm}$$

Problem 4.57 : 300 V, 15 kW shunt motor has a maximum efficiency 90% and a speed of 800 rpm, when delivering 85% of its rated output. The resistance of its shunt field is 150 Ω. Determine the efficiency and speed when motor draws a current of 72 A from mains.

Solution :

Given, $\eta_m = 0.90$, $N_1 = 800$ rpm, $r_f = 150\,\Omega$,

$V_t = 300$ V, $I_{L2} = 72$ A

Rated output = 15 kW,

We know that, efficiency $= \dfrac{\text{output}}{\text{output + lossess}}$

$$\therefore \text{ Total losses } = \left(\frac{1}{\eta_m} - 1\right) \times \text{output} = \left(\frac{1}{0.9} - 1\right) \times 0.85 \times 15000$$

$$= 1416.67 \text{ W}$$

At maximum efficiency, constant loss = variable armature circuit losses

$$= \frac{1416.67}{2} = 708.34 \text{ W}$$

$\therefore$ Input to motor at maximum efficiency

$$= 0.85 \times 15000 + 1416.67 = 14166.7 \text{ W}$$

Line current $I_{L1} = \dfrac{14166.7}{300} = 47.22$ A

$\therefore$ Armature currrent $I_{a1} = (47.22 - 2)\text{ A} = 45.22$ A

But, armature copper loss = 708.34 W

or $(I_{a1})^2 \times r_a = 708.34$

$\therefore$ $r_a = 0.346\,\Omega$

When motor input current is 72 A

i.e. $I_{L2} = 72$ A since $I_f = \dfrac{300}{150} = 2$ A

Armature current $I_{a2} = 72 - 2 = 70$ A

Total Copper loss $= (I_{a2})^2 \times r_a = (70)^2 \times 0.346 = 1695.4$ W

Total power input $= 72 \times 300 = 21600$ W

Total losses = 1695.4 + 708.34 = 2403.74 Watts

Now Motor efficiency at 72 A is

$$= \left(1 - \frac{2403.74}{21600}\right) \times 100\% = 88.87\% \textbf{ Ans.}$$

We have, $E_{b1} = 300 - 45.22 \times .346 = 284.35 \text{ V} \propto \phi_1 N_1$

$$E_{b2} = 300 - 70 \times .346 = 275.78 \text{ V} \propto \phi_2 N_2$$

Since field current is constant, $\phi_1 = \phi_2 = \phi$

Now, $$\frac{E_{b2}}{E_{b1}} = \frac{\phi N_2}{\phi N_1} \quad \therefore \; N_2 = \frac{E_{b2}}{E_{b1}} \times N_1$$

$$= \frac{275.78}{284.35} \times 800 = 775.9 \text{ r.p.m.}$$

Thus, Speed = 775.9 rpm **Ans.**

> **Problem 4.58 :** A 250 V, 10 kW dc shunt motor draws a line current of 5 A while running at no-load speed of 1000 rpm dc supply. It has an armature resistance of 0.5 Ω and field resistance of 250 Ω. Estimate the efficiency of the motor it delivers rated load.

Solution :

Given $V_t = 250$ V, $V_a = 0.5$ Ω

$I_{Lo} = 5$ A, $N_o = 1000$ rpm

$r_f = 250\ \Omega$

Field current $$I_f = \frac{250}{250} = 1 \text{ A}$$

No-load armature current $I_{ao} = 5 - 1 = 4$ A

Constant no-load rotational loss $= V_t \times I_{Lo} - I_{ao}^2 r_a$

$$= 250 \times 5 - (4)^2 \times 0.5$$

$$= 1250 - 8 = 1242 \text{ W}$$

Let I_a be the armature current at load

Now, $$E_b = 250 - 0.5 \times I_a$$

Now, Electromagnetic power = Shaft power + rotational loss

or $$E_b \times I_a = 10000 + 1242$$

or $$(250 - 0.5\, I_a) \times I_a = 11242$$

or $$500\, I_a - I_a^2 = 22484$$

$$I_a^2 - 500\, I_a + 22484 = 0$$

This yields $$I_a = \frac{500 \pm \sqrt{(500)^2 - 4 \times 22484}}{2} = \frac{500 \pm 400}{2}$$

Neglecting higher value, $I_a = 50$ A

$$\text{Power input to the motor} = E_b I_a + \text{full-load } I_a^2 r_a \text{ loss + shunt field loss}$$
$$= 11242 + (50)^2 \times 0.5 + 250 \times 1$$
$$= 11242 + 1250 + 250 = 11242 + 1500$$
$$= 12742 \text{ W}$$

$$\text{Efficiency at rated load} = \left(\frac{10000}{12742}\right) \times 100\% = 78.48\% \quad \textbf{Ans.}$$

Problem 4.59 : A 420 V dc shunt motor has no-load current zero at speed of 1500 rpm. It is running at 1000 rpm at full load torque, reduced armature voltage and full field. If load torque is reduced to 50% of rated value with armature voltage and field voltage held constant at previous values, the speed increases to 1150 rpm. Find armature voltage drop at full load. Neglect armature reaction.

Solution :

Given, $V_t = 420$ V, $N_1 = 1000$ rpm,
$N_2 = 1150$ rpm

Let Ia_1 be the armature current at full load. Back emf at full load torque

$$E_{b1} = 420 - I_{a1} r_a \alpha N_1 \phi_1 \quad (1)$$

Motor torque $$T_{e1} = k_a \phi_1 I_{a1} \quad (2)$$

Let I_{a2} be the armature current when full load torque is reduced to half

$$\therefore \quad \frac{T_{e1}}{2} = k_a \phi_2 I_{a2} \quad (3)$$

As the field constant is held constant, $\phi_1 = \phi_1 = \phi$

From (2) and (3)

$$\frac{T_{e1}}{T_{e1}/2} = \frac{k_a \phi_1 I_{a1}}{k_a \phi_2 I_{a2}} \quad \text{or} \quad \frac{I_{a1}}{I_{a2}} = 2$$

$$\therefore \quad I_{a2} = \frac{I_{a1}}{2} \quad (4)$$

Now, Back emf at half of the rated torque

$$E_{b2} = \left(420 - \frac{I_{a1}}{2} \times r_a\right) \propto \phi_2 N_2 \quad (5)$$

From (1) and (5) $$\frac{E_{b2}}{E_{b1}} = \frac{\phi_2 N_2}{\phi_1 N_1}$$

or $$\frac{420 - \frac{I_{a1} r_a}{2}}{420 - I_{a1} r_a} = \frac{1150}{1000}$$

or $$420 - \frac{I_{a1} r_a}{2} = 1.15(420 - I_{a1} r_a)$$

or $$420 - 0.5 I_{a1} r_a = 1.15 \times 420 - 1.15 I_a r_a$$

or $$(1.15 - 0.5) I_{a1} r_a = 420(1.15 - 1)$$

$\therefore$ $$I_{a1} r_a = \frac{420}{0.65} \times .15 = 96.92 \text{ V}$$

Thus, Armature drop at full load = 96.92 V **Ans.**

> **Problem 4.60 :** A 420 V dc shunt motor takes 6 A at no-load. Its armature resistance (including brushes) is 1 Ω and shunt field resistance is 210 ohm. Estimate the kW output and efficiency when motor takes 52 A on full load. Also calculate the percentage change in speed from No-load to full load.

Solution :

Given, $V_t = 420$ V, $r_f = 210$ Ω, $r_a = 1$ Ω,

$I_{Lo} = 6$ A, $I_{L1} = 52$ A

Field current $= \frac{420}{210} = 2$ A, $I_{ao} = 6 - 2 = 4$ A

At No-load Back emf developed,

$$E_{b0} = V_t - I_{a0} \times r_a \propto \phi_0 N_0$$

$$E_{b0} = 420 - (6-2) \times 1 \propto \phi_0 N_0 \quad (1)$$

$$E_{b0} = 416 \propto \phi_0 N_0 \quad \textbf{Ans.}$$

Constant no-load rotational loss $= V_t \times I_{a0} - I_{a0}^2 r_a$

$$= 420 \times 6 - (4)^2 \times 1$$

$$= 1680 - 16 = 1664 \text{ W}$$

Constant shunt field loss $= I_a^2 r_a = (50)^2 \times 1 = 2500$ W

Full load armature circuit loss $= V_t \times I_f = 420 \times 2 = 840$ W

$\therefore$ Total losses at full load = 2500 + 840 + 1664 = 5004 W

Power input at full load $= 52 \times 420 = 21840$ W

Motor efficiency $= \left(1 - \dfrac{5004}{21840}\right) \times 100\% = 77.09\%$ **Ans.**

Back emf at full load

$$E_{b1} = 420 - 50 \times 1 \alpha \phi_1 N_1 = 370 \text{ V} \propto \phi_1 N_1 \quad (2)$$

From (1) and (2)

$$\frac{\phi_1 N_1}{\phi_0 N_0} = \frac{370}{416}$$

Since field current is constant, $\phi_1 = \phi_2 = \phi$,

$$\frac{N_1}{N_0} = 0.889$$

$$\therefore \quad N_1 = 0.889 \ N_o$$

Now, % change in speed

$$= \left(\frac{N_0 - N_1}{N_0}\right) \times 100\% = \left(\frac{N_0 - 0.889 N_0}{N_0}\right) \times 100\%$$

$$= 0.111 \times 100\% = 11.1\% \text{ **Ans.**}$$

Problem 4. 61 : A 420 V, 20 kW dc shunt motor takes 3 A when running light. For an armature resistance of 0.5 Ω, field resistance is 420 Ω and brush drop of 2 V. Find the full load efficiency.

Solution :

Given, $V_t = 420$ V, $I_{Lo} = 3$ A, $r_f = 420\ \Omega$,

$r_a = 0.5\ \Omega$, $V_b = 2$ V

Now, Field current $I_f = \dfrac{420}{420} = 1$ A

No-load current $I_{a0} = 3 - 1 = 2$ A

Constant no-load rotational loss

$$W_0 = V_t I_{a0} - I_{a0}^2 r_a - V_b I_{a0}$$
$$= 420 \times 2 - 4 \times 0.5 - 2 \times 2$$
$$= 840 - 2 - 4 = 834 \text{ W}$$

∴ Total constant loss

$$W_c = 834 + V_t I_f = 834 + 420 \times 1 = 1254 \text{ W}$$

Let I_a be the full load armature current

$$\text{Power input} = 20000 + I_a^2 r_a + V_b I_a + W_c$$
$$= 20000 + I_a^2 \times 0.5 + 2I_a + 1254$$

or $$V_t \times I_L = 21254 + 0.5 I_a^2 + 2I_a$$

or $$420 \times (I_a + 1.0) = 21254 + 0.5 I_a^2 + 2I_a$$

or $$420 \times I_a + 420 = 21254 + 0.5 I_a^2 + 2I_a$$

or $$840 I_a + 840 = 42508 + I_a^2 + 4I_a$$

or $$I_a^2 - 836 I_a + 41668 = 0$$

This yields $$I_a = \frac{836 \pm \sqrt{(836)^2 - 4 \times 41668}}{2} = \frac{836 \pm 729.5}{2}$$

Neglecting higher value $$I_a = \frac{106.5}{2} = 53.25 \text{ A}$$

∴ Power input $$= 420 \times (53.25 + 1.0) = 22785 \text{ W}$$

Full load efficiency of the motor

$$\left(\frac{P_{out}}{P_{in}}\right) \times 100\% = \left(\frac{20000}{22785}\right) \times 100\% = 87.78\% \quad \textbf{Ans.}$$

Problem 4.62 : Find an expression for the maximum steady state power output of a d.c. shunt motor with an armature resistance r_a, if the applied voltage V_t and the shunt field current. If are kept constant. Neglect rotational loss and assume that the theoretical maximum power output will not overload the motor.

Calculate the maximum power output of a dc shunt of armature resistance of 0.5 Ω and applied votage is 400 V. Neglect brush drop voltage.

Solution :

For dc shunt motor $I_L = I_a + I_f$

$\therefore$ $I_a = I_L - I_f$

Now, Power output (P) = Power input – armature copper loss – Shunt field loss – rotational loss – brush drop loss.

Let I_a be the armature currrent,

$$p = V_t I_L - I_a^2 r_a - V_t I_f - V_b I_a - \text{rotational losses}$$

It is given that rotational losses are neglected,

$$p = V_t I_L - I_a^2 r_a - V_t I_f - V_b I_a$$

or $$p = V_t (I_L - I_f) - I_a^2 r_a - V_b I_a$$

or $$p = V_t I_a - I_a^2 r_a - V_b I_a \qquad (1)$$

For maximum power output

$$\frac{dp}{dI_a} = 0$$

or $$V_t - 2I_a r_a - V_b = 0$$

or $$I_a = \frac{V_t - V_b}{2r_a}$$

Substituting $I_a = \dfrac{V_t - V_b}{2r_a}$ in (1)

$$P_{\max} = V_t \times \frac{(V_t - V_b)}{2r_a} - \frac{(V_t - V_b)^2}{4r_a^2} \times r_a - V_b \times \frac{(V_t - V_b)}{2r_a}$$

$$= \frac{V_t^2 + V_b^2 - 2V_t V_b}{4r_a} = \frac{(V_t - V_b)^2}{4r_a}$$

$$P_{\max} = \frac{(V_t - V_b)^2}{4r_a}$$

If the brush drop is neglected

$$P_{\max} = \frac{V_t^2}{4r_a}$$

Also, given $V_t = 400$ V, $r_a = 0.5\,\Omega$

$$\therefore \qquad P_{max} = \frac{400 \times 400}{4 \times 0.5} = \frac{400 \times 400}{2} = 80 \text{ kW} \quad \textbf{Ans.}$$

Problem 4.63 : Two identical dc shunt machines when tested by Hopkinson's method gave the following data:

Line voltage – 220 V, Line current excluding both the field currents – 30 A, motor armature current – 220 A, field currents 4 A and 3 A.

If the armature resistance of each machine (including brustes) is 0.02 Ω. Calculate efficiency of both machines.

Solution :

The diagram of Hopkinson test can be drawn as below.

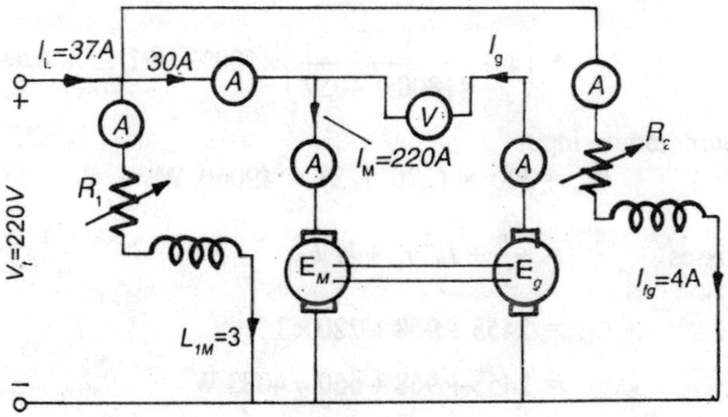

Fig. P. 4.63

Given, $\quad I_{fm} = 3$ A, $\quad I_{fg} = 4$ A,

$I'_L = 30$ A, $\quad I_L = 30 + 4 + 3 = 37$ A,

$V_t = 220$ V, $\quad I_m = 220$ A

Generator armature current

$$I_G = \text{motor armature current} - \text{line current}$$
$$= 220 - 30 = 190 \text{ A}$$

Armature circuit loss in generator $= (I_G)^2 \times r_a$

$$= (190)^2 \times 0.02 = 722 \text{ W}$$

Armature circuit loss in motor $= (I_m)^2 \times r_a = (220)^2 \times 0.02 = 968$ W

Power drawn from the supply excluding field losses of both machines

$$= 220 \times 30 = 6600 \text{ W}$$

No-load rotational loss in both machines

$$W_0 = V_t I'_L - I_a^2 r_a + I_m^2 r_a)$$

$$= 6600 - (722 + 968) = 6600 - 1690 = 4910 \text{ W}$$

No-load rotational loss for each machine

$$\frac{W_0}{2} = \frac{4910}{2} = 2455 \text{ W}$$

For generator, Generator output $= V_t \times I_g = 220 \times 190 = 41800$ W

Total losses $\quad \frac{W_0}{2} = V_t I_{f_g} + I_a^2 r_a = 2455 + 220 \times 4 + 722$

$$= 2455 + 880 + 722 = 4057 \text{ W}$$

Efficiency of generator

$$= \left(1 - \frac{4057}{41800 + 4057}\right) \times 100\% = 91.15\% \textbf{ Ans.}$$

For motor, motor input

$$P_{im} = 200 \times (220 + 3) = 49060 \text{ W}$$

Total losses $\quad = \frac{W_0}{2} + I_a{}^2 r_a + V_t I_{fm}$

$$= 2455 + 968 + 220 \times 3$$

$$= 2455 + 968 + 660 = 4083 \text{ W}$$

$\therefore$ Efficiency of motor $= \left(1 - \frac{4083}{49060}\right) \times 100\% = 91.68\%$ **Ans.**

Problem 4.64 : A Field's test on two similar series machines gave the following data:

Motor: armature current	=	50 A
Voltage across armature	=	400 V
Voltage across field	=	25 V
Generator: terminal voltage	=	350 V
Output current	=	40 A
Voltage across field	=	25 V

Armature resistance (including brushes) of each machine is 0.5 Ω. Calculate efficiency of both machines.

Solution :

Given, $V_G = 350$ V, $I_G = 40$ A,

$I_m = 50$ A $r_a = 40\ \Omega$

Power input to the whole set $= (400 + 25 + 24) \times 50$

$= 450 \times 50 = 22500$ W

Generator output $= V_G\ I_G = 350 \times 40 = 14000$ W

Total losses in whole set $= 22500 - 14000 = 8500$ W

Total ohmic losses $= (50)^2 \times 0.5 + 50\,(25 + 25) + (40)^2 \times 0.5$

$= 4550$ W

No-load rotational loss of each machine $= \dfrac{8500 - 4550}{2} = 1975$ W

Motor power input $= (400 + 25) \times 50 = 425 \times 50 = 21250$ W

Total Motor losses = Armature current loss + field circuit loss + No-load rotational loss

$= (50)^2 \times 0.25 + 50 \times 25 + 1975 = 3850$ W

Motor efficiency $= \left(1 - \dfrac{3850}{21250}\right) \times 100\% = 81.88\%$ **Ans.**

Total generator circuit loss = Armature circuit loss + field loss + No-load rotational loss

$= (40)^2 \times 0.5 + 25 \times 40 + 1975 = 3775$ W

Generator input = P_G out + Total loss = 14000 + 3775 = 17775 W

Generator efficiency $= \left(1 - \dfrac{3775}{17775}\right) \times 100\% = 78.76\%$ **Ans.**

Problem 4.65 : A 4kW, 400 V, 300 rad/sec amplidyne has the following constants,

$r_f = 60\ \Omega$, $r_d = 6\ \Omega$, $r_c = 1\ \Omega$

Voltage constants $k_{qf} = 300$ V/field amp

$k_{dq} = 150$ V/amp

$k_{qd} = 100$ V/amp

Calculate field current and power gain at rated output. Also obtain these value when compensation is zero.

Solution :

For an amplidyne, $V_d = AV_f - I_d\,(r_d + r_c)$ (1)

At rated voltage of 400 V, $I_d = \dfrac{4000}{400} = 10\text{ A}$

Voltage amplification $A = \dfrac{k_{dq} \times k_{qf}}{r_a\, r_f} = \dfrac{150 \times 300}{60 \times 62} = 125$

Substituting the values of V and I_d in eq (1)

$$400 = 125\,V_f - 10\,(6+1)$$

or

$$400 = 125\,V_f - 70$$

$$V_f = \frac{400+70}{125} = \frac{470}{125} = 3.8\text{ V}$$

$\therefore$ Field current $(I_f) = \dfrac{V_f}{r_f} = \dfrac{3.8}{60} = 63\text{ mA}$

$\therefore$ Power gain $= \dfrac{V_d\, I_d}{V_f\, I_f} = \dfrac{400 \times 10}{3.8 \times 63 \times 10^{-3}} = 16708$ **Ans.**

For an amplidyne with zero compensation

$$k = \frac{k_{da} \times k_{qd}}{r_d} = \frac{150 \times 100}{6} = 2500$$

Voltage equation is $V_d = AV_f - I_d\,(k + r_d)$

or

$$400 = 125\,V_f - 10\,(2500 + 6)$$

or

$$V_f = 203.7\text{ V}$$

$\therefore$ Field current $I_f = \dfrac{203.7}{60} = 3.4\text{ A}$ **Ans.**

Now, Power gain $= \dfrac{V_d\, I_d}{V_f\, I_f} = \dfrac{400 \times 10}{203.7 \times 3.4} = 5.78$ **Ans.**

Problem 4.66 : A PMDC motor has armature resistance of $1\,\Omega$. When fed from 40 V dc source it runs at a speed of 2000 rpm. While taking 0.5 A. Determine *(a)* no-load rotational losses of the motor *(b)* the motor output when running at a speed of 1500 rpm and with voltage source of 32 V dc and *(c)* its stall torque when the source voltage is 20 v d.c.

Solution :

At no-load $V_t = 40$ V, $I_{a0} = 0.5$ A $r_a = 1\Omega$

$\therefore$ $E_b = 40 - 0.5 \times 1 = 39.5$ V

(a) As we know that at no-load all the electromagnetic power developed is used to supply rotational losses.

$\therefore$ No-load rotational losses $= E_b \times I_a = 39.5 \times .5 = 19.75$ W **Ans.**

(b) At speed 1500 rpm $E_b = k_m \omega_{m0}$

$$\therefore \text{ Speed-voltage costant } = \frac{39.5 \times 60}{2 \times \pi \times 2000} = 0.189 \text{ V-sec/rad}$$

Generated emf at speed 1500 rpm

$$E_{b1} = k_m \omega_{m1} = 0.189 \times 2\pi \times \frac{1500}{60}$$

$$= 29.67 \text{ V}$$

New armature current

$$I_{a1} = \frac{V_t - E_{b1}}{r_a} = \frac{32 - 29.67}{1} = 2.33 \text{ A}$$

Electromagnetic power developed

$$= E_{b1} \times I_{a1} = 29.67 \times 2.33 = 69.13 \text{ W}$$

Motor output $= E_{b1} I_{a1}$ – rotational losses
$= 69.13 - 19.75 = 49.38$ W **Ans.**

(c) When motor stalls, $E_b = 0$ and $V_t = I_a r_a$

$$\therefore \quad \text{Stall current } I_a = \frac{20}{1} = 20 \text{ A}$$

Stall torque $= K_m I_a = 0.189 \times 20 = 3.78$ Nm **Ans.**

Problem 4.67 : A 20 kW, 250 V, 2000 rpm d.c. motor has the following constants:

$r_a = 1.00\,\Omega$, $L_a = 0.20$ H,

$k_m = M_d I_f = 5.00$ Nm/armature amp

The load coupled with the motor has its inertia equal to 2.00 kg-m^2. If the load torque varies linearly with speed, then calculate ω_n and investigate its dynamic behaviour. Neglect rotational losses.

Solution :

Given, $P_{rated} = 20$ kW, $V_t = 250$ V,
$N_r = 2000$ rpm, $J = 2.00$ kg-m²

Here $$\omega_r = \frac{2 \times \pi \times 2000}{60} = \text{rad/sec} = 209.3 \text{ rad/sec}$$

Rated load torque,

$$T_L = \frac{p}{\omega_r} = \frac{20 \times 1000}{209.3} \text{ Nm} = 95.57 \text{ Nm}$$

It is given that torque-speed characteristic is a straigth line through origin, the slop of the torque speed curve of the load is

$$B = \frac{T_L}{\omega_r} = \frac{95.57}{209.3} = 0.457$$

Now $$\frac{B}{J} = \frac{0.457}{2} = 0.229, \; \tau_a = \frac{L_a}{r_a} = \frac{0.20}{1.0} = 0.2$$

$$T_m = \frac{Tr_a}{k_m^2} = \frac{2 \times 1.0}{(5)^2} = 0.08$$

We know that $$\omega_n = \sqrt{\frac{1}{T_a}\left(\frac{B}{J} + \frac{1}{T_m}\right)} = \sqrt{\frac{1}{0.2}\left(0.229 + \frac{1}{0.08}\right)}$$

$$= \sqrt{63.65} = 7.98 \text{ rad/sec } \textbf{Ans.}$$

Also, we have $$\alpha = \frac{1}{2}\left(\frac{B}{J} + \frac{1}{\tau_a}\right) = \frac{1}{2}(0.229 + 5)$$

$$= \frac{5.229}{2} = 2.615 \textbf{ Ans.}$$

and $$E_b = \frac{\alpha}{\omega_n} = \frac{2.615}{7.98} = 0.33 \textbf{ Ans.}$$

Problem 4.68 : An ideal d.c. motor with negligible internal power loss operates at no load with fixed field excitation supplied from a separate source. Write down an expression for the current drawn when the armature is subjected to a varying voltage V_t.

Determine the constant in this expression if the moment of inertia of the armature is 0.6 kgm² and the armature develops 110 V at a speed of 1100 rpm.

Solution :

The voltage equation for the armature circuit is

$$V_t = (r_a + L_a P)\, i_a + E_a \tag{1}$$

When armature resistance is zero and $E_a = k_m\, \omega_r$

Then, the equation becomes

$$V_t = L_a P i_a + k_m\, \omega_r$$

The torque balance equation for the dc motor is

$$T_e = k_m\, i_a = jp\omega_r$$

$$\therefore \quad i_a = \frac{jp\omega_r}{k_m} = \frac{J}{k_m} pd\omega_r = \frac{J}{k_m} p\left(\frac{V_t - L_a\, Pi_a}{k_m}\right)$$

$$\therefore \quad i_a = \frac{J}{(k_m)^2} p\,(V_t - L_a p i_a)$$

or

$$i_a = \frac{\left(\dfrac{J}{k_m^2}\right) p\, V_a}{1 + \left(\dfrac{J}{k_m^2}\right) p^2\, L_a}$$

The constant in above equation is $= \dfrac{J}{k_m}$

$$V_t = 110 \text{ volts at no-load} = k_m\, \omega_r$$

or

$$k_m = \frac{110 \times 60}{2\,\pi \times 1100} = \frac{3}{\pi}$$

$$\therefore \quad \text{Constant} = \frac{J}{k_m^2} = 0.6 \times \left(\frac{\pi}{3}\right)^2 = 0.657 \quad \textbf{Ans.}$$

Problem 4.69 : A cumulatively compounded generator, with its shunt field separately excited, has the following parameters.

$r_f = 100\,\Omega$, $\quad L_f = 50$ H

$r_a = 6\,\Omega$, $\quad L_a = 0.03$

$r_s = 0.4\,\Omega$, $\quad L_s = 0.02$

M_d = mutual inductance between series and shunt field = 1 H

Generated emf / shunt field ampere = k_g =100 at 100 rad/sec.

Generated emf / series field ampere = k_s = 0.5 at 100 rad/sec.

(a) The generator is running at constant speed of radians/second and is open circuited. Shunt field is suddenly switched across 200 volts. Find the rise of the armature voltage.
(b) After steady state has reached in part (a) the amature is suddenly short-circuited. Find an expression for the armature short circuit current.

Solution :

We have $$V_f = (r_f + L_f\, p)\, i_f + M_d\, p i_a$$

(a) When voltage to the shunt field is suddenly applied

$$V_t = V_g.1 \text{ and } i_a = 0$$

$$V_f.1 = (r_f + L_f p)\, i_f$$

and $$V_t = k_g i_f - M_d p_{if}$$

or $$V_t = \frac{k_g \times V_f\,.L}{r_f\,(1+\tau_f p)} - \frac{M_d\, V_f\,.L}{r_f\,(1+\tau_f p)}$$

$$= \frac{k_g i_f\,.1}{r_f\,(1+\tau_f p)} - \frac{M_d V_f}{r_f \tau_f} \cdot \frac{T_f p \cdot 1}{1+\tau_f p}$$

or $$V_t\,(t) = \frac{k_g\, V_f}{r_f}\left(1 - e^{-\frac{t}{\tau_f}}\right) - \frac{M_d\, V_f}{I_f}\, e^{-\frac{t}{\tau_f}}$$

Substituting the given values

$$V_t\,(t) = 200\,(1 - e^{-2t}) - 4e^{-2t} = 200 - 204\, e^{-2t}$$

(b) $t = \infty$, $V_f = 200$ V

Applying the relation

$$i_{at} = \frac{(r_f + L_f p)\, V_t\,.1}{(k_g - M_d p)\, M_d p + (R + L_P - k_s)\,(r_f + L_{fP})}$$

Substituting given values

$$i_{at} = \frac{200\,(100 + 50\,P)\,1}{(100 - P)\,P + [1 + 0.05P - 0.5)\,(100 + 50\,P)}$$

or $$i_{at} = \frac{10000\,(P+2)\,1}{1.5\,P^2 + 130\,P + 50}$$

This expression in s-domain is

$$I_{at}(s) = \frac{10000\,(s+2)}{s\,(1.5s^2+120\,s+50)} = \frac{10000\,(s+2)}{s\,(s+0.386)\,(s+86.28)}$$

Taking inverse Laplace transform,

$$i_{at}(t) = 600.5 - 486.8\,e^{-0.386\,t} - 113.72\,e^{-86.28\,t} \quad \textbf{Ans.}$$

> **Problem 4.70 :** The open circuit voltage of a metadyne generator running at 100 rad/sec is 200 V. When the field voltage is 5 V. Its other parameters are
>
> $$r_a = 1.00\,\Omega, \quad L_a = 0.03\,\text{H}, \quad r_c = 0.4\,\Omega$$
>
> Deduce the output voltage and output power versus load current characteristics of 100%, 50% and zero percent compensations. Armature leakage flux may be Neglected.

Solution :

Here, $E_d = 200$ V, $V_f = 5$V

The voltage amplification for a metadyne is given by

$$\frac{E_d}{V_f} = \frac{L_a\,M_d\,\omega_r^2}{r_a\,r_f} = \frac{200}{5} = 40 = \text{A (say)}$$

The load voltage is given by

$$V_L = AV_f - \left[\frac{L_a\omega_r^2\,(L_a - kM_d)}{r_a} + kr_c + r_a\right] I_L$$

For 100% compensation $K = 1$

$\therefore\ (L_a - M_d) \propto$ Armature leakage flux $(L_a - M_d) \propto 0$, i.e.

$$L_a = M_d$$

$$\therefore \quad V_L = AV_f - \left[\frac{L_a^2\,\omega_r^2\,(1-k)}{r_a} + kr_c + r_a\right] I_L$$

Substituting the given values and for 100% compensation, $k = 1$

$$V_L = AV_f - (r_c + r_a)\,I_L$$

$$V_L = 40\,V_f - (r_c + r_a)\,I_L$$

$$= 40 \times 5 - (1.00 + 0.4)\,I_L$$

$$V_L = 200 - 1.4\,I_L$$

Power $= V_L\,I_L = (200 - 1.4\,I_L)\,I_L$ **Ans.**

When the compensation is 50%, $K = 0.5$

$$V_L = 200 - \left[\frac{(0.03)^2 \times (100)^2 \times 0.5}{1} + 0.5 \times 0.4 + 1\right] I_L$$

$$= 200 - 5.7\, I_L$$

Power = $V_L I_L = (200 - 5.7\, I_L)\, I_L$ **Ans.**

For zero compensation $K = 0$

$$V_L = 200 - \left[\frac{(0.03)^2 \times (100)^2}{1} + 1\right] I_L$$

$$V_L = 200 - 10\, I_L$$

Power $V_L I_L = (200 - 10\, I_L)\, I_L$ **Ans.**

Problem 4.71 : Two d.c. shunt generators, operating in parallel, share a load current of 300 A. Each generator has armature resistance 0.2 Ω and field resistance of 150 Ω. Their no-load generated emf's are 275 and 270 volts. Calculate the bus bar voltage and the power output of each machine.

Solution :

Let the generators A and B supply load currents I_A and I_B respectively at a common bus voltage V_b.

$$I_{LA} + I_{LB} = 300 \quad (1)$$

Shunt field current for both generators $= \dfrac{V_b}{150}$

For generator A,

Armature current $I_{aA} = I_{LA} + \dfrac{V_b}{150}$

For generator B,

Armature current $I_{aB} = I_{LB} + \dfrac{V_b}{150}$

For generator A $E_{bA} = V_b + I_{aA} \times r_A$

or $275 = V_b + 0.2\, I_{aA}$

or $275 = V_b + 0.2\left(I_{LA} + \dfrac{V_b}{150}\right)$ (2)

For generator B $E_{bB} = V_b + I_{aB} \times r_B$

or $$270 = V_b + 0.2 \times \left(I_{LB} + \frac{V_b}{150} \right) \quad (3)$$

Adding (2) and (3) $$545 = 2\,V_b + 0.2 \times \left(I_{LA} + I_{LB} + \frac{2\,V_b}{150} \right)$$

or $$545 = 2\,V_b + 0.2 \times \left(300 + \frac{2\,V_b}{150} \right)$$

or $$545 = 2\,V_b + 60 + 0.4\,\frac{V_b}{150}$$

or $$2.003\,V_b = 545 - 60 = 485$$

$\therefore$ $$V_b = \frac{485}{2.003} = 242 \text{ V}$$

From equation (2) $$275 = 242 + 0.2 \left(I_{LA} + \frac{242}{150} \right)$$

or $$I_{LA} = \frac{33}{0.2} - \frac{242}{150} = 165 - 1.6 = 163.4 \text{ A}$$

From equation (2) $$270 = 242 + 0.2 \left(I_{LB} + \frac{242}{150} \right)$$

$\therefore$ $$I_{LB} = \frac{28}{0.2} - \frac{242}{150} = 140 - 1.6 = 138.4 \text{ A}$$

Power output of machine A

$= V_b\, I_{LA} = 242 \times 163.4 = 39542.8 \text{ W} = 39.543 \text{ kW}$ **Ans.**

Power output of machine B

$= V_b\, I_{LB} = 242 \times 138.4 = 33.5 \text{ kW}$ **Ans.**

Problem 4.72 : Two dc generators operating in parallel, supply a total output current of 100 A. The terminal voltage of one machine falls from 270 at no load to 250 V at an output current of 60 A. The terminal voltage of second machine is 280 V at no-load to 250 V at an output current of 80 A. Their external characteristics are linear.

(a) The current output of each machine

(b) Find the line voltage and total load in kW when one generator is floating.

Solution :

Given, $I_{LA} + I_{LB} = 100 \text{ A} \quad (1)$

$V_{AO} = 270 \text{ V}, \quad V_{AL} = 250 \text{ V}, \quad I_{AO} = 60 \text{ A}$

$V_{BO} = 280 \text{ V}, \quad V_{BL} = 250 \text{ V}, \quad I_{BO} = 80 \text{ A}$

(a) For an output current of 60 A

Voltage drop = 270 – 250 = 20 V

Now, Voltage drop per output ampere = $\frac{20}{60}$ V/A

Since their external characteristics is linear.

Let I_{LA} be the load shared by generator A.

$$\text{Voltage drop} = \frac{2}{6} \times I_{LA}$$

Back emf at load $V_{AO} = V_t + \frac{2}{6} I_{LA}$

or $$270 = V_t + \frac{2}{6} I_{LA} \quad (2)$$

For generator B, Voltage drop per output amp = $\frac{30}{80} = \frac{3}{8}$ V / amp

Back emf at load, $280 = V_t + \frac{3}{8} I_{LB}$ (3)

From (2) and (3) $280 = 270 - \frac{2}{6} I_{LA} + \frac{3}{8} I_{LB}$

or $$20 = \frac{-2}{6}(100 - I_{LB}) + \frac{3}{8} I_{LB}$$

or $$20 = \frac{-200}{6} + \frac{1}{3} I_{LB} + \frac{3}{8} I_{LB}$$

or $$20 + \frac{200}{6} = \frac{(8+9)\, I_{LB}}{24}$$

or $$\frac{320}{6} = \frac{17}{24} \times I_{LB}$$

$\therefore$ $$I_{LB} = \frac{320 \times 24}{6 \times 17} = 75.3 \text{ A} \quad \textbf{Ans.}$$

From (1) $I_{LA} = 100 - 75.3 = 24.7$ A **Ans.**

From equation (2) $V_t = 270 - \frac{2}{6} \times 24.7 = 232$ V

(b) When the generator A is floating, the line voltage is 270 V since No-load emf of generator B is 280 A, A line voltage of 270 V corresponds to voltage drop of 10 V in its armature. But the voltage drop of this generator is $\frac{3}{8}$ V/amp.

$\therefore$ Load carried by generator

$$B = \frac{8}{3} \times 10 = \frac{80}{3} = 26.67 \text{ A}$$

Now, Power delivered to the load

$$= 26.67 \times 270 \times \frac{1}{1000} \text{ kW} = 7.2 \text{ kW} \quad \textbf{Ans.}$$

Problem 4.73 : Two d.c. shunt generators, each with an armature resistance of 0.01 Ω, have no load emf's of 250 and 240 V. Field resistances of 60 Ω and 50 Ω respectively. They are in parallel and supply a total load of 50 kW. How do they share the total load.

Solution :

For generator A, $r_A = 0.10\ \Omega$, $E_{AO} = 250$ V, $r_{Af} = 60\ \Omega$

For generator B, $r_B = 0.01\ \Omega$, $E_{BO} = 240$ V, $r_{Bf} = 50\ \Omega$

Let V_t be the terminal voltage

$$\therefore \quad V_t\,(I_A + I_B) = 500 \times 1000 \tag{1}$$

For Generator A, $$250 = V_t + \left(I_A + \frac{V_t}{60}\right) \times 0.01 \tag{2}$$

For Generator B, $$240 = V_t + \left(I_B + \frac{V_t}{50}\right) \times 0.01 \tag{3}$$

Adding (2) and (3) $$490 = 2\,V_t + 0.01\,(I_A + I_B) + 0.01\left(\frac{V_t}{60} + \frac{V_t}{50}\right)$$

or $$490 = 2\,V_t + 0.01\,(I_A + I_B) + 0.0002\,V_t + 0.0002\,V_t$$

or $$490 = 2.0004\,V_t + 0.01\,(I_A + I_B)$$

or $$I_A + I_B = \frac{490}{0.01} - \frac{2.0004}{0.01} V_t$$

$$I_A + I_B = 49000 - 200.04\,V_t \tag{4}$$

Substituting this in (1)

or $$(49000 - 200.04\,V_t) \times V_t = 50000$$

or $$49000\,V_t - 200.04\,V_t^2 = 50000$$

or $$200.04\,V_t^2 - 49000\,V_t + 50000 = 0$$

or $$V_t^2 - 245\,V_t + 249.95 = 0$$

This yields, $V_t = \dfrac{245 \pm \sqrt{(245)^2 - 4 \times 249.95}}{2} = \dfrac{245 \pm 242.95}{2}$

Neglecting lower value, $V_t = 244$ V

Putting $V_t = 244$ V in (2)

$$250 = 244 + \left(I_A + \frac{244}{60} \right) \times 0.01$$

or $$\frac{250 - 244}{0.01} = I_A + \frac{244}{60}$$

This yealds $I_A = 596$ A **Ans.**

Since $E_{B0} < V$ the machine B is opteraing as motor

Again, from equation (3),

$$240 = 244 + \left(I_B - \frac{244}{50} \right) \times 0.01$$

This yealds $I_B = -\ 395.12$ A

This negative sign shows that Machine B is operating as motor.

Output of Machine $A = V_t \times I_A = 244 \times 596 = 145.42$ kW **Ans.**

Power input to Machine $B = V_t \times I_B = 395.12 \times 244$

$= 96.41$ kW **Ans.**

Problem 4.74 : A d.c. series motor has a magnetization curve given by a straight line $E_a = (4\ I_a + 200)$ at a speed of 1000 r.p.m. and for I_a between 50 A and 75 A. The series motor shaft is fitted with a pulley of diameter 50 cms and a load of 100 kgs is to be lowered. In order to limit the descending speed to a safe value, reheostatic braking is employed. Find the breaking resistance to be connected in series with the motor armature circuit for limiting the downward speed to 500 rpm. The total motor resistance is 1 Ω. Neglect rotational loss.

Solution :

Given, At $N = 1000$ rpm, $E_a = 4\ I_a + 200$, $r_a = 1\Omega$ (1)

At $N_1 = 500$ rpm, $E_{a1} = (4\ I_{a1} + 200) \times \dfrac{500}{1000}$

$E_{a1} = 2\ I_{a1} + 100$ (2)

Now, Power developed in armature $= E_{a1} \times I_{a1} = (2\ I_{a1} + 100) \times I_{a1}$ watts

Armature terminal power $= E_{a1} \times I_{a1} - I_{a1}^2 \times r_a$

$= E_{a1} \times I_{a1} - I_{a1}^2 \times 1$

Now, Descending load torque $= rF = r$ mg

$$= \frac{0.50}{2} \times 100 \times 9.8 = 245 \text{ Nm}$$

For torque balance, Dynamic breaking torque = Descending torque

or
$$\frac{(2\,I_{a1} + 100) \times I_{a1} - I_{al}^2}{2 \times \pi \times \frac{500}{60}} = 245$$

or
$$2\,I_{a1}^{\;2} + 100\,I_{a1} - I_{a1}^{\;2} = \frac{2\,\pi \times 500 \times 245}{60} = 12821.7$$

or
$$I_{a1}^{\;2} + 100\,I_{a1} - 12821.7 = 0$$

This yields
$$I_{a1} = \frac{-100 \pm \sqrt{(100)^2 + 4 \times 12821.7}}{2} = \frac{-100 \pm 247.6}{2}$$

Neglecting Negative value, $I_{a1} = \frac{147.6}{2} = 73.8 \text{ A}$

At 500 r.p.m. Back emf $= 2 \times 73.8 + 100 = 147.6 + 100 = 247.6$ V

Let R_b be the breaking resistance to be connected in series with armature, then

$$I_{a1} = \frac{E_{a1}}{R_b + r_a} \quad \text{or} \quad R_b + r_a = \frac{E_{a1}}{I_{a1}}$$

or
$$R_b = \frac{E_{a1}}{I_{a1}} - r_a = \frac{247.6}{73.8} - 1 = 2.36\ \Omega \quad \textbf{Ans.}$$

> **Problem 4.75 :** A 200 HP, 1200 rpm dc shunt motor has stored energy per HP of 8000 Nm. Determine the time taken to start the motor if the load torque is equal to the full load torque and the current is limited to 2 times the full load current.

Solution :

Given, $N = 1200$ rpm

$$\therefore \quad \omega = \frac{2\,\pi \times 1200}{60} = 40\,\pi \text{ rad / sec}$$

Here, energy stored $= \frac{1}{2} J\,\omega^2 = 200 \times 8000$

or $$J = \frac{2 \times 200 \times 8000}{(40 \times \pi)^2} = 202.8 \text{ kgm}^2$$

For d.c. motor, the dynamic equation is $T_e = \frac{jd\omega}{dt} + T_L$

or $$T_e - T_L = \frac{Jd\omega}{dt}$$

or $$dt = \frac{J}{T_e - T_L} d\omega$$

Integrating both sides, $$t = \frac{J}{T_e - T_L} \times \omega \quad (1)$$

According to the question

$$T_L = T_{fl}, \quad T_L = \frac{200 \times 746}{40\,\pi} = T_{fl}$$

Again, during starting, the current is limited to 2 times the full load current.

Starting torque $T_e = T_{fl}$

Now, from (1) $$t = \frac{J \times \omega}{2\,T_{fl} - T_{fl}} = \frac{J \times \omega}{T_{fl}}$$

$$= \frac{202.8 \times 40\,\pi \times \pi}{5 \times 746} \text{ sec} = 21.44 \text{ sec} \quad \textbf{Ans.}$$

Thus, Time taken to start the motor = 21.44 sec **Ans.**

Problem 4.76 : Two series motors are connected to the same supply system and coupled to the same load. The one when running at 600 rpm normally takes a current of 60 A; the other when taking 60 A, runs at 1200 rpm; in each case the supply being same 600 V. Neglecting the effect of saturation, at what speed will the motors run when the load current is 120 A? Each motor has a resistance of 0.5 Ω. Calculate also the value of current in each motor under these conditions.

Solution :

If the motors are disignated by A and B, then for Motor A,

At $N_A = 600$ rpm, $I_A = 60$A

For motor B At $I_B = 60$ A, $N_B = 1200$ rpm

Given, $V_t = 600\,V,$ $r_A = r_B = 0.5\,\Omega$

At $I_L = 120$ A, $N = ?$

When saturation effects are neglected, $\phi = C\,I_a$

For motor A, $E_{bA} = V_t - I_A\,(r_A + r_S) = K_{aA}\,C_A\,I_A\,N_A$

$$= 600 - 60 \times 0.5 = K_{aA}\,C_A\,I_A\,N_A \quad (1)$$

$$K_{aA} \times C_A = \frac{570}{60 \times 600} = \frac{570}{36000}$$

For motor B, $E_{bB} = V_t - I_B\,(r_B + r_S) = K_{aB}\,C_B\,N_B\,I_B$

or $600 - 60 \times 0.5 = K_{aB}\,C_B\,N_B\,I_B \quad (2)$

From (1) and (2)

$$\frac{K_{aA} \times C_A \times N_A \times I_A}{K_{aB} \times C_B \times N_B \times I_B} = 1$$

or
$$\frac{K_{aA} \times C_A}{K_{aB} \times C_B} = \frac{N_B \times I_B}{N_A \times I_A} = \frac{1200 \times 60}{60 \times 600} = 2$$

When motors are connected in parallel, and coupled to the same load, then let the current in motor A be I_{LA}.

In this case, Back emf of the motor A is given by

$$600 - 0.5 \times I_{LA} = K_{aA}\,C_A \times I_{LA} \times N \quad (3)$$

For motor B, Back emf of the motor B is given by

$$600 - 0.5\,(120 - I_{LA}) = K_{aB} C_B \times (120 - I_{LA}) \times N \quad (4)$$

From (3) and (4)

$$\frac{K_{aA}\,C_A \times I_{LA}}{K_{aB}\,C_B \times (120 - I_{LA})} = \frac{600 - 0.5 \times I_{LA}}{600 - 0.5\,(120 - I_{LA})}$$

or
$$2 \times \frac{I_{LA}}{120 - I_{LA}} = \frac{1200 - I_{LA}}{1200 - 120 + I_{LA}}$$

or
$$2\,I_{LA} \times (1080 + I_{LA}) = (120 - I_{LA})\,(1200 - I_{LA})$$

or
$$2160\,I_{LA} + 2\,I_{LA}^2 = 144000 - 120\,I_{LA} - 1200\,I_{LA} + I_{LA}^2$$

or
$$I_{LA}^{\;2} + (2160 + 120 + 1200)\,I_{LA} - 144000 = 0$$

or
$$I_{LA}^{\;2} + 3480\,I_{LA} - 144000 = 0$$

This yields $$I_{LA} = \frac{-3480 \pm \sqrt{(3480)^2 + 4 \times 144000}}{2}$$

$$= \frac{-3480 \pm 3562}{2}$$

Neglecting Negative value $I_{LA} = 41$ A

$\therefore$ Current in motor $A = 41$ A **Ans.**

Current in motor $B = 120 - 41 = 79$ A **Ans.**

The speed can be calculated by

$$N = \frac{(600 - 0.5 \times 41)}{41 \times 570} \times 36000 = 893 \text{ rpm}$$ **Ans.**

Chapter 5

Polyphase Synchronous Machines

EMF EQUATION OF ALTERNATOR

The r.m.s. value of the induced emf per phase is given by

$$E/\text{phase} = 4.44\ fN\phi$$

Where f = Frequency of rotor rotation

N = No. of turms per phase

ϕ = Flux per pole

The equivalent circuit of armature circuit is given as

In figure

E_f = Excitation emf per phase

V_t = Terminal voltage/phase

I_a = Armature current

X_s = Synchronous reactance

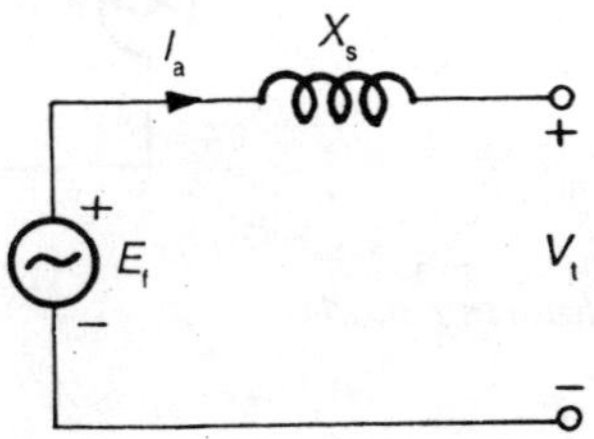

Fig. 5.1

$$\overline{E}_f = \overline{V}_t + j\, I_a\, X_s$$

Synchronous reactance

$$X_s = \frac{V_{oc}}{I_{sc}}$$

I_f = Constant

Where V_{oc} = Open circuit voltage at I_f

I_{sc} = Short circuit armature current at same field current, I_f

The equivalent circuit under armature-reaction condition can be drawn as

From circuit diagram

$\overline{E}_f = \overline{V}_t + \overline{I}_a (r_a + X_s)$

Where

r_a = Armature resistance

$X_s = X_a + X_l$ = Synchronous reactance

X_a = Armature leakage reactance

X_l = Armature reaction leakage reactance

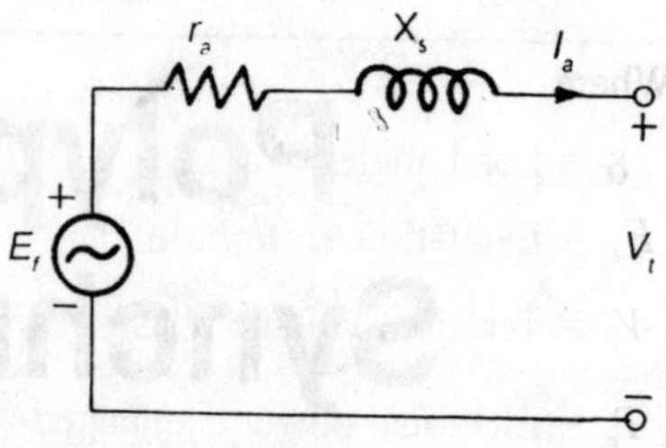

Fig. 5.2

Its phasor diagram can be drawn as

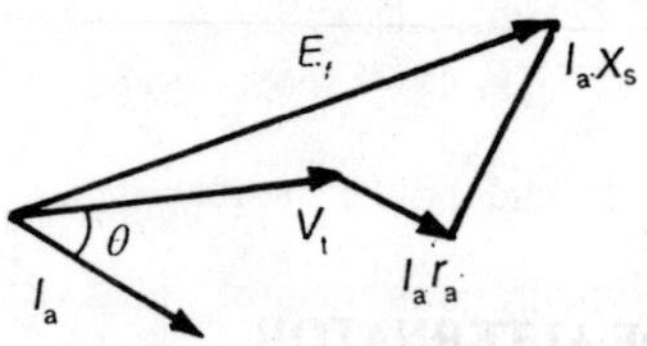

Fig. 5.3

SCR of an alternator SCR $= \dfrac{1}{X_s(\text{adjusted})\ Pu}$

Generating mode: Let as assume armature resistance negligible.

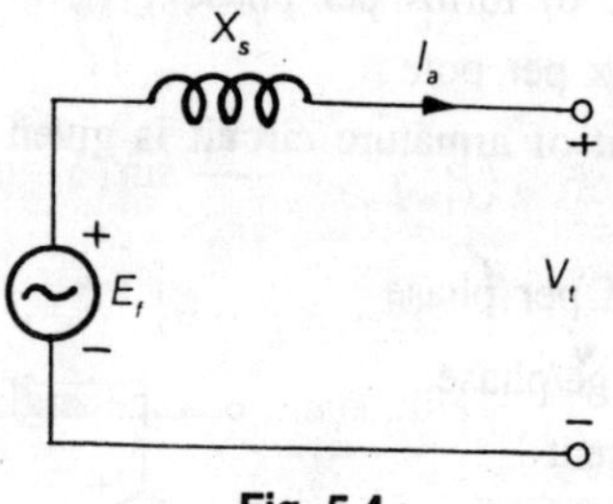

Fig. 5.4

Motoring mode

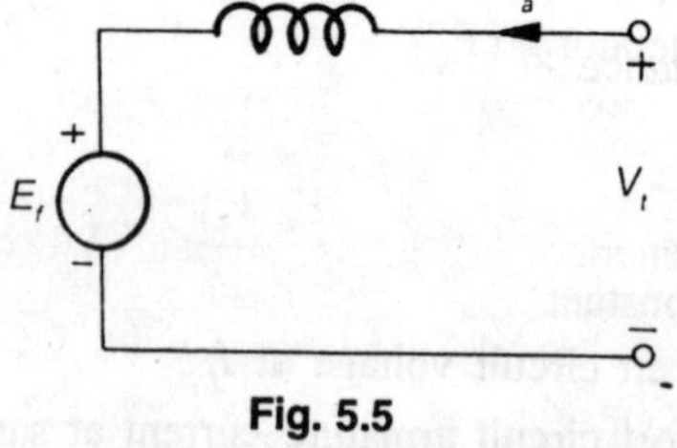

Fig. 5.5

Power-Angle characteristics: The power-angle characteristics of a Synchronous machine is given by

$$P_e = \frac{V_t E_f}{X_s} \sin \delta$$

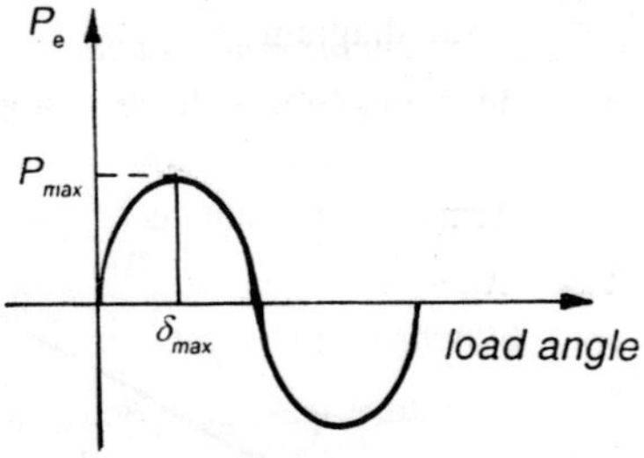

Fig. 5.6

Where

δ = Load angle

E_f = Excitation emf/phase

V_t = Terminal voltage/phase

P_e = Electrical power exchanged with bus bar.

- The excitation emf/phase can be calculated as

$$E_f = \sqrt{(V_t \cos\theta + I_a r_a)^2 + (V_t \sin\theta + I_a X_s)^2}$$

Where $\cos\theta$ = load power factor

- The total drop in synchronous machin is given by

$$I_a Z_s = \sqrt{E_f^2 + V_t^2 - 2 E_f V_t \cos\delta}$$

- For Motoring Mode

Power input to Motor $(P_{im}) = \dfrac{E_f V_t}{Z_s} \sin(\delta - \alpha_z) + \dfrac{V_t^2}{Z_s^2} r_a$

Power output of Motor $(P_{om}) = \dfrac{E_f V_t}{Z_s} \sin(\delta - \alpha_z) - \dfrac{E_t^2}{Z_s^2} r_a$

Where $\alpha_z = 90° - \theta_z$

θ_z = impedance angle

For maximum power output $\delta = \theta_z$

- For Generating Mode

Power input to generator $(P_{ig}) = \dfrac{E_f V_t}{Z_s} \sin(\delta - \alpha_z) + \dfrac{E_f^2}{Z_s^2} r_a$

Power output of generator $(P_{og}) = \dfrac{E_f V_t}{Z_s} \sin(\delta - \alpha_z) - \dfrac{V_t^2}{Z_s^2} r_a$

For maximum power output

Load angle = impedance angle $\delta = \theta_z$

Salient pole Synchronous machine: The phasor diagram of salient pole synchronous machine can be drawn as

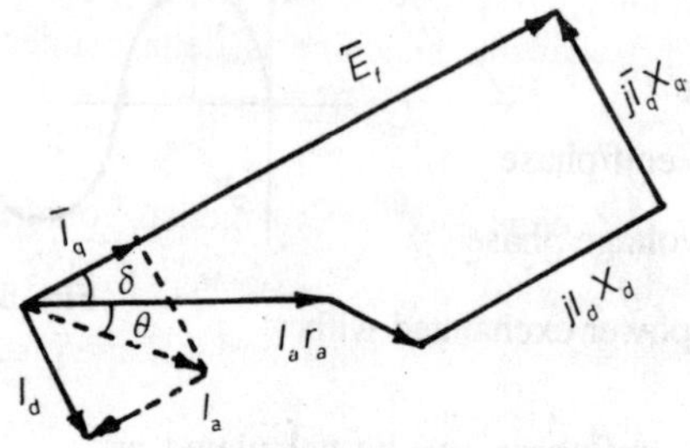

Fig. 5.7

The excitation voltage equation for generating mode is

$$\bar{E}_f = \bar{V}_t + \bar{I}_a r_a + j x_a \bar{I}_a + j x_d \bar{I}_d$$

For motoring mode $\bar{E}_f = \bar{V}_t - \bar{I}_a r_a - j x_a \bar{I}_a + j x_d \bar{I}_d$

Where X_q = Synchronous reactance along q-axis

X_d = Synchronous reactance along d-axis

I_q = Component of armature current along q-axis

I_d = Component of armature current along q-axis

Also $I_d = I_a \sin\psi$

$I_q = I_a \cos\psi$

Where $\tan\psi = \dfrac{V_t \sin\theta - I_a x_q}{V_t \cos\theta - I_a r_a}$ for leading p.f.

Power exchanged with bus bar is given by

$$P = \frac{V_t E_f}{X_d} \sin\delta + \frac{V_t^2}{2}\left(\frac{1}{x_q} - \frac{1}{x_d}\right) \sin 2\delta$$

Voltage-regulation: The voltage regulation of synchronous machine is given by% voltage regulation

$$= \frac{E_f - V_t}{V_t} \times 100\%,$$

where E_f = Induced emf

V_t = Terminal voltage

> **Problem 5.1 :** The phase emf of a 3-phase, 50 Hz alternator consists of a fundamental, a 20% third harmonic and a 10% fifth harmonic. The maximum amplitude of the fundamental voltage is 1000 V. Calculate the rms line voltage when the alternator windings are in *(i)* star and *(ii)* delta.
>
> If the reactance per phase at 50 Hz is 12 Ω, calculate the circulating current when the machine is delta connected.

Solution :

Given, Maximum amplitude of the fundamental voltage/phase

$$E_{ph1} = 1000 \text{ V}$$

Maximum amplitude of the third harmonic voltage/phase

$$E_{ph3} = 1000 \times \frac{20}{100} = 200 \text{ V}$$

Maximum amplitude of the fifth harmonic voltage

$$E_{ph5} = 1000 \times \frac{10}{100} = 100 \text{ V}$$

(i) When alternator windings are star connected third harmonic emf is absent in the lines. Maximum value of line voltage,

$$E_L = \sqrt{3} \times E_{ph} = \sqrt{3} \times \sqrt{E_{ph1}^2 + E_{ph5}^2}$$

∴ R.M.S. value of line voltage

$$= \frac{E_L}{\sqrt{2}} = \frac{\sqrt{3}}{\sqrt{2}} \times \sqrt{(1000)^2 + (100)^2}$$

$$= 1230.85 \text{ V } \textbf{Ans.}$$

(ii) When alternator is delta connected, the third harmonic emf is absent in both phase and line.

$$\therefore \quad E_{ph} = E_{line} = \sqrt{E_{ph_1}^2 + E_{ph_5}^2}$$

∴ R.M.S. value of line voltage

$$= \frac{\sqrt{E_{ph_1}^2 + E_{ph_5}^2}}{\sqrt{2}} = \frac{\sqrt{(1000)^2 + (100)^2}}{\sqrt{2}} = 710.63 \text{ V } \textbf{Ans.}$$

Also, given reactance per phase = 12 Ω.

When the alternator is delta connected, the resultant of fundamental and fifth harmonic emf's in the closed delta is zero but the resultant of third harmonic is 3 E_{ph3}. This resultant third harmonic emf rise to circulating current.

Now, Reactance of the line $= 12 \times 3 = 36\,\Omega$

Resultant of third harmonic emf

$$(E_L) = \frac{3\,E_{ph3}}{\sqrt{2}}\,(r.m.s) = \frac{3 \times 200}{\sqrt{2}}$$

$\therefore$ Circulating current

$$(I_c) = \frac{E_L}{X} = \frac{3 \times 200}{\sqrt{2} \times 36} = 11.785\,\Omega \quad \textbf{Ans.}$$

Problem 5.2 : A 3-phase alternator has 2 slots per pole per phase and coil span of 5 slot pitches. The flux density wave of alternator consists of a fundamental and a 25% third harmonic. Calculate the percentage increase in the phase voltage due to the harmonic.

Solution :

Given, Slot per pole per phase $(q) = 2$

No. of phases $(m) = 3$

Now, Slots per pole $(Q) = m \times q = 2 \times 3 = 6$

Angular slot pitch or slot angle

$$(\gamma) = \frac{180^\circ}{\text{slots per pole}} = \frac{180^\circ}{6} = 30^\circ$$

Coil span $= 5$ slot pitches $= 5 \times 30^\circ = 150^\circ$

Chording angle $(\varepsilon) = 180^\circ -$ Coil span $= 180^\circ - 150^\circ = 30^\circ$

We know that, distribution factor

$$(k_d) = \frac{\sin \frac{qn\gamma}{2}}{q \sin \frac{n\gamma}{2}}$$

$$\text{Pitch factor } (k_p) = \cos \frac{n \in}{2}$$

Fundamental distribution factor

$$k_{d1} = \frac{\sin \frac{1 \times 2 \times 30^\circ}{2}}{2 \sin \frac{30^\circ}{2}} = \frac{1}{4 \sin 15^\circ}$$

Fundamental pitch factor $k_{p1} = \cos \frac{\varepsilon}{2} = \cos \frac{30^\circ}{2} = \cos 15^\circ$

Third harmonic distribution factor

$$k_{d3} = \frac{\sin\dfrac{3\times 2\times 30^\circ}{2}}{2\sin\dfrac{3\times 30^\circ}{2}} = \frac{1}{2\sin 45^\circ}$$

Third harmonic pitch factor

$$k_{p3} = \cos\frac{3\varepsilon}{2} = \cos\frac{3\times 30^\circ}{2} = \cos 45^\circ$$

Applying the relation,

$$E_{Ph} = \sqrt{2}\,\pi f\, k_d\, k_p\, N_{ph}\, B_{\max}\, A$$

$$\therefore \qquad \frac{E_{ph3}}{E_{ph1}} = \frac{k_{d3}\,k_{p3}}{k_{d1}\,k_{p1}} \times \frac{B_{\max 2}}{B_{\max 1}}$$

$$\therefore \qquad E_{ph3} = \frac{k_{d3}\times k_{p3}}{k_{d1}\times k_{p1}} \times \frac{B_{\max 2}}{B_{\max 1}} \times E_{ph1}$$

$$= \frac{4\sin 15^\circ \times \cos 45^\circ}{2\sin 45^\circ \times \cos 15^\circ} \times \frac{0.25\,B_{\max 1}}{B_{\max 1}} \times E_{ph1}$$

$$= 0.134\,E_{ph1}$$

Resultant phase emf $E_R = \sqrt{E_{ph1}^2 + E_{ph3}^2}$

$$= \sqrt{E_{ph1}^2 + (0.134\,E_{ph1})^2} = 1.00894\,E_{ph1}$$

Now, Percentage increase in phase voltage due to third harmonic emf

$$= \frac{E_R - E_{ph1}}{E_{ph1}} \times 100\% = \frac{1.00894\,E_{ph1} - E_{ph1}}{E_{ph1}} \times 100\%$$

$= 0.894\%$ **Ans.**

Problem 5.3 : A 3-phase star-connected alternator is delivering 20 MW and 8 MVAr to an infinite bus at 11 kV. The alternator has synchronous impedance of $0 + j\,3\,\Omega$. Determine the load angle and the excitation emf of the alternator.

Solution :

Given, Real power (P) = 20 MW

Reactive power (Q) = 8 MVAr

Complex power delivered

$$\overline{S}_L = 20 \text{ MW} - j\, 8 \text{ MVAr}$$

$$\overline{S}_L = 21.541 \angle -21.8° \text{ MVA}$$

Now, Load current $(\overline{I}_L) = \dfrac{\overline{S}_L}{\sqrt{3} \times V_L} = \dfrac{21.541 \angle -21.8°}{\sqrt{3} \times 1000} \times 10^6$

$$= 1130.61 \angle -21.8° \text{ A}$$

Bus terminal voltage per phase

$$V_{ph} = \frac{V_t}{\sqrt{3}} = \frac{11 \times 1000}{\sqrt{3}} = 6351 \text{ V}$$

$$\overline{Z}_L = 3 \angle 90° \ \Omega$$

Let us take bus terminal voltage as reference phasor

$$\overline{V}_{ph} = 6351 \angle 0°$$

Now, Excitation emf per phase of the alternator

$$\overline{E}_{ph} = \overline{V}_{ph} + \overline{I}_L \cdot \overline{Z}_L$$

$$= 6351\,(1 + j\,0) + 1130.61 \angle -21.8° \times 3 \times \angle 90°$$

$$= 6351 + 3391.83 \angle 68.2°$$

$$= 7610.617 + j\ 3151.46 = 8237.305 \angle 22.49°$$

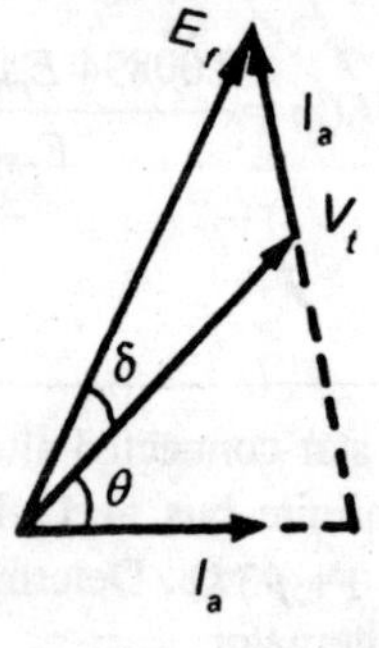

Fig. P. 5.3

Load angle $(\delta)^{\bullet} = 22.49°$ **Ans.**

Line excitation emf of the alternator

$$E_L = \sqrt{3} \times E_{ph} = \sqrt{3} \times 8237.305 = 14267.431 \text{ V } \textbf{Ans.}$$

> **Problem 5.4 :** A 3-phase, 17.32 kVA, 400 V, star-connected alternator is delivering rated load at 400 V and at pf 0.8 lag. Its synchronous impedance is $0.2 + j\,2\ \Omega$ per phase. Find the load angle at which it is operating.
>
> Now, with the magnitude of excitation voltage held constant, the nature of rated load p.f. is altered to p.f. 0.8 leading. Determine the new value of terminal voltage and load angle.

Solution :

Given, $V_t = 400$ V (L – L), p.f. = 0.8 (lagging)

$\therefore$ $\theta = \cos^{-1} 0.8 = 36.9°$

$$\text{Rated load current } (I_L) = \frac{\text{Rated power}}{\sqrt{3} \times V_t} = \frac{17.32 \times 1000}{\sqrt{3} \times 400}$$

$$= 24.999 \text{ A}$$

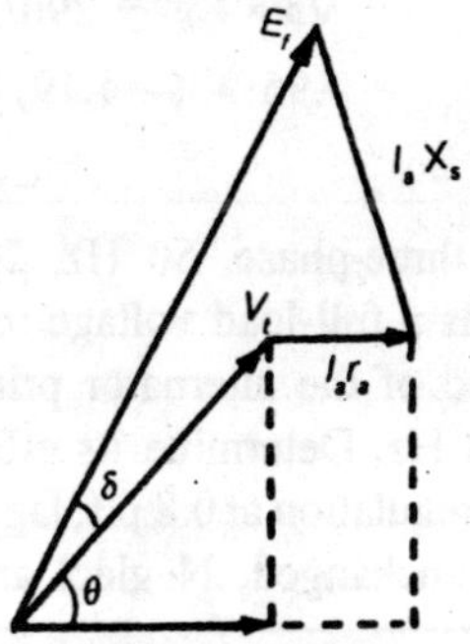

Fig. P. 5.4

In phasor form, $\bar{I}_L = 24.999 \angle -36.9°$

$$\bar{Z}_s = 0.2 + j\,2\ \Omega = 2.01 \angle 84.29°$$

$$\text{Terminal voltage per phase } V_{ph} = \frac{V_t}{\sqrt{3}} = \frac{400}{\sqrt{3}} = 230.94 \text{ V}$$

Let us take terminal voltage as reference phasor

$\therefore$ $\bar{V}_{ph} = 230.94 \angle 0°$

Now, Excitation emf per phase,

$$\overline{E}_{ph} = \overline{V}_{ph} + \overline{I}_L \overline{Z}_s$$

$$= 230.94 \angle 0° + 24.999 \angle -36.9° \times 2.01 \times \angle 84.29°$$

$$= 230.94 + 50.248 \angle 47.39° = 264.958 + j\,36.98$$

$$= 267.526 \angle 7.95°$$

$\therefore$ Load angle = 7.95° **Ans.**

When the load is operated at leading p.f.

$$\overline{I}_L = 24.999 \angle 36.9°$$

Applying the relation, $\overline{E}_{ph} = \overline{V}_{ph} + \overline{I}_L \overline{Z}_s$

or $\quad 264.958 + j\,36.98 = \overline{V}_{ph} + (24.999 \angle 36.9) \times 2.01 \angle 84.29°$

or $\quad 264.958 + j\,36.98 = \overline{V}_{ph} + (50.248) \angle 121.19°$

or $\quad 264.958 + j\,36.98 = \overline{V}_{ph} + 26.022 + j\,42.985$

or $\quad \overline{V}_{ph} = 290.022 - j\,6.005 = 290.084 \angle -1.19°$

Now, Terminal voltage at p.f. 0.8 leading

$$= \sqrt{3} \times V_{ph} = 290.04 \times \sqrt{3} = 502.44 \text{ V}$$ **Ans.**

New load angle $= 7.95 - (-1.19) = 9.14°3$ **Ans.**

> **Problem 5.5 :** A three-phase, 50 Hz, 2000 kVA, 11 kV star-connected alternator has a full-load voltage regulation of 10% at 0.8 p.f. lag. Now the speed of the alternator prime-mover decreases to give a frequency of 48 Hz. Determine its effect on alternator rating and also on the voltage regulation at 0.8 p.f. lag. The field and armature currents are assumed unchanged. Neglect armature resistance.

Solution :

We know that the excitation emf per phase

$$E_{ph} = \sqrt{2}\, \pi\, f\, N_{ph}\, \phi$$

When field current and no. of series turns remain constant

$$\frac{E_{f1}}{E_{f2}} = \frac{f_1}{f_2}$$

Given $f_1 = 50\text{ Hz}, \quad f_2 = 48\text{ Hz}, \quad V_{t1} = 11\text{ kV}$

$$\therefore \quad E_{f2} = \frac{48}{50} E_{f1} = 0.96\, E_{f1}$$

Hence $X_{s2} = 0.96\, X_{s1}$

Given, at 50 Hz and p.f. 0.8 lagging

$$\text{Voltage regulation} = 0.01, \frac{E_{f1} - V_{t1}}{V_{t1}} = 0.01 \quad (1)$$

For negligible armature resistance

$$V_{t2} = 0.96\, V_{t1} = 0.96 \times 11\,\text{kV} = 10.56 \text{ kV}$$

kVA rating at 48 Hz $= 0.96 \times 2000\,\text{kVA} = 1920$ kVA **Ans.**

Since power factor is same

$$\therefore \text{ voltage regulation at 48 Hz} = \left(\frac{E_{f2} - V_{t2}}{V_{t2}}\right) = \frac{0.96\, E_{f1} - 0.96\, V_{t1}}{0.96\, V_{t1}}$$

$$= \frac{E_{f1} - V_{t1}}{V_{t1}} = 0.01 = 10\% \text{ Ans.}$$

> **Problem 5.6 :** An alternator has a synchronous reactance of 20% and negligible resistance. Calculate its voltage regulation when working at full-load *(i)* 0.8 p.f. lag *(ii)* unity pf and *(iii)* 0.8 p.f. lead.

Solution :

We have, $$E_f^2 = (V_t \cos\theta + I_a\, r_a)^2 + (V_t \sin\theta + I_a\, x_s)^2$$

For negligible resistacne $r_a = 0$

$$\therefore \quad E_f^2 = (V_t \cos\theta)^2 + (V_t \sin\theta + I_a\, x_s)^2$$

or $$E_f^2 = V_t^2 \left[\cos^2\theta + \left(\sin\theta + \frac{I_a\, x_s}{V_t}\right)^2\right]$$

(i) Given synchronous reactance in p.u. $\left(\dfrac{I_a X_s}{V_t}\right) = 0.20$

$$\text{p.f.} = 0.8 \text{ (lagging)}$$

$$\therefore \quad \sin\theta = 0.6$$

Now, $$E_f^2 = V_t^2 \left[(0.8)^2 + (0.6 + 0.2)^2\right]$$

$$E_f = 1.13\, V_t$$

$$\text{Now, Voltage regulation} = \left(\frac{E_f - V_t}{V_t}\right) = \frac{1.13\,V_t - V_t}{V_t}$$

$$= 0.13 = 13\% \textbf{ Ans.}$$

(ii) At unity p.f. $\cos\theta = 1, \quad \sin\theta = 0$

$$\therefore \qquad E_f^2 = V_t^2\left[1^2 + (0 + 0.2)^2\right] = V_t^2\,[1.04]$$

$$\therefore \qquad E_f = 1.0198\,V_t$$

$$\text{Now, voltage regulation} = \frac{E_f - V_t}{V_t} = \frac{1.0198\,V_t - V_t}{V_t}$$

$$= 0.0198 = 1.98\% \textbf{ Ans.}$$

(iii) At 0.8 leading, $\cos\theta = 0.8, \quad \sin\theta = 0.6$

$$E_f^2 = V_t^2\left[\cos^2\theta + \left(\sin\theta - \frac{I_a X_s}{V_t}\right)^2\right]$$

$$\text{or} \qquad E_f^2 = V_t^2\left[(0.8)^2 + (0.6 - 0.2)^2\right]$$

$$\text{or} \qquad E_f^2 = V_t^2 \times 0.8$$

$$\therefore \qquad E_f = 0.894\,V_t$$

$$\text{Voltage regulation} = \frac{E_f - V_t}{V_t} = \frac{0.894\,V_t - V_t}{V_t}$$

$$= -\,0.106 = -\,10.6\% \textbf{ Ans.}$$

Problems 5.7 : A 3-phase star-connected alternator is rated 1600 kVA, 13,500 V. The armature effective resistance and synchronous reactance are 1.5 Ω and 30 Ω respectively per phase. Calculate the percentage regulation for a load of 1280 kW at power factors of *(a)* 0.8 leading, *(b)* unity and *(c)* 0.8 lagging

Solution :

Given, $r_a = 1.5\,\Omega\,/\,\text{phase}, \; x_s = 30\,\Omega\,/\,\text{phase}$

Load Power $(P_L) = 1280\ \text{kW}$

(a) For load of 1280 kW at p.f. 0.8 (leading)

$$\text{Load kVA} = \frac{P_L}{p.f.} = \frac{(1280)\times(1000)}{0.8} = 1600\ \text{kVA}$$

Load current $I_L = \frac{kVA}{\sqrt{3} \times V_L} = \frac{1600 \times 1000}{\sqrt{3} \times 13500} = 68.427$ A

Load terminal voltage/phase

$$V_{ph} = \frac{V_t}{\sqrt{3}} = \frac{13.5 \times 1000}{\sqrt{3}} = 7794.229 \text{ V}$$

To calculate excitation emf per phase for leading p.f.

$$E_{ph}^2 = (V_{ph} \cos\theta + I_a r_a)^2 + (V_{ph} \sin\theta - I_a x_s)^2$$

$$= (7794.229 \times 0.8 + 68.427 \times 1.5)^2 + (7794.229 \times 0.6 - 68.427 \times 30)^2$$

$$= 47054489.89$$

$\therefore$ $E_{ph} = 6859.628$ V

Now, Voltage regulation

$$= \frac{E_{ph} - V_{ph}}{V_{ph}} = \frac{6859.628 - 7794.229}{7794.229}$$

$= -\ 0.11991 = -\ 11.991\%$ **Ans.**

(b) At unity p.f. $\cos\theta = 1, \quad \sin\theta = 0$

$$\text{Load kVA} = \frac{P_L}{1} = 1280 \text{ kVA}$$

Load current

$$(I_L = I_a) = \frac{\text{Load kVA}}{\sqrt{3} \times V_t} = \frac{1280}{\sqrt{3} \times 13.5} = 54.741 \text{ A}$$

Terminal voltage per phase,

$$V_{ph} = \frac{13500}{\sqrt{3}} = 7794.229 \text{ V}$$

To calculate excitation emf

$$E_{ph}^2 = (V_{ph} \cos\theta + I_a r_a)^2 + (V_t \sin\theta - I_a X_s)^2$$

$$= (7794.229 \times 1 + 54.741 \times 1.5)^2 + (0 - 30 \times 54.741)^2$$

$$= 64733659.04$$

$\therefore$ $E_{ph} = 8045.723$ V

Now, voltage regulation $= \frac{E_{ph} - V_{ph}}{V_{ph}} = \frac{8045.723 - 7794.229}{7794.229}$

$= 0.03227 = 3.227\%$ **Ans.**

(c) At p.f. 0.8 lagging $\cos\theta = 0.8$, $\sin\theta = 0.6$

$$E_{ph}^2 = (V_{ph}\cos\theta + I_a r_a)^2 + (V_{ph}\sin\theta + I_a X_s)^2$$

$$= (7794.229 \times 0.8 + 68.427 \times 1.5)^2 + (7794.229 \times 0.6 + 68.427 \times 30)^2$$

$$= 85454660.85$$

$$\therefore \quad E_{ph} = 9244.169$$

$$\text{Voltage regulation} = \frac{E_{ph} - V_{ph}}{V_{ph}} = \frac{9244.169 - 7794.229}{7794.229}$$

$$= 0.18603 = 18.603\% \textbf{ Ans.}$$

> **Problem 5.8 :** A 3-phase star-connected alternator has synchronous impedance of $1 + j\,10\,\Omega$ per phase. O is operating at a constant voltage of 6.6 kV and its field current is adjusted to give an excitation voltage of 6.4 kV. Find the power output, armature current and pf under the conditions of *(i)* maximum power output and *(ii)* maximum power input.

Solution :

Given, $Z_s = 1 + j\ 10 = 10.05\angle 84.29°$

$\theta_z = 84.29°$

Excitation emf per phase $E_{ph} = \dfrac{E_f}{\sqrt{3}} = \dfrac{6400}{\sqrt{3}} = 3695.042$ V

Terminal voltage per phase $V_{ph} = \dfrac{V_t}{\sqrt{3}} = \dfrac{6600}{\sqrt{3}} = 3810.512$ V

(i) For maximum power output

Load angle (δ) = impedance angle $(\theta_z) = 84.29°$

Power output/phase

$$= \frac{E_{ph} \times V_{ph}}{Z_s} - \frac{V_{ph}^2}{Z_s^2} \times r_a$$

$$= \frac{3695.042 \times 3810.512}{10.05} - \left(\frac{3810.512}{10.1}\right) \times 1$$

$$= 1257.236 \text{ kW}$$

Total power output $= 3 \times P_{ph} = 3 \times 1257.236$

$= 3771.71$ kW **Ans.**

To calculate armature current

$$I_a Z_s = \sqrt{E_{ph}^2 + V_{ph}^2 - 2\, E_{ph} \times V_{ph} \times \cos\delta}$$

$$I_a = \frac{\sqrt{(3695.042)^2 + (3810.512)^2 - 2 \times 3810.512 \times 3695.042 \times \cos 84.29^\circ}}{10.05}$$

$\therefore \quad = 501.20$ A **Ans.**

To calculate p.f.,

$$\cos\theta = \frac{(V_{ph}\, Z_s - E_{ph}\, r_a)}{I_a Z_s^2}$$

$$= \frac{3695.042 \times 10.05 - 3810.512 \times 1}{501.20 \times 101} = 0.6583$$

Since alternator is underexcited, p.f. is leading

$\therefore \quad$ p.f. = 0.6583 (leading) **Ans.**

(*ii*) For maximum input power load angle

$$(\delta) = 180^\circ - \theta_z = 180^\circ - 84.29^\circ = 95.71^\circ$$

$$\alpha_z = 90^\circ - \theta_z = 5.71^\circ$$

$$\text{Per phase power output} = \frac{E_{ph}\, V_{ph}}{z_s} \sin(\delta + \alpha_z) - \left(\frac{V_{ph}}{Z_s}\right)^2 \times r_a$$

$$= \frac{3695.042 \times 3810.512}{10.05} \sin(5.71^\circ + 95.71^\circ) - \left(\frac{3810.512}{10.05}\right)^2 \times 1$$

$= 1229.47$ kW

Total power output, $P_o = (1229.47 \times 3) = 3688.41$ kW **Ans.**

Now, Armature current,

$$I_a = \frac{\sqrt{E_{ph}^2 + V_{ph}^2 - 2\, E_{ph}\, V_{ph} \cos\delta}}{Z_s}$$

$$= \frac{\sqrt{(3695.042)^2 + (3810.512)^2 - 2 \times 3810.512 \times 3695.042 \times \cos 95.71^\circ}}{10.05}$$

$= 553.784$ A **Ans.**

$$\text{p.f.} = \frac{E_f\, Z_s - V_t\, V_a}{I_a\, Z_s^2} = \frac{3695.042 \times 10.05 - 3810.512 \times 1.0}{(10.05)^2 \times 553.784}$$

$= 0.5958$

Since $E_{ph} \cos\delta < V_{ph}$, p.f = 0.5958 (leading) **Ans.**

Problems 5.9 : A 3-phase star-connected non-salient pole synchronous motor connected to 6.6 kV mains has an armature impedance of $2.5 + j\,15.0\ \Omega$ per phase. The excitation of the machine gives a generated emf of 7.0 kV. The iron and friction losses amount to 10 kW. Determine the output of the motor when operating at load angle of 31° (electrical)

Solution :

Given $\quad \overline{Z}_{ph} = 2.5 + j\ 15\ \Omega$

$$= 15.207 \angle 80.54°,\ \theta_z = 80.54°,\ \delta = 31°$$

$$\therefore \quad \delta + \alpha_Z = \delta + (90° - \theta_z) = 31° + 90° - 80.54° = 40.46°$$

Iron and friction loss $\quad (W_o) = 10$ kW

Terminal voltage per phase, $V_{ph} = \dfrac{V_t}{\sqrt{3}} = \dfrac{6.6}{\sqrt{3}}$ kV

Excitation emf per phase, $E_{ph} = \dfrac{E_t}{\sqrt{3}} = \dfrac{7}{\sqrt{3}}$ kV.

Now, Motor output power developed/phase

$$= \frac{E_{ph} V_{ph}}{Z_s} \sin(\delta + \alpha_Z) - \frac{E_f^2}{Z_e^2} \times r_a$$

$$= \frac{6.6 \times 7}{3 \times 15.207} \sin 40.46° - \frac{49}{3 \times (15.207)^2} \times 2.5 = 480.578 \text{ kW}$$

Total output power developed = $480.578 \times 3 = 1441.734$ kW

Net power output = Total output power developed – iron and friction losses

$$= (144.734 - 10)\text{ kW} = 1431.734 \text{ kW} \quad \textbf{Ans.}$$

Problem 5.10 : A 3300-V, star-connected synchronous motor works at constant terminal voltage and constant excitation. Its synchronous impedance is $1 + j\ 10\ \Omega$ per phase. It operates at a pf of 0.8 leading when taking 600 kW from the mains. Find the p.f. when the input is increased to 900 kW.

Solution :

Given, $\bar{Z} = 1 + j\ 10 = 10.05\ \angle 84.29°$

$$\alpha_Z = 90° - \theta_z = 90° - 84.29° = 5.71°$$

Terminal voltage per phase

$$V_{ph} = \frac{V_t}{\sqrt{3}} = \frac{3300}{\sqrt{3}} = 1905.256 \text{ V}$$

p.f. = 0.8 (leading), Load power = 600 kW.

Load curent $I_L = \dfrac{\text{Load power}}{\sqrt{3} \times V_L \times \text{p.f.}} = \dfrac{600 \times 1000}{\sqrt{3} \times 3300 \times 0.8}$

$= 131.216$ A

For motor operating at leading p.f.

$$E_{ph} = \sqrt{(V_{ph} \cos\theta - I_a r_a)^2 + (V_{ph} \sin\theta + I_a x_s)^2}$$

$$= \sqrt{(1905.256 \times 0.8 - 131.216 \times 1)^2 + (1905.256 \times 0.6 + 131.216 \times 10)^2}$$

$= 2822.939$ V

When input power is changed, load angle also changes. For motor,

$$\text{Power input per phase} = \frac{V_{ph} E_{ph}}{Z_s} \sin(\delta - \alpha_z) + \frac{V_t^2}{Z_s^2} \times r_a$$

or $$\frac{900 \times 1000}{3} = \frac{2822.939 \times 1905.256}{10.05} \sin(\delta - \alpha_z) + \left(\frac{1905.256}{10.05}\right)^2 \times 1$$

or $\sin(\delta - \alpha_z) = 0.4934$

$\therefore$ $\delta - \alpha_z = 29.56°$

$\therefore$ $\delta = 29.56° + 5.71° = 35.27°$

Now, Armature current,

$$I_a = \frac{\sqrt{E_{ph}^2 + V_{ph}^2 - 2 \times E_{ph} \times V_{ph} \cos 35.27°}}{Z_s}$$

$$= \frac{\sqrt{(2822.939)^2 + (1905.250)^2 - 2 \times 1905.250 \times 2822.939 \cos 35.27°}}{10.05}$$

$= 166.994$ A

Now, power input per phase = 300 × 1000

or, $I_a \times V_{ph} \times \text{p.f.} = 300 \times 1000$

$$\therefore \qquad \text{p.f.} = \frac{300 \times 1000}{166.994 \times 1905.256} = 0.9429$$

Since $E_f \cos\delta = 2822.939 \cos 35.27 > 1905.256$ V.

p.f. = 0.9429 (leading) **Ans.**

Problems 5.11 : The synchronous reactance per phase of a 3-phase star-connected 6600 V synchronous motor is 20 Ω. For a certain load, the input is 915 kW at normal voltage and the induced e.m.f. is 8942 V. Determine the line current and the p.f. Neglect armature resistance.

Solution :

Given, $\bar{Z} = 0 + j\ 20\ \Omega = 20\ \angle 90°\ \Omega$

Here $\alpha_z = 90° - \theta_z = 90° - 90° = 0°$

Power input $= 915$ kW

Terminal voltage per phase $V_{ph} = \dfrac{V_t}{\sqrt{3}} = \dfrac{6600}{\sqrt{3}} = 3810.512$ V

Excitation emf per phase, $E_{ph} = \dfrac{E_f}{\sqrt{3}} = \dfrac{8942}{\sqrt{3}} = 5162.666$ V

Now, for motor, power input per phase $= \dfrac{915 \times 1000}{3}$

or $$\frac{E_{ph}\, V_{ph}}{Z_s} \sin(\delta - \alpha_z) = 305 \times 1000$$

or $$\frac{5162.666 \times 3810.512}{20} \sin\delta = 305 \times 1000$$

or $$\sin\delta = 0.31$$

$\therefore$ Load angle $(\delta) = 18.06°$

Now, Armature curent

$$I_a = \frac{\sqrt{V_{ph}^2 + E_{ph}^2 - 2\, V_{ph}\, E_{ph} \cos\delta}}{Z_s}$$

$$= \frac{\sqrt{(5162.666)^2 + (3810.512)^2 - 2 \times 3810.512 \times 5162.666 \cos 18.06°}}{20}$$

$= 97.04$ A

To calculate p.f., power input per phase = 305 × 1000

or, $V_{ph} \times I_a \times p.f. = 305 \times 1000$

$$\therefore \quad \text{p.f.} = \frac{305 \times 1000}{3810.512 \times 97.04} = 0.8248$$

Since $E_f \cos\delta > V_t$, p.f. = 0.8248 (leading) **Ans.**

> **Problems 5.12 :** A 2200-V, 400 kW, 3-phase, star-connected synchronous motor has a synchronous impedance of $0.3 + j\,3\ \Omega$. Determine the excitation e.m.f. and the power angle if the motor works at rated load at 0.8 p.f. leading and with an efficiency of 94%.

Solution :

Given, efficiency = 0.94

or, $$\frac{p_o}{p_{in}} = 0.94$$

$$\therefore \quad P_{in} = \frac{p_o}{0.94} = \frac{400}{0.94}\ \text{kW} = 425.532\ \text{kW}$$

Load curent $$(I_L) = \frac{425.532 \times 1000}{\sqrt{3} \times 2200 \times 0.8} = 139.591\ \text{A}$$

$$\text{p.f.} = 0.8,\ \theta = \cos^{-1} 0.8 = 36.9°$$

Armature current in phasor form,

$$\bar{I}_a = 139.591\ \angle 36.9° \text{ (for leading p.f.)}$$

$$\bar{Z}_s = 0.3 + j\,3 = 3.015\ \angle 84.29°$$

Terminal voltage per phase,

$$V_{ph} = \frac{V_t}{\sqrt{3}} = \frac{2200}{\sqrt{3}} = 1270.171\ \text{V}$$

Now, for motor, $\bar{E}_{ph} = \bar{V}_{ph} - \bar{I}_a\,\bar{Z}_s$

Let us take terminal voltage as reference phasor.

$$\bar{E}_{ph} = 1270.171\angle 0° - 3.015\angle 84.29° \times 139.591\angle 36.9°$$

$$= 1487.958 - j\,360.033 = 1530.896\angle -13.602°$$

$\therefore$ Power angle $(\delta) = 13.602°$ **Ans.**

Excitation emf $= \sqrt{3} \times E_{ph} = \sqrt{3} \times 1530.896 = 2651.59$ V **Ans.**

Problems 5.13 : A 2000 V, 3-phase, star-connected synchronous motor has synchronous impedance of 0.5 + *j* 5 Ω per phase. For an excitation voltage of 3000 V, the motor takes an input of 900 kW at rated voltage. Compute the line current and p.f.

Solution :

Given, $p_{in} = 900$ kW

Power input per phase $= \dfrac{p_{in}}{3} = \dfrac{900}{3} = 300$ kW

$$\bar{Z} = 0.5 + j5 = 5.025\angle 84.29°\ \Omega,$$

$$\alpha_z = 90° - 84.29° = 5.71°$$

Terminal voltage per phase, $V_{ph} = \dfrac{V_t}{\sqrt{3}} = \dfrac{2000}{\sqrt{3}} = 1154.701$ V

Excitation voltage per phase, $E_{ph} = \dfrac{E_f}{\sqrt{3}} = \dfrac{3000}{\sqrt{3}} = 1732.051$ V

For motor, power input/phase

$$= \frac{E_{ph} \times V_{ph}}{Z_s} \sin(\delta - \alpha_z) + \frac{V_{ph}^2}{Z_s^2} \times V_a$$

or $300 \times 1000 = \dfrac{1732.051 \times 1154.701}{5.025} \sin(\delta - \alpha_z) + \left(\dfrac{1154.701}{5.025}\right)^2 \times 0.5$

or $\sin(\delta - \alpha_z) = 0.6874$

$\therefore \quad \delta - \alpha_z = \sin^{-1}(0.6874) = 43.42$

$\therefore \quad \delta = 43.42° + 5.71° = 49.13°$

Let us terminal voltage as reference phasor.

$$\bar{E}_{ph} = \bar{V}_{ph} - \bar{I}_a \bar{Z}_s$$

or, $\bar{I}_a = \dfrac{\bar{V}_{ph} - \bar{E}_{ph}}{\bar{Z}_s} = \dfrac{1154.701\angle 0°}{5.025\angle 84.29°} - \dfrac{1732.051\angle 49.13°}{5.025\angle 84.29°}$

$$= 229.791\angle -84.29° - 344.687\angle -35.16°$$

$$= -258.935 - j30.159 = 260.69\angle 6.64°$$

Armature curent $= 260.69$ A,

p.f. $= \cos 6.64° = 0.9933$ (leading) **Ans.**

Problems 5.14 : A 11-kV, 3-phase, mesh-connected synchronous motor has synchronous reactance of 10 Ω per phase. Its friction, windage and iron losses are 1000 kW.

(*a*) The motor excitation is so adjusted that it takes u.p.f. current at a shaft load of 10 MW. Calculate the excitation voltage.

(*b*) Now the excitation e.m.f. is increased to 12.5 kV and the shaft load is adjusted so that motor again operates at u.p.f. Calculate the shaft load output.

Solution :

Given, $\bar{Z} = 0 + j\,10 = 10\angle 90°\ \Omega$

Windage, friction and iron loss $(W_o) = 1000$ kW

Shaft power $(p_{sh}) = 10$ MW

Output power developed $= p_{sh} + w_o = 11\,\text{MW}$

Neglecting copper loss $(r_a = 0)$

∴ Power input to motor = power developed = 11 MW

Line load current $= \dfrac{p_{in}}{\sqrt{3} \times V_L \times pf}$, $(I_L) = \dfrac{11 \times 10^6}{\sqrt{3} \times 11 \times 10^3 \times 1}$

Phase load curent $= \dfrac{I_L}{\sqrt{3}} = \dfrac{11 \times 10^6}{3 \times 11 \times 10^3} = \dfrac{1000}{33} = 333.33$ A

At unity p.f. $\bar{I}_a = 333.33\ \angle 0°$

Now, let us take terminal voltage as reference phasor,

For dalta connection $\bar{V}_{ph} = \bar{V}_L = 11000\ \angle 0°$

For Motor, $\bar{E}_{ph} = \bar{V}_{ph} - \bar{I}_a\,\bar{Z}_s$

$$\bar{E}_{ph} = 11000\ \angle 0° - 333.33\ \angle 0° \times 10\angle 90°$$

$$= 11000\,(1 + j\,0) - j\,333.33 = 11000 - j\,3333.3$$

$$= 11493.95\ \angle -16.86°$$

∴ Motor excitation = 11493.95 V **Ans.**

(*a*) When the excitation is changed to 12.5 kV.

$$E_L = E_{ph} = 12500\ \text{V}$$

Let I_a be the per phase armature current. Applying the relation,

$$E_{ph}^2 = (V_{ph}\cos\theta)^2 + (I_a\,X_s)^2$$

or, $I_a = \frac{\sqrt{E_{ph}{}^2 - V_{ph}{}^2}}{X_s} = \frac{\sqrt{(12500)^2 - (11000)^2}}{10}$

$= 593.717$ A

Line curent $(I_L) = \sqrt{3} \times I_a = \sqrt{3} \times 593.717$

Since there is no i_r^2 loss

Gross power output = power input to the motor =

$\sqrt{3} \times V_L \times I_L \times \text{p.f.} = \sqrt{3} \times 11000 \times \sqrt{3} \times 593.717 \times 1$

$= 19592.661$ kW

Now, shaft load = Gross power output – No. load loss

$= (19592.661 - 1000)$

$= 18592.661$ kW **Ans.**

Problem 5.15 : A 3-phase star-connected alternator with synchronous impedance of $0 + j\,5\ \Omega$ per phase is connected to an 11- kV system. The alternator power output is found to be 10 MW and reactive power output as 3 MVA. Compute the magnitude of excitation voltage, load angle, line current and power factor.

Solution :

Given,

Real power $(P) = 10$ MW

Reactive power $(Q) = 3$ MVAr

$\therefore$ MVA rating of alternator $= \sqrt{P^2 + Q^2} = \sqrt{100 + 9}$

$= 10.44$ MVA

$$\text{p.f. angle} = \tan^{-1}\left(-\frac{3}{10}\right) = -16.7°$$

$$Z_s = 5\angle 90°\ \Omega$$

Now, Line curent $(I_L) = \frac{\text{Load power in MVA}}{\sqrt{3} \times V_L}$

$$= \frac{10.44 \times 10^6}{\sqrt{3} \times 11 \times 10^3} = 547.96 \text{ A } \textbf{Ans.}$$

In phasor form, $\bar{I}_L = 547.96\angle -16.7°$

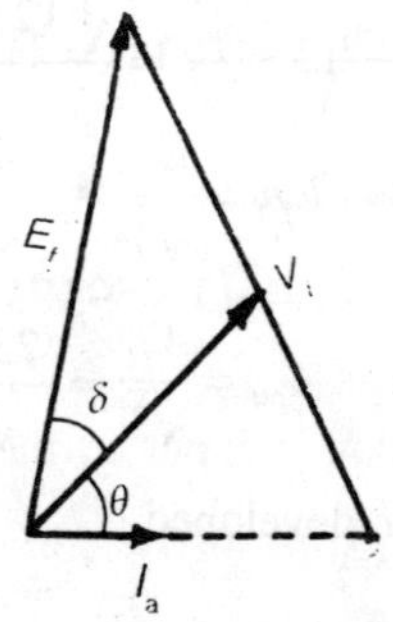

Fig. P. 5.15

Terminal voltage per phase $V_{ph} = \dfrac{V_t}{\sqrt{3}} = \dfrac{11 \times 1000}{\sqrt{3}} = 6350.853$ V

Let us take terminal voltage as reference phasor

$$\overline{V}_{ph} = 6350.852 \angle 0°$$

To calculate excitation,

$$\overline{E}_{ph} = \overline{V}_t + \overline{I}_a\, \overline{Z}_s$$

$$= 6350.853 + (1 + j0) + 547.96 \angle -16.7° \times 5 \angle 90°$$

$$= 6350.853 + j\, 2739.8 \angle 73.3° = 7138.163 + j\, 2624.242$$

$$= 7605.262 \;\; \angle 20.185°$$

Hence, Load angle (δ) = (20.185°)

Now, Line excitation voltage =

$$\sqrt{3} \times E_{ph} = \sqrt{3} \times 7605.262 = 13172.7 \text{ V } \textbf{Ans.}$$

We have, $\overline{I}_L = 546.97 \angle -16.7°$

Since $E_f \cos\delta > V_t$,

The p.f. will be leading

$\therefore$ p.f. = cos 16.7° = 0.9578 (lead) **Ans.**

Problems 5.16 : A 2300-V, 3-phase, star-connected, 50 Hz synchronous motor operating at normal voltage is excited to give an excitation voltage of 2400 V. Determine the maximum power developed, armature current and power factor under this excitation. Per-phase synchronous impedance is $1.5 + j\, 21\ \Omega$.

Solution :

Given, $\bar{Z}\text{/phase} = 1.5 + j\,21\,\Omega = 21.054\,\angle 85.9°\,\Omega$

Excitation emf per phase, $E_{ph} = \dfrac{E_f}{\sqrt{3}} = \dfrac{2400}{\sqrt{3}} = 1385.641\text{ V}$

Terminal voltage per phase, $V_{ph} = \dfrac{V_t}{\sqrt{3}} = \dfrac{2300}{\sqrt{3}} = 1327.906\text{ V}$

In motor, Maximum power developed,

$$p_{o\,max} = \frac{E_{ph}\,V_{ph}}{Z_s} - \frac{E_{ph}^{\;2}}{Z_s^{\;2}} \times r_a$$

$$= \frac{1327.906 \times 1385.641}{21.054} - \left(\frac{1385.641}{21.054}\right)^2 \times 1.5 = 80.897\text{ kW}$$

Total power developed $= 3 \times$ power output per phase

$= 242.691$ kW **Ans.**

For maximum power developed in motor,

Load angle (δ) = impedance angle

$\therefore \qquad \delta = 85.91°$

Let us take terminal voltage as reference phasor. For motor,

$$\bar{E}_{ph} = \bar{V}_{ph} - \bar{I}_a\,\bar{Z}_s \text{ or } \bar{I}_a = \frac{-\bar{E}_{ph} + \bar{V}_{ph}}{\bar{Z}_s}$$

$$= \frac{-1385.641\,\angle 85.91° + 1327.906\,\angle 0°}{21.054\,\angle 85.91}$$

$$= -65.814 + 63.071\,\angle -85.91° = 61.316 - j\,62.91$$

$$= 87.848\,\angle -45.74°$$

$\therefore$ Armature current (I_a) = 87.848 A **Ans.**

p.f. = cos 45.74° = 0.6979 (lagg) **Ans.**

Problem 5.17 : A 10 kVA, 380 V, 4-pole, 50 Hz, star-connected cylindrical rotor alternator has a stator resistance and synchronous reactance of 1 ohm and 15 ohms respectively. It supplies a load of 8 kW at rated voltage and 0.8 power factor lagging.

(*a*) Draw a phasor diagram of operation.

(*b*) Express the resistance and synchronous reactance in per unit values with the machine rating as the base.

(*c*) Calculate the percentage regulation.

(*d*) What is the terminal voltage if the load is suddenly removed (with the speed and excitation unaltered) ?

Solution :

(*a*) $r_a = 1\ \Omega,\ X_s = 15\ \Omega$

$$\overline{Z} = 1 + j15 = 15.033\angle 86.19°\ \Omega$$

Load power = 8 kW, p.f. = 0.8 lagging.

Terminal voltage = 380 V

$$\text{Load curent} \quad (I_a) = \frac{\text{load power}}{\sqrt{3} \times V_L \times p.f.}$$

$$= \frac{8 \times 1000}{\sqrt{3} \times 380 \times 0.8} = 15.193\ \text{A}$$

Now, $I_a r_a$ drop = 15.193 × 1 = 15.193 V

$I_a X_s$ drop = 15.193 × 15 = 227.895 V

Let us take terminal voltage as reference phasor

$$\overline{E}_f = \overline{V}_t + \overline{I}_a\, \overline{Z}_s$$

$$= 380\angle 0° + 15.193\angle -36.9° \times 15.033\angle 86.19°$$

$$= 380 + 228.396\angle 49.29°$$

$$= 528.967 + j\,173.129 = 556.579\angle 18.12°$$

Now, The phasor diagram can be drawn as

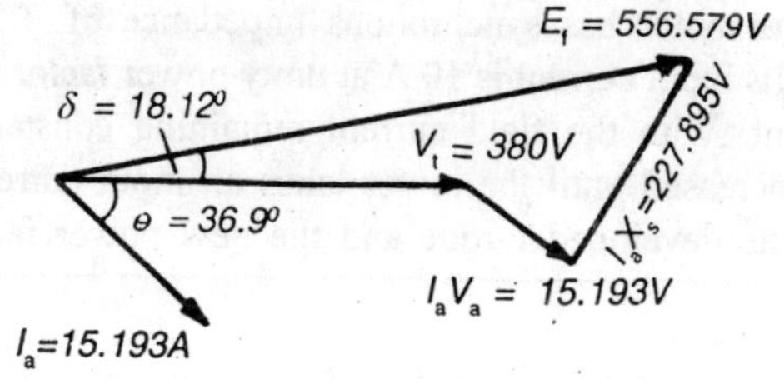

Fig. P. 5.17

(*b*) Base impedance, $Z_{base} = \dfrac{(V_t)^2}{Rating} = \dfrac{(380)^2}{10 \times 1000} = 14.44\ \Omega.$

p.u. resistance $= \dfrac{r_a}{Z_{base}} = \dfrac{1}{14.44} = 0.0639$ p.u. **Ans.**

p.u. synchronous reactance

$$= \frac{x_s}{Z_{base}} = \frac{15}{14.44} = 1.039\ \text{p.u.} \quad \textbf{Ans.}$$

(*c*) For p.f. 0.8 leading, $\cos\theta = 0.8, \sin\theta = 0.6$

$$\text{Now, } E_f = \sqrt{(V_t \cos\theta + I_a r_a)^2 + (V_t \sin\theta + I_a X_s)^2}$$

$$\text{or, } E_f = V_t\sqrt{\left(\cos\theta + \frac{I_a r_a}{V_t}\right)^2 + \left(\sin\theta + \frac{I_a X_s}{V_t}\right)^2}$$

$$r = V_t\sqrt{(\cos\theta + r_a \text{ in p.u.})^2 + (\sin\theta + X_s \text{ in p.u.})^2}$$

$$= V_t\sqrt{(0.8 + 0.0693)^2 + (0.6 + 1.039)^2} = 1.8553\,V_t$$

Now, Voltage regulation

$$= \frac{E_f - V_t}{V_t} = \frac{1.8553\,V_t - V_t}{V_t} = 0.8553 = 85.53\% \textbf{ Ans.}$$

(*d*) When the load is suddenly removed, terminal voltage is equal to excitation voltage.

We have, $E_f = 1.8553\,V_t$

For $V_t = 380 \text{ V} = E_f$ at No-load.

$\therefore$ $E_f = 1.8553 \times 380 = 705.01$ V **Ans.**

Problem 5.18 : A 3-phase, 400 V, 6-pole, 50 Hz, star-connected synchronous motor has synchronous impedance of $0.5 + j\,8\ \Omega$ per phase. Its input current is 10 A at unity power factor for a certain field current. With the field current remaining constant, the load torque is increased until the motor takes an input current of 35 A. Calculate the developed torque and the new power factor.

Solution :

Given $\bar{Z} = 0.5 + j8 = 8.0156\angle 86.42°\ \Omega$

$\alpha_Z = 90° - \theta_z = 90 - 86.42° = 3.58^c$

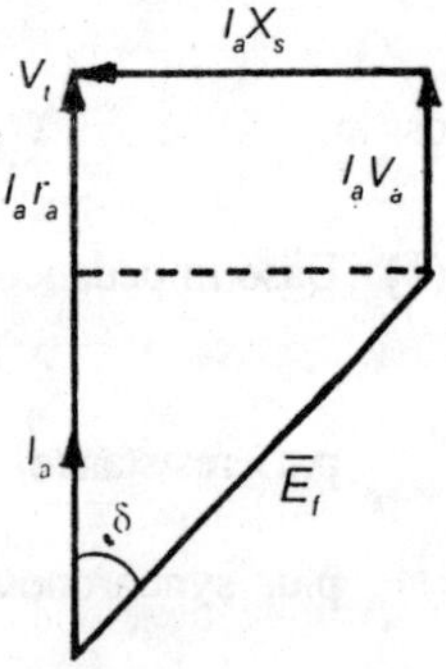

Fig. P. 5.18

Terminal voltage per phase

$$V_{ph} = \frac{V_t}{\sqrt{3}} = \frac{400}{\sqrt{3}}$$

$$= 230.94 \text{ V}$$

$$I_a = 10 \text{ A}$$

Now, Excitation emf at unity p.f. is given by

$$E_f = \sqrt{(V_t - I_a\, r_a)^2 + (I_a\, X_s)^2}$$

$$= \sqrt{(230.94 - 10 \times 0.5)^2 + (10 \times 8)^2} = 239.685 \text{ V}$$

When the load torque is increased keeping field currnet i.e. excitation constant, load angle increases. This load angle at current of 35 A is calculated by the relation

$$I_a\, Z_s = \sqrt{V_{ph}^2 + E_{ph}^2 - 2E_{ph}\, V_{ph} \cos\delta}$$

or, $$\cos\delta = \frac{V_{ph}^2 + E_{ph}^2 - I_a^2\, Z_s^2}{2\, E_{ph}\, V_{ph}}$$

$$= \frac{(230.94)^2 + (239.685)^2 - (35)^2 \times 64.25}{2 \times 230.94 \times 239.685}$$

$$= 0.2896$$

$\therefore$ $$\delta = \cos^{-1}(0.2896) = 73.17°$$

Now, power input per phase

$$= \frac{E_{ph}\, V_{ph}}{Z_s} \sin(\delta - \alpha_z) + \frac{V_t^2}{Z_s^2} \times V_a$$

$$= \frac{230.94 \times 239.68}{8.01} \sin(73.17° - 3.58°) + \frac{(230.94)^2}{64.25} \times 0.5$$

$$= 6887.151 \text{ W}$$

Total power input p_{in} = 3 × power input per phase

= 3 × 6887.15 = 20661.47 W

Total armature copper loss

$$p_c = 3\, I_a^2\, r_a = 3 \times (35)^2 \times 0.5 = 1837.5 \text{ W}$$

Gross power output

$$p_o = p_{in} - p_c = 20661.471 - 1837.5 = 18823.97 \text{ W}$$

$$\text{Speed} = \frac{120\, f}{p} = \frac{120 \times 50}{6} = 1000 \text{ rpm}$$

$$= \frac{1000 \times 2\pi}{60} \text{ rad/sec.} = 104.67 \text{ rad/sec.}$$

Gross torque developed

$$= \frac{p_o}{w_s} = \frac{18823.97 \times 60}{1000 \times 2\pi} = 179.76 \text{ Nm}$$ **Ans.**

Power input $= 20661.471$ or $\sqrt{3} \times V_L \times I_L \times \cos\phi = 20661.471$

$$\therefore \quad \text{p.f.} = \frac{20661.471}{\sqrt{3} \times 400 \times 35} = 0.852$$

Since $E_f \cos\delta < v_t$ P.f. = 0.852 (logging) **Ans.**

> **Problem 5.19 :** An alternator, with a synchronous impedance of $(0 + j1.25)$ p.u., delivers rated current to infinite busbars at pf 0.8 lagging. For the same excitation, find the current and power factor just before falling out to step.

Solution :

Given, $\bar{Z}_{pu} = 0 + j1.25 = 1.25\angle 90°$ p.u.

p.f. = 0.8 (lagging)

Let us take terminal voltage as reference phasor.

In p.u. system, $\bar{V}_t = 1.00\angle 0°$ p.u., $\bar{I}_a = 1.00\angle 0°$ p.u.

To Calculate excitation at p.f. 0.8,

$$\bar{E}_f = \bar{V}_t + \bar{I}_a\,\bar{Z}$$

$$\bar{E}_f = 1.00\angle 0° + 1\angle -36.9° \times 1.25\angle 90°$$

$$= 1\angle 0° + 1.25\angle 53.1° = 1.75 + j0.99 = 2.01\angle 29.5°$$

$$\therefore \quad (E_g) = 2.01 \text{ p.u.}$$

When the motor just falls out of step, load angle will be nearly equal to $90°, \delta \cong 90°$

Now, Let I_a be the armature current in this case.

$$\therefore \quad \bar{E}_g = \bar{V}_t + \bar{I}_a \times \bar{Z}_s$$

or $$2.01\angle 90° = 1\angle 0° + \bar{I}_a \times 1.25\angle 90°$$

$$\bar{I}_a = \frac{2.0\angle 90° - 1\angle 0°}{1.25\angle 90°} = 1.61 + j0.8 = 1.8\angle 26.42° \text{ p.u.}$$

Armature current $(I_a) = 1.80$ p.f. **Ans.**

$$\text{p.f.} = \cos 26.42° = 0.8956$$

Since, $E_g \cos\delta = 2.01 > V_t$

$\therefore$ p.f. will be leading Hence, p.f. = 0.8956 (lead) **Ans.**

Problem 5.20 : A turbo-alternator having a reactance of 10 Ω has an armature current of 200 A at unity power factor, when running on 11,000 V, constant frequency bus-bars. If the prime-mover input is not changed but the emf is raised by 20%, find the value of armature current and power factor. (*I.A.S., 1994*)

Solution :

Given, $V_t = 11000\text{ V}, X_s = 10\,\Omega,\ I_a = 200\text{ A}, \text{p.f.} = 1$

$\therefore \quad \bar{Z}_s = 10\angle 90°\,\Omega$

At unity p.f. $\quad I_a = 200\angle 0°$

Terminal voltage per phase, $V_{ph} = \dfrac{V_t}{\sqrt{3}} = \dfrac{11000}{\sqrt{3}} = 6350.853\text{ V}$

Let us take terminal voltage as reference phasor.

Now, Excitation emf per phase, $\bar{E}_{ph} = \bar{V}_{pn} + \bar{I}_a\,\bar{Z}_s$

$$\bar{E}_{ph} = 6350.853\angle 0° + (200)\times(10\angle 90°)$$
$$= 6350.853 + j2000 = 6658.328\angle 17.48°$$

Thus, Load angle $(\delta) = 17.48°$

When the excitation is increased by 20%

New excitation emf per phase

$$E'_{ph} = 1.2\times E_{ph} = 1.2\times 6658.328 = 7989.994\text{ V}$$

As the excitation is changed, load angle changes,

Since input remains constant

$$\frac{E_{ph}V_{ph}}{Z_s}\sin\delta_o = \frac{E'_{ph}\times V_{ph}}{Z_s}\sin\delta'$$

or
$$\frac{E_{ph}\times V_{ph}}{Z_s}\sin 17.48° = \frac{1.2\,E_{ph}\times V_{ph}}{Z_s}\sin\delta'$$

$$\therefore \quad \sin\delta' = \frac{\sin 17.48°}{1.2} = 0.25$$

$$\therefore \quad \delta' = \sin^{-1}(0.25) = 14.48°$$

Now, Armature current, $I'_a = \dfrac{\sqrt{E_{ph}{}^2 + V_{ph}{}^2 - 2\times E_{ph}V_{ph}\cos\delta'}}{Z_s s}$

$$= \frac{\sqrt{(7989.994)^2 + (6350.853)^2 - 2\times 6350.853\times 7989.994\cos(14.48)}}{10}$$

$= 243.12$ A **Ans.**

As here $E_f \cos\delta' = 7989.994 \cos 14.48 > V_t$

The p.f. will be lagging

Now, to calculate p.f.

$$\bar{E}'_{ph} = \bar{V}_{ph} + \bar{I}_a \bar{Z}_s$$

or
$$\bar{I}_a = \frac{\bar{E}'_{ph} - \bar{V}_{ph}}{\bar{Z}_s} = \frac{7989.994\angle 14.48° - 6350.853\angle 0°}{10\angle 90°}$$

$$= 789.99\angle -75.52° + j635.08$$

$$= 197.53 - j129.81 = 236.37\angle -33.31°$$

$$\text{p.f.} = \cos 33.31° = 0.8374 \text{ (lagging)} \quad \textbf{Ans.}$$

> **Problem 5.21 :** A 3-phase, turbo-alternator with a synchronous reactance of 10 Ω per phase and negligible armature resistance is connected to 11 kV constant voltage constant-frequency bus-bars and supplies 100 A at unity pf to the system. If the turbine power is kept constant and the excitation of the alternator is increased by 25%, what would be the new current and p.f.?
> Make appropriate assumptions. *(I.A.S., 1994)*

Solution :

Given,
$$V_t = 11\text{ kV} = 11000\text{ V}, \; X_s = 10\,\Omega$$
$$\bar{Z}_s = 10\angle 90°\,\Omega, \; \bar{I}_a = 100\angle 0° \text{ at UPF}$$

Terminal voltage per phase,

$$\bar{V}_{ph} = \frac{11\times 1000}{\sqrt{3}} = 6350.853\text{ V}$$

Let us take terminal voltage as reference phasor.

$$\bar{V}_{ph} = 6350.853\angle 0°$$

Now, Excitation emf per phase,

$$\bar{E}_{ph} = \bar{V}_{ph} + \bar{I}_a \bar{Z}_s = 6350.853\angle 0° + j1000$$

$$= 6350.853 + j1000 = 6429.101\angle 8.95°$$

Here load angle $(\delta) = 8.95$, $E_{ph} = 6429.101$ V

When the excitation is increased by 25%,
New excitation emf per phase,

$$E'_{ph} = (1+.25)\times E_{ph},\ 1.25\times 6429.101 = 8036.376 \text{ V}$$

Since input power is constant, therefore,

$$\frac{E_{ph}V_{ph}}{Z_s}\sin\delta_o = \frac{E'_{ph}\times V_{ph}}{Z_s}\sin\delta'$$

or $$\frac{E_{ph}V_{ph}}{Z_s}\sin\delta = \frac{1.25\,E_{ph}\times V_{ph}}{Z_s}\sin\delta'$$

or $$\sin\delta' = \frac{\sin 8.95^\circ}{1.25} = 0.1245$$

$$\therefore \qquad \delta' = \sin^{-1}(0.1245) = 7.15^\circ$$

Let I'_a be the armature current when the excitation is increased by 25%.

$$\therefore \qquad E'_{ph} = \bar{V}'_{ph} + \bar{I}'_a \times \bar{Z}_s$$

or
$$\bar{I}'_a = \frac{\bar{E}'_{ph} = \bar{V}_{ph}}{\bar{Z}_s} = \frac{8036.376\angle 7.15^\circ - 6350.853\angle 0^\circ}{10\angle 90^\circ}$$
$$= 803.638\angle -82.9^\circ - 635.085\angle -90^\circ$$
$$= 803.638\angle -82.9^\circ + j\,635.085$$
$$= 99.331 - j\,162.391 = 190.36\angle -58.55^\circ$$

Armature current $= 190.36$ A **Ans.**

$$\text{p.f.} = \cos 58.55^\circ = 0.5218\,(\text{lag}) \quad \textbf{Ans.}$$

Problem 5.22 : A 433-V, 3-phase, star-connected synchronous motor has a synchronous reactance of 5 Ω per phase. For a power output of 15 kW, find its minimum armature current, excitation voltage and the power angle. Armature resistance is negligible.

Solution :

Given, $V_L = 433\text{ V}, X_s = 5\ \Omega/\text{phase}$

As we know that, Power input – Armature copper loss = Power output

If armature resistance in zero, armature loss = 0

∴ Power input = Power output

or $\sqrt{3}\times V_L \times I_L \times \text{p.f.} = 15\times 1000$

or $$I_L = \frac{15 \times 1000}{\sqrt{3} \times 433 \times \text{p.f.}}$$

For fixed power output and line voltage, armature current will be minimum when p.f. is maximum, i.e. unity.

Then, $\cos\theta = 1$

and, $$\text{Armature current} = \frac{15 \times 1000}{\sqrt{3} \times 433} = 20\text{ A} \quad \textbf{Ans.}$$

Now, Terminal voltage per phase,

$$V_{ph} = \frac{V_L}{\sqrt{3}} = \frac{433}{\sqrt{3}} = 250\text{ V}$$

Let us take terminal voltage as reference phasor,

$$\therefore \quad \overline{V}_{ph} = 250\angle 0°, \overline{I}_a = 20\angle 0° \text{ at UPF}$$

For motor, Now, excitation emf per phase,

$$\overline{E}_{ph} = \overline{V}_{ph} - \overline{I}_a \overline{Z}_s$$

or $$= 250 - j\,20 \times 5$$

$$= 250 - j100 = 269.258\angle -21.8°$$

Excitation emf $\sqrt{3} \times E_{ph} = 269.258 \times \sqrt{3} = 466.369$ V **Ans.**

Load angle $(\delta) = 21.8°$ **Ans.**

Problem 5.23 : *(a)* A salient-pole synchronous generator with negligible resistance, has the following per unit parameters:

$$X_d = 0.8, \; X_q = 0.5$$

If this generator is supplying rated kVA at rated voltage and at 0.8 p.f. lagging, compute the load angle and the excitation e.m.f.

(b) Now cylindrical-rotor theory is applied to salient-pole machine of part *(a)* by assuming it to have only one reactance equal to X_d. Compute the load angle and excitation voltage by using cylindrical rotor theory and compare the results.

Solution :

(a) Given $X_d = 0.8\,\Omega, \; X_a = 0.5\,\Omega, \; r_a = 0$

Terminal voltage in p.u. = 1 p.u.

Rated armature current in p.u.= 1 p.u.

Let us take terminal voltage as reference phasor.

$$\therefore \quad \bar{V}_t = 1.00\angle 0° \text{ p.u.}$$

$$\bar{I}_a = 1.00\angle -36.9° \text{ at p.f. 0.8 lagging}$$

Now, Excitation emf,

$$\bar{E}'_f = \bar{V}_t + j\,\bar{I}_a\,\bar{X}_a$$

$$= 1.00\angle 0° + j \times 1.00 \times 0.5\left(0.8 - j0.6\right)$$

$$= 1.00 + j\,0.4 + 0.3$$

$$= 1.30 + j\,0.4 = 1.36\angle 17.1°$$

Here, $\delta = 17.1°$ and $\angle\delta + \theta_z = 17.1° + 36.9° = 54°$

Component of armature current along d-axis,

$$I_d = I_a \sin\left(\delta + \theta_z\right)$$

$$= 1.00 \sin 54° = 0.81 \text{ p.u.}$$

Component of armature current along q-axis

$$I_a = I_a \cos\left(\delta + \theta_z\right)$$

$$= 1.00 \cos 54° = 0.59 \text{ p.u.}$$

The phasor diagram can be drawn as

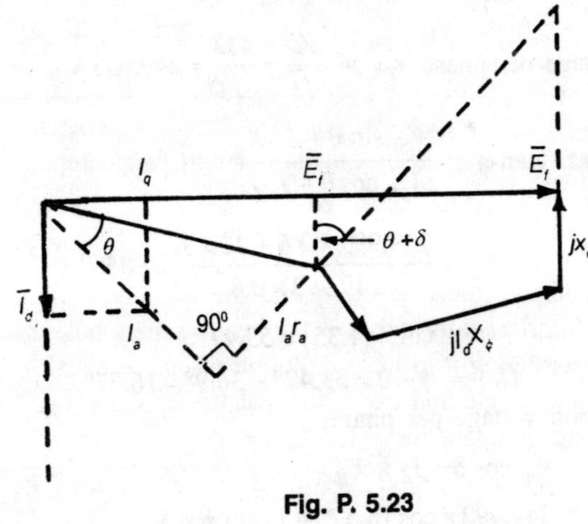

Fig. P. 5.23

Now, $E_f = E'_f + I_a\left(X_d - X_a\right)$

$$= 1.36 + 0.81(0.8 - 0.5) = 1.603 \text{ p.u.}$$

Load angle $\left(\delta\right) = 17.1°$

(b) Let E_{f1} be the excitation voltage with cylindrical rotor theory

Such that $\bar{E}_{f1} = \bar{V}_t + j\, I_a\, X_d$

$$= 1.00 \angle 0° + j \times 0.8 \times 1.00\,(0.8 - j0.6)$$

$$= 1.00 \angle 0° + j0.64 + 0.48$$

$$= 1.48 + j\,0.64 = 1.612 \angle 23.4°$$

Thus, Excitation Voltage = 1.612 p.u.

Load angle = 23.4°

> **Problem 5.24 :** A 3-phase, 30 kW, 433 volts, 50 Hz, star-connected salient-pole synchronous motor is operating at 0.8 p.f. leading and taking 40 A from the mains. If the direct and quadrature axes reactances of the machine are 5 Ω and 3 Ω respectively, calculate the maximum power the motor can develop if its excitation is maintained constant.

Solution :

Given, $V_t = 433$ V, $X_d = 5\,\Omega$, $X_q = 3\,\Omega$, $I_a = 40$

In phasor form, $\bar{I}_a = 40 \angle +36.9°$ at p.f. 0.8 leading

Terminal Voltage per phase $V_{ph} = \dfrac{V_t}{\sqrt{3}} = \dfrac{433}{\sqrt{3}} = 249.993$ V

For leading p.f. $\tan\varphi = \dfrac{V_{ph}\sin\theta + I_a X_q}{V_{ph}\cos\theta - I_a r_a}$

$$= \frac{249.993 \times 0.6 + 40 \times 3}{249.993 \times 0.8} = 1.35$$

$\therefore$ $\varphi = \tan^{-1}(1.35) = 53.47°$

Load angle $(\delta) = \varphi - \theta = 53.47° - 36.9° = 16.57°$

Now, Excitation voltage per phase,

$$E'_{ph} = V_{ph}\cos\delta + I_d \times X_d$$

$$= 249.993 \times \cos 16.57° + I_a \sin\phi \times X_d$$

$$= 249.993 \times \cos 16.57° + 40 \times 5 \times \sin 53.47° = 400.32 \text{ V}$$

Now, Power output of the motor, at any Load angle δ,

$$P_{om} = \frac{E_{ph} V_{ph}}{X_d}\sin\delta + \frac{V_{ph}^2}{2}\left(\frac{X_d - X_q}{X_d X_q}\right)\sin 2\delta$$

$$= \frac{400.32 \times 249.993}{5} \sin\delta + \frac{(249.993)^2}{2}\left(\frac{5-3}{5\times 3}\right)\sin 2\delta$$

$$= 20015.44 \sin\delta + 4166.43 \sin 2\delta$$

For maximum power output $\frac{d\,P_{om}}{d\,\delta} = 0$

or $\quad 20015.44 \cos\delta + 2\times 4166.43 \cos 2\delta = 0$

or $\quad 20015.44 \cos\delta + 8332.86\left(2\cos^2\delta - 1\right) = 0$

or $\quad 2.402 \cos\delta + 2\cos^2\delta - 1 = 0$

or $\quad \cos^2\delta + 1.201 \cos\delta - 0.3 = 0$

This yields $\quad \cos\delta = \frac{-1.201 \pm \sqrt{(1.201)^2 + 4\times 0.5}}{-\,2}$

$$= \frac{-1.201 \pm 1.8554}{2} = 0.3272 \text{ or } -1.5282$$

Since $\cos\theta \nless -1 \;\therefore\; \cos\delta_{max} = 0.3272$

or $\quad \delta_{max} = \cos^{-1}(0.3272) = 70.90°$

$\therefore$ Maximum power output per phase.

$$P_{o\max} = 20015.44 \sin 70.90° + 41.66\,43 \sin 191.80°$$
$$= 21.49 \text{ kW}$$

Total Maximum power output $= (21.49\times 3)\text{ kW} = 64.47 \text{ kW}$ **Ans.**

Problem 5.25 : A salient-pole synchronous generator has the following per unit, parameters:

$$X_d = 1.00,\; X_q = 0.60,\; r_a = 0.02$$

If this generator is delivering rated kVA at rated voltage and at 0.8 pf leading, compute the power angle and the excitation emf. Draw the phasor diagram also.

Solution :

Given, $\quad r_a = 0.02$ p.u., $X_d = 1.00$ p.u. $= X_q = 0.60$ p.u.

Terminal voltage in p.u. $= 1.00$ p.u.

rated current in p.u. $= 1.00$ p.u.

$$\bar{I}_a = 1.00\angle 36.9° \text{ at p.f. } 0.8 \text{ leading}$$

Take terminal voltage as reference phasor

$\therefore \qquad \bar{V}_t = 1.00 \angle 0°$

Now, $\qquad E'_f = \bar{V}_t + j\,\bar{I}_a\,X_q + \bar{I}_a\,r_a$

$$= 1.00 \angle 0° + j \times 0.60 \times 1.00\,(0.8 + j0.6) + (0.8 + j0.6) \times 1.00 \times 0.02$$

$$= 1.00 + j\,0.48 - 0.36 + 0.016 + j\,0.012$$

$$= 0.656 + j0.492 = 0.82 \angle 36.87°$$

Load angle $(\delta) = 36.87°$

Internal p.f. angle, $\varphi = \delta - \theta = 36.87 - 36.9° = -0.03°$

Now, $\quad I_d = I_a \sin\varphi = 1.00 \sin(-0.03°) = -0.0005$ p.u.

$\qquad\quad I_a = I_a \cos\varphi = 1.00 \cos(-0.03°) = -0.99999$ p.u.

$\therefore$ excitation voltage in p.u. $= E'_g + I_d\,(X_d - X_a)$

$$= 0.82 - 0.0005\,(1.00 - 0.60) = 0.8198 \quad \textbf{Ans.}$$

Load angle $(\delta) = 36.87°$ **Ans.**

The phasor diagram can be drawn ,as

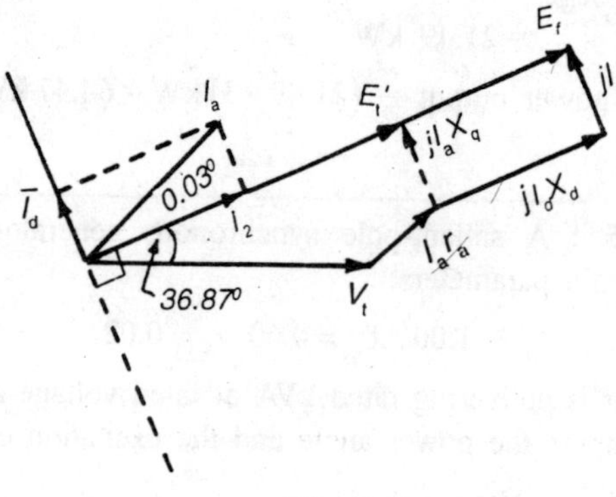

Fig. P. 5.25

Problem 5.26 : A salient-pole synchronous motor has the following per unit constants:

$$X_d = 1.25\,,\; X_q = 1.00$$

Find the excitation voltage when the motor takes rated current (leading) at rated voltage, delivering 0.5 p.u. mechanical power. Ignore all losses.

Solution :

Given, $X_d = 1.25$ p.u., $X_q = 1.00$ p.u.

Terminal voltage in p.u. = 1.00 p.u.

Rated current in p.u. = 1.00 p.u.

At any power factor, $\bar{I}_a = 1.00 \angle \cos\theta$ p.u.

Let us take terminal voltage as reference phasor.

$$\bar{V}_t = 1.00 \angle 0° \text{ p.u.}$$

We know that, Power input – losses = power output

Neglecting losses, Power input = Power output

$$V_t I_a \cos\theta = 0.5$$

$$\therefore \quad \text{p.f.} = \frac{0.5}{1 \times 1} = 0.5$$

$$\therefore \quad \theta = \cos^{-1} 0.5 = 60° \text{ (leading)}$$

Let δ be the load angle,

We have, $X_q I_q = V_t \sin\delta$, $1 \times I_q = 1 \times \sin\delta$

$$\therefore \quad I_q = \sin\delta \qquad (1)$$

Also, we have $I_q = I_a \cos(\theta + \delta) = I_a \cos\varphi$

From (1) $\sin\delta = 1 \times \cos(\theta + \delta)$

or $\cos(90° + \delta) = \cos(\theta + \delta)$

or $90° - \delta = \theta + \delta$

or $2\delta = 90° - \theta = 90° - 60° = 30°$

Load angle, $\delta = 15°$

Applying the relation, $I_d = I_a \sin\varphi = I_a \sin(\theta + \delta)$

$$= I_a \sin(60 + 15°)$$

$$= 1.00 \sin 75° = 0.9659 \text{ p.u.}$$

$$I_q = \sin 15° = 0.2588 \text{ p.u.}$$

Now, Excitation voltage for leading p.f.

$$E_f = V_t \cos\delta + I_d X_q$$

$$= 1.00 \cos 15° + 0.9659 \times 1.25$$

$$= 2.173 \text{ p.u.} \quad \textbf{Ans.}$$

Problem 5.27 : A 5 MVA, 11 kV, 50 Hz, 4 pole, star-connected synchronous generator with synchronous reactance of 0.7 p.u. is connected to an infinite bus. Find synchronizing power and the corresponding torque per unit of mechanical angle displacement
(a) At no load and
(b) At full load of 0.8 p.f. lag.

Solution :

Given, $X_s = 0.7$ p.u., $V_t = 11$ kV,

$f = 50$ Hz, $P = 4$

$$\text{Synchronous speed } (N_s) = \frac{120 f}{P} = \frac{120 \times 50}{4} = 1500 \text{ rpm}$$

$$\text{Base impedance, } Z_{base} = \frac{(V_t)^2}{\text{rating}} = \frac{\left(11 \times 10^3\right)^2}{5 \times 10^6} = \frac{121}{5} = 24.2\ \Omega$$

$\therefore$ Synchronous reactance in ohm $= X_{pu} \times Z_{base} = 0.7 \times 24.2 = 16.94\ \Omega$

(a) At No-load, $\delta = 0$, $E_f = V_t$

Now, Synchronising power per unit mechanical displacement,

$$P_{sy} = \frac{4\pi E_f V_t}{360 X_s} = \frac{2\pi \times 11 \times 11 \times 1000}{360 \times 16.94} = 249.333 \text{ **Ans.**}$$

Now, Synchronising torque,

$$T_s = \frac{P_{sy}}{\omega_s} = \frac{P_{sy}}{2\pi N_s}$$

$$= \frac{249.333 \times 1000 \times 60}{2\pi \times 1500} = 1587.303 \text{ Nm **Ans.**}$$

(b) At full load of 0.8 p.f. lagging, Rated armature current

$$I_a = \frac{\text{Rating}}{\sqrt{3} \times V_L} = \frac{5 \times 10^6}{\sqrt{3} \times 11 \times 1000} = 262.432 \text{ A}$$

In phasor form, $\bar{I}_a = 262.432\angle -36.9°$

$\bar{Z}_s = 16.94\angle 90°\ \Omega$

Terminal voltage per phase

$$V_{ph} = \frac{V_t}{\sqrt{3}} = \frac{11 \times 1000}{\sqrt{3}} = 6350.853 \text{ V}$$

Let us take terminal voltage as reference phasor.

$$\bar{V}_{ph} = 6350.853\angle 0°$$

Now, Excitation voltage per phase,

$$\bar{E}_{ph} = \bar{V}_{ph} + \bar{I}_a\,\bar{Z}_s$$

$$= 6350.853\angle 0° + 262.432\angle -36.9° \times 16.94\angle 90°$$

$$= 6350.853 + 4445.598\angle 53.1°$$

$$= 9020.08 + j3555.077 = 9695.38\angle 21.51°$$

Here $\bar{E}_{ph} = 9695.38$ V,

Load angle $(\delta) = 21.51°$

Now, Synchronising power per phase

$$= \frac{4\pi\, E_{ph}\, V_{ph} \cos\delta}{360\, X_s}$$

$$= \frac{4\pi \times 9695.38 \times 6350.853 \cos 21.51°}{360 \times 16.94} = 118.043 \text{ kW}$$

Total synchronising power,

$$P_{sy} = 3 \times \text{synchronising power per phase}$$

$$= (3 \times 118.043)\text{ kW} = 354.129 \text{ kW} \quad \textbf{Ans.}$$

Synchronising torque,

$$T_{sy} = \frac{P_{sy}}{\omega_s} = \frac{P_{sy}}{2\pi\, N}$$

$$= \frac{354.129 \times 1000 \times 60}{2\pi \times 1500} = 2254.46 \text{ Nm} \quad \textbf{Ans.}$$

Problem 5.28 : On no load, the emf of a 2.2 kV, star-connected, 3-phase, 4-pole, synchronous motor may be considered equal to and in phase with the terminal voltage. If, on the application of load, the emf induced is retarded by 1 mechanical degree, calculate the armature current and power developed by the motor, taking the synchronous reactance to be 3 Ω per phase and the stator resistance to be negligible.

(I.A.S., 1989)

Solution :

Given, $V_t = 2.2$ kV, $X_s = 3\ \Omega$

At no-load, $E_f = V_t = 2200$ V

Now, 1 mechanical degree reduced means $\frac{4 \times 1}{2} = 2$ electrical degree reduced. Now excitation voltage in the case.

$$E'_f = \frac{2\,E_f}{\sqrt{3}} \sin \frac{\Delta\delta}{2} \text{ per phase}$$

$$= 2 \times \frac{2200}{\sqrt{3}} \sin \frac{2°}{2} = 2 \times \frac{2200}{\sqrt{3}} \times \sin 1° = \frac{76.791}{\sqrt{3}} \text{ V}$$

Now, Armature current $= \frac{E'_f}{X_s} = \frac{76.791}{\sqrt{3} \times 3} = 14.7784$ A **Ans.**

$$\text{Motor power developed} = \frac{3 \times E'_f \times \frac{V_t}{\sqrt{3}} \times \cos \frac{\Delta\delta}{2}}{\text{reactance per phase}} = \frac{E'_f\, V_t \cos \frac{\Delta\delta}{2}}{\sqrt{3}}$$

$$= \frac{76.791 \times 2200 \cos 1°}{3} = 56304.82 \text{ W}$$ **Ans.**

Problem 5.29 : An industrial plant takes a load of 5 MW at p.f. 0.8 lagging from a power system network. It has been decided to replace one 500 kW induction motor of the same rating in order to raise the plant p.f. The induction motor and synchronous motor both have efficiency of 90%. The induction motor operates at 0.9 lagging but synchronous motor is designed to operate at p.f. 0.8 leading. Find the new p.f. of the plant. Calculate also percentage reduction in line current.

Solution :

Let us take terminal voltage as reference phasor.
Initial complex power of the plant.

$$A = 5000 - j500 \tan(\cos^{-1} 0.8)$$
$$= (5000 - j\,3750) \text{ kVA}$$

Given, Efficiency of the inductor motor = 0.9

or $\frac{\text{output}}{\text{input}} = 0.9$

$\therefore$ Power input to the induction motor = $\dfrac{500}{0.9}$ kW

Complex power requirement of induction motor

$$B = \frac{500}{0.9} - j\frac{500}{0.9}\tan\cos^{-1}(0.9) \qquad [\because p.f. = 0.9]$$

$$= 555.56 - j\ 2690.7 \text{ kVA}$$

Complex power requirement of synchronous motor

$$C = \frac{500}{0.9} + j\frac{500}{0.9}\tan\cos^{-1}(0.8)$$

$$= 555.56 + j\ 416.67$$

Now requirement of the plant

$$= A - B + C$$

$$= 5000 - j\ 3750 - 555.50 + j\,2690.07 + 555.50 + j\ 416.67$$

$$= 5000 - j\ 30\ 64.32 = 5864.3\angle -31.50°$$

$\therefore$ New p.f. = 0.8526 (lagging) **Ans.**

New line current $I_2 = \dfrac{5864.3}{\sqrt{3}\,V_t}$ kA

Original line current $= \dfrac{5000 \times 10^3}{\sqrt{3} \times V_t \times 0.8}$, $I_1 = \dfrac{6250}{\sqrt{3}\,V_t}$ kA

% Reduction in line current= $\left(\dfrac{I_1 - I_2}{I_1}\right) \times 100\%$

$$= \left(\frac{6250 - 5864.3}{6250}\right) \times 100\%$$

$$= 6.171\% \textbf{ Ans.}$$

Problem 5.30 : A factory has a total load of 2000 kW at a pf of 0.8 lagging. It is desired to improve the p.f. to 0.96 lagging with the installation of a synchronous condenser, then calculate

(a) The kVA rating of the synchronous condenser

(b) Total kVA of the factory

Solution :

This problem can be easily analyzed by drawing phasor diagram as below
Initial kVA load

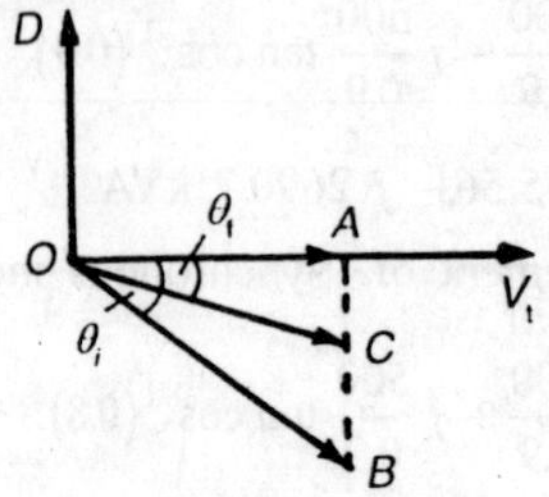

Fig. P. 5.30

$$OB = \frac{2000}{0.8} \text{ kVA} = 2500 \text{ kVA}$$

$$\text{p.f. angle} = \angle AOB = \cos^{-1} 0.8 = 36.9° = \theta_i$$

Improved power factor angle

$$\angle AOC = \cos^{-1} 0.96 = 16.26° = \theta_f$$

Initial load in kVAr = AB = $OB \sin \theta_2 = 2500 \sin 36.9°$

$= 2500 \times 0.6 = 1600$ kVAr

Combined kVAr = AC = $OA \tan \theta_f = OB \cos \theta_i \times \tan \theta_f$

$= 2500 \times 0.8 \times \tan 16.26° = 583.33$ kVAr

(a) Synchronous condenser rating

= OD = BC = AB – AC

= (1600 – 583.33) kVA = 1016.67 kVA **Ans.**

(b) Total kVA of the factory

$$= \sqrt{(OB \cos \theta_i)^2 + (AC)^2}$$

$$= \sqrt{(2000)^2 + (583.33)^2} = 2083.33 \text{ kVA} \quad \textbf{Ans.}$$

Problem 5.31 : A 150 kVA, 500 V, 3 phase, star connected alternator has the following data.

Friction and windage loss	=	350 watts
Open circuit core loss	=	500 watts
Field winding resistance at 75°C	=	200 Ω
Effective armature resistance/phase	=	0.03 Ω.

The voltage applied to field winding is 250 V. Calculate the alternator efficiency at 0.8 p.f. and at
(a) Half full load and
(b) Full load

Solution :

Rated armature current/phase $= \dfrac{150 \times 1000}{\sqrt{3} \times 500} = 173.21$ A

(a) Field loss $= \dfrac{V_f^{\,2}}{r_f} = \dfrac{(250)^2}{200} = 312.5$ W

Short-circuit load loss

$$= 3 \times \left(\frac{I_a}{2}\right)^2 \times (\text{effective armature resistance / phase})$$

$$= 3 \times \left(\frac{173.4}{2}\right)^2 \times 0.03 = 675.04 \text{ W}$$

$\therefore$ Total losses = 675.04 + 312.5 + 350 + 500
= 1837.54 W

Power output at half full load at 0.8 p.f.

$$= 150 \times 1000 \times \frac{1}{2} \times 0.8 = 60000 \text{ W}$$

$\therefore$ Efficiency of the alternator $= \left(1 - \dfrac{1837.54}{60000 + 1837.54}\right) \times 100\%$

$= 97.03\%$ **Ans.**

Short circuit load loss at full load $= 3 \times (173.21)^2 \times 0.03$
$= 2700.15$ W

$\therefore$ Total losses = 350 + 500 + 315.5 + 2700.15 = 3865.65 W

(b) Power out at full load at 0.8 p.f. $= 150 \times 0.8 \times 1000$
= 120,000 W

Efficiency $= \left(1 - \dfrac{3865.65}{123865.65}\right) \times 100\% =$ 46.88% **Ans.**

Problem 5.32 : A 50 kVA, 500 V, star connected synchronous machine has short circuit load loss of 1.6 kW at rated armature current and at 30°C. At this temperature, the d.c. resistance of the armature is 0.120 Ω per phase. Calculate the effective armature resistance in per unit and also in ohms per phase at 30°C. Find also the ac to dc ratio.

Solution :

Rated armature current $= \dfrac{50 \times 1000}{\sqrt{3} \times 500} = 57.74 \text{ A}$

Given, Short circuit load loss = 1.6 kW

Short circuit load loss in p.u. $= \dfrac{1.6}{50} \text{ p.u.} = 0.032 \text{ p.u.}$

Effective armature resistacne is given by

$$r_a \cdot \text{eff} = \frac{S_c \text{ load loss in p.u.}}{(\text{armature current in p.u.})^2} = \frac{0.032}{\left(\dfrac{57.74}{57.74}\right)^2} = 0.032 \text{ p.u.}$$

Short circuit load loss/phase $= \dfrac{1.6 \times 1000}{3} = 533.33 \text{ W} = 533.33 \text{ W}$

Effective armature resistance

$$r_a \cdot \text{eff} = \frac{533.33}{(57.74)^2} = 0.16\ \Omega / \text{phase} \quad \textbf{Ans.}$$

Ratio of ac to dc resistance $= \dfrac{0.16}{0.12} = 1.333$ **Ans.**

Problem 5.33 : A star connected 10 pole, 50 Hz synchronos motor has synchronous reactance of 10 Ω/phase. At no-load, both the excitation and the terminal voltages are equal to 4 kV. Neglect losses.

(a) For a sudden disturbance in the shaft load torque, the rotor falls back in space by two mechanical degree.Calculate the synchronizing current, power and synchronising torque tending to restore the rotor to its previous position.

(b) With gradual increase in shaft load, the rotor slides back in space by two mechanical degree. Calculate the armature current. For an armature current of 40 A. Find the angle in Mehanical degrees through which the rotor slips back.

Solution :

We have $\theta_m = \dfrac{P}{2}\theta_e$, Given $p = 10$

Two mechanical degree in space $= \dfrac{10}{2} \times 2$ electrical degree in space

$= 10$ electrical degree is space

At No-load, $\delta \cong 0, \quad V_t = E_f = 4000 \text{ V}$

(a) Synchronising voltage is given by,

$$E_s = 2\,E_f \sin\frac{\Delta\,\delta}{2} = 2 \times 4000 \times \sin\frac{10^\circ}{2} = 697.24 \text{ V}$$

Synchronising current in armature

$$= \frac{E_s}{X_s} = \frac{697.24}{10} = 69.72 \text{ A} \quad \textbf{Ans.}$$

Synchronising power

$$= 3\,V_t\,I_s \cos\left(\delta + \frac{\Delta\,\delta}{2}\right) = 3 \times 4000 \times 69.72 \cos 5^\circ$$

$$= 833.46 \text{ kW} \quad \textbf{Ans.}$$

$$\text{Synchronous speed} = \frac{120\,f}{P} = \frac{120 \times 50}{10} = 600 \text{ rpm}$$

$$= \frac{600 \times 2\,\pi}{60} = 20\,\pi \text{ rad / sec} = 62.8 \text{ rad / sec}$$

$$\therefore \text{ Synchronous torque} = \frac{833460}{20\,\pi} = 13.27 \text{ kNm} \quad \textbf{Ans.}$$

(b) When the rotor slides back by two mechanical degree, then $I_a\,X_s$ drop becomes equal to 697.2 and armature current increases from zero to 697.2 A

For armature current 40 A,

We have $$I_a\,X_s = 2\,E_f \sin\frac{\Delta\,\delta}{2}$$

or $$\sin\frac{\Delta\,\delta}{2} = \frac{40 \times 10}{2 \times 4000} = \frac{1}{20} = 0.05$$

or $$\frac{\Delta\,\delta}{2} = 2.866 \;\therefore\; \Delta\,\delta = 5.732^\circ$$

In mechanical degree

$$\Delta\,\delta = 5.732 \times \frac{1}{5} = 1.1464^\circ \text{ mechanical} \quad \textbf{Ans.}$$

Problem 5.34 : A 2000 kVA, 3-phase, star connected, 5000 V, 10 pole, 50 Hz alternator has a synchronous reactance of 10% and is running in parallel with infinite bus. Calculate the synchronising power and the corresponding synchronising torque per mechanical degree of phase displacement *(a)* at no-load, and *(b)* at full load, 0.8 power factor lagging.

Solution :

Given, $V_t = 5000$ V, $P = 10, f = 50$ Hz

Terminal Voltage/phase $= \dfrac{5000}{\sqrt{3}} = 2886.75$ V

per phase armature current $I_a = \dfrac{2000 \times 1000}{\sqrt{3} \times 5000} = 231$ A

percentage reactance $= 0.10, \dfrac{(X_s \text{ in ohm}) I_a}{V_t} = 0.10$

$$X_\delta \text{ (in ohms)} = \frac{0.10 \times 2886.75}{231} = 1.25\,\Omega$$

(a) At no load, $\delta = 0$ and $V_t = E_f$

Synchronising power/mechanical degree

$$= 3 \times \frac{dP}{d\delta} \times \frac{\pi P}{360} = 3 \times \frac{V_t \times E_f}{X_s} \cos\delta \frac{\pi P}{360}$$

$$= 3 \times \frac{(2886.75)^2}{1.25} \times \frac{\pi \times 10}{360} = 1744.44 \text{ kW } \textbf{Ans.}$$

Synchronous speed

$$= \frac{120 \times f}{P} = \frac{120 \times 50}{10} = \frac{600 \times 2\pi}{60} \text{ rad / sec} = 62.8 \text{ rad / sec}$$

The corresponding synchronising torque is given by

$$= \frac{1744440}{62.8} = 27{,}777.7 \text{ Nm } \textbf{Ans.}$$

(b) At full load p.f. = 0.8 (lag)

From phasor equation

$$E_f = \bar{V}_t + \bar{I}_a \bar{X}_s$$

$$= 2886.75 + j \times 231 \times 1.25(0.8 - j\,0.6)$$

$$= 3060 + j\,289.6 = 3073.6 \angle 5.41°$$

Now, Synchronising power

$$= 3 \times \frac{3073.6 \times 2886.75}{1.25} \times \cos 5.41° \times \frac{\pi \times 10}{360°}$$

$$= 1849.9 \text{ kW } \textbf{Ans.}$$

$$\text{Synchronising torque} = \frac{P_{sy}}{\omega_s}$$

$$= \frac{1849.9 \times 1000}{6.28} = 29{,}456.8 \text{ N - m } \textbf{Ans.}$$

Problem 5.35 : A three phase synchronous generator is delivering a power of 0.8 p.u. to an infinite bus at rated voltage and at p.f. 0.8 lagging. The generator has $X_d = 1.0$ p.u. and $X_q = 0.5$ p.u. Determine the load angle and the excitation voltage.

In case loss of excitation takes place, will the generator remain in synchronism.

Solution :

In p.u. system, Terminal voltage = 1.00 p.u.

Given, $X_d = 1.0 \text{ p.u.}, \quad X_q = 0.5 \text{ p.u.}$

Power $= 0.8 \text{ p.u.}, \; V_t \times I_a \times \cos\theta = 0.8$

or $1 \times I_a \times 0.8 = 0.8, \; I_a = 1 \text{ p.u.}$

Here $r_a = 0$

To calculate load angle, applying the relation

$$\tan(\delta + \theta) = \frac{I_a X_a + V_t \sin\theta}{V_t \cos\theta} = \frac{1 \times 0.5 + 1 \times 0.6}{1.00 \times 0.8} = \frac{1.1}{0.8} = 1.375$$

or $\delta + \theta = \tan^{-1}(1.375) = 53.97°$

$$\text{since } \theta = \cos^{-1}(0.8) = 36.9°$$

or $\delta = 53.97° - 36.9°$

Load angle $= 17.07°$ **Ans.**

Also, $I_d = I_a \sin(\delta + \theta) = 1.00 \sin(17.07 + 36.9)$

$$= 1.00 \sin(53.97°) = 0.9976 \text{ p.u.}$$

Applying the relation $E_f = V_t \cos\delta + I_d X_d$

$$= 1.00 \cos 17.07° + 0.9976 \times 1 = 1.954 \text{ p.u.}$$

Excitation $= 1.954$ p.u. **Ans.**

When loss of excitation takes place.

$E_f = 0$ and the maximum power is given by

$$P_{\max} = \frac{1}{2} V_t^2 \left(\frac{1}{X_a} - \frac{1}{X_d} \right) \sin 2\delta$$

$$= \frac{1}{2} \left(\frac{1}{0.5} - \frac{1}{1} \right) \times \sin 34.14° = \sin 34.14 = 0.5612 \text{ p.u.}$$

With the loss of excitation, the maximum power delivered to the infinite bus is less than the 0.8 p.u. Hence the generator will loss the synchronism.

> **Problem 5.36 :** A synchronous generator is running over excited with $E_f = 150$ p.u. This machine with a synchronous reactance of 1.30 p.u. is delivered a synchronous power of 0.60 p.u. to the bus. If the prime mover is increased by 2%, by how much will the synchronous power p and the reactive power Q change?

Solution :

Given, $E_f = 1.50$ p.u., $X_s = 1.30$ p.u.

$P_s = 0.60$ p.u., $V_t = 1.00$ p.u.

The synchronous power of a cylindrical-rotor synchronous machine is

given by $$P = \frac{E_f V_t}{X_s} \sin\delta$$

or $$0.60 = \frac{1.50 \times 1}{1.30} \times \sin\delta$$

or $$\sin\delta = 0.52$$

∴ $$\delta = \sin^{-1}(0.52) = 31.33°$$

Now, 2% increase in torque means 2% increase in real power.

∴ $$dP = 2\% \text{ of its previous value}$$

$$= \frac{2}{100} \times 0.6 = 0.012 \text{ p.u.} \qquad (1)$$

For cylindrical rotor machine, the reactive power is given by

$$Q = \frac{E_f V_t}{X_s} \cos\delta - \frac{V_t^2}{X_s}$$

$$\therefore \qquad \frac{dQ}{d\delta} = \frac{-E_f V_t \times \sin\delta}{X_s}$$

But
$$\frac{dP}{d\delta} = \frac{E_f V_t}{X_s}\cos\delta$$

$$\therefore \qquad \frac{dQ}{dp} = -\tan\delta = -\tan 31.33° = -0.6087$$

$$dQ = -0.6087 \times (2)\% = -1.2174\% \qquad (2)$$

From (1) and (2) shows that with 2% increases in prime-mover torque, the active power is increased by 2% while reactive power is decreased by 1.217% of its previous value.

Problem 5.37 : A three phase 440 V, star connected synchronous machine is synchronized with an infinite bus at rated voltage. The synchronous machine is now made to deliver a shaft load of 9 kW. The machine resistance is negligible and $X_d = 6\,\Omega$ per phase and $X_a = 4$ per phase. Friction, windage and core loss total 500 watts. For this shaft output. Calculate

(a) Power angle, armature current, p.f. and draw the phasor diagram

(b) Maximum power output and the corresponding power angle.

Solution :

Given,
$$X_d = 6\,\Omega / \text{phase}, \quad X_q = 4\,\Omega / \text{phase}, \quad r_a = 0$$

$$P_{wf} + P_c = 500\text{ W}, \quad V_t = 440\,V\ (L-L)$$

(a) Power developed in the synchronous motor

$$= P_{wf} + P_c + P_{sh} = 9000 + 500 = 9500\text{ W}$$

Applying the relation

$$P = 3\left[\frac{E_f V_t}{X_d}\sin\delta + \frac{1}{2}V_t^2\left(\frac{1}{X_a} - \frac{1}{X_d}\right)\sin 2\delta\right]$$

$$10{,}500 = 3\left[\left(\frac{440}{\sqrt{3}}\right)^2 \times \frac{1}{6}\sin\delta + \frac{1}{2}\left(\frac{440}{\sqrt{3}}\right)^2 \sin 2\delta \times \frac{2}{24}\right]$$

or
$$3500 = \frac{440 \times 440}{3 \times 6}[\sin\delta + 0.25\sin 2\delta]$$

or $\quad \sin\delta + 0.25\sin 2\delta = \dfrac{3500\times 2\times 6}{440\times 440} = 0.217$

or $\quad \sin\delta + 0.25\sin 2\delta = 0.217$

By hit & trial $\quad \delta = 8.4°$ **Ans.**

with $\quad E_f = V_t = \dfrac{440}{\sqrt{3}} = 254\ \text{V / phase}$

The synchromous motor would be operating at leading p.f. with $r_{a=0}$, we have

$$I_d X_d = E_f - V_t\cos\delta$$

$$I_d = \frac{E_f - V_t\cos\delta}{X_d} = \frac{254 - 254\cos 8.4°}{6} = 0.454\ \text{A}$$

and
$$I_q = \frac{V_t\sin\delta}{X_q} = \frac{254\sin 8.4°}{4} = 9.28\ \text{A}$$

$\therefore$ Armature current,

$$I_a = \sqrt{I_d^{\,2} + I_q^{\,2}} = \sqrt{(9.28)^2 + (0.454)^2} = 9.29\ \text{A}\ \textbf{Ans.}$$

Applying $\quad I_q = I_a\cos(\delta+\theta)$

or $\quad 9.28 = 9.29\cos(\delta+\theta)$

or $\quad \cos(\delta+\theta) = \dfrac{9.28}{9.28}$

or $\quad \delta+\theta = \cos^{-1}\left(\dfrac{9.28}{9.28}\right) = 2.659°$

$\therefore \quad \theta = 2.659° - 8.4° = -5.74°$

As θ turns out to be negative for a leading p.f. phasor diagram of figure shown, power factor is actually lagging

p.f. = cos (5.74°) = 0.995 (lagging) **Ans.**

The phasor diagram can be drawn as

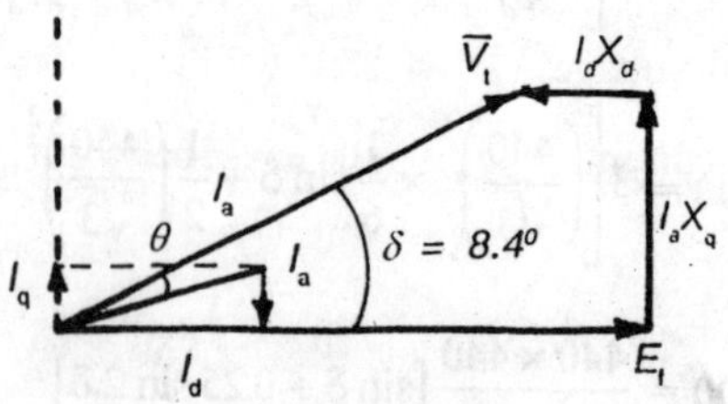

Fig. P. 5.37

(b) For maximum power

$$\cos\delta = \frac{E_f X_a}{4V_t(x_d - x_q)} \pm \sqrt{\frac{1}{2} + \left(\frac{E_f x_q}{4V_t(x_d - x_q)}\right)^2}$$

$$= -\frac{4}{4(6-4)} \pm \sqrt{\frac{1}{2} + \left(\frac{4}{4\times 2}\right)^2}$$

$$= -\frac{1}{2} \pm \sqrt{\frac{1}{2} + \frac{1}{4}} = -0.5 \pm \sqrt{\frac{2+1}{4}}$$

$$= -0.5 \pm \frac{\sqrt{3}}{2} = -0.5 \pm 0.866$$

Since $|\cos\delta| \leq 1$

$\therefore \qquad \cos\delta = 0.366$

$\therefore \qquad \delta = 68.53°$

Maximum power developed

$$= 3\times\left(\frac{400}{\sqrt{3}}\right)^2 \times \frac{1}{6}\sin 68.53° + 3\left(\frac{400}{\sqrt{3}}\right)^2 \times \frac{2}{4\times 6}\sin 137.06°$$

$$= \frac{400\times 400}{6}[\sin 68.53 + 0.5\sin 137.06] = 33807 \text{ W}$$

$\therefore$ Maximum power output= 33807 – 500 = 33.307 kW **Ans.**

Problem 5.38 : A salient-pole synchronous generator has the following p.u. parameters

$$X_d = 1.00 \text{ p.u.}, \quad X_q = 0.8 \text{ p.u.}, \quad r_a = 0.02 \text{ p.u.}$$

Calculate the excitation E_f on a per unit basis, when generator is delivering kVA load at rated voltage and a p.f. of 0.8 lagging.

Solution :

In p.u., Terminal voltage $\bar{V}_t = 1.00\angle 0°$

Rated armature current $\bar{I}_a = 1.00\angle -36.9°$, p.f. = 0.8

Rated kVA = $1.00\angle 0°$

Now, $j\,I_a\,X_q = j(0.8 - j\,0.6)\times 0.8 = 0.48 + j\,0.64$

$$\bar{I}_a r_a = 0.02\ (0.8 - j\ 0.6)$$

$$\bar{I}_a r_a = 0.016 - j\ \ 0.012$$

Applying the relation. $E'_f = \bar{V}_t + j\,\bar{I}_a\,X_q - \bar{I}_a r_a$

$$= 1 + 0.48 + j\ 0.64 + 0.016 - j\ 0.012$$

$$= 1.496 + j\ 0.532 = 1.588 \angle 19.58$$

$\therefore$ Load angle = 19.58°

$\therefore$ $\delta + \theta = 19.58° + 36.9° = 56.48°$

We have $I_d = I_a \sin(\delta + \theta) = 1.00 \sin 56.48° = 0.834$ pu

$$I_q = I_a \cos(\delta + \theta) = 1.00 \cos 56.48 = 0.552 \text{ pu}$$

To calculate excitation $E_f = E'_f + I_d\,(x_d - x_q)$

$$= 1.588 + 0.834 \times 0.2 = 1.755 \textbf{ Ans.}$$

$$\bar{E}_f = 1.755 \angle 19.58° \text{ pu} \quad \textbf{Ans.}$$

Problem 5.39 : A 25 MVA, 3-phase star-connected alternator, with an impedance of 6 Ω and a resistance of 0.4 Ω is operating in parallel with constant voltage 10 kV bus bars. If its field is adjusted to give an excitation voltage of 11 kV. Calculate

(a) The maximum power output from the alternator

(b) The armature current and power factor under maximum power condition.

Solution :

Given, $E_f = 11\,\text{kV},\ V_t = 10\,\text{kV},\ Z_s = 6\,\Omega,\ r_a = 0.4\,\Omega$

Alternator excitation/phase $= \dfrac{11000}{\sqrt{3}} = 6351$ V

Alternator terminal voltage/phase $= \dfrac{10000}{\sqrt{3}} = 5774$ V

(a) To calculate maximum power, Applying the relation

$$P_{o\max} = \frac{E_f\,V_t}{Z_s} - \frac{V_t^2}{Z_s^2} \times r_a = \frac{5774 \times 6351}{6} - \frac{(5774)^2}{36} \times 0.4$$

$$= 5741344.8 = 5.7 \text{ MW/phase}$$

$$= 17.1 \text{ MW for line } \textbf{Ans.}$$

(b) Here impedance power factor $\cos\theta_z = \frac{0.4}{6} = 0.067$

To calculate armature current

Applying the relation

$$I_a Z_s = \sqrt{E_f^2 + V_t^2 - 2E_f V_t \cos\theta_z}$$

or $$= \sqrt{(5774)^2 + (6351)^2 - 2 \times 6351 \times 5774 \times 0.067}$$

$$= 8292.2 \qquad \therefore I_a = 1382 \text{ A } \textbf{Ans.}$$

To calculate p.f. under maximum condition

Applying $$\cos\theta = \frac{E_f Z_s - V_t r_a}{I_a Z_s^2} = \frac{6351 \times 6 - 5774 \times 0.4}{1382 \times 36}$$

$$= 0.7195 \text{ (leading) } \textbf{Ans}$$

Problem 5.40 : A 2.5 kV, 3-phase, star connected cylindrical rotor synchronous motor with synchronous reactance $X_{sn} = 3\,\Omega$ is connected to a turbogenerator with $X_{sg} = 4\,\Omega$, through the transmission line of reactance $2\,\Omega$. The synchronous motor is drawing 120 A at rated terminal voltage and at unity p.f. Calculate

(i) the alternator and motor excitation voltage

(ii) the power transfer between motor and alternator

Solution :

Let us draw the equivalent circuit for the given situation

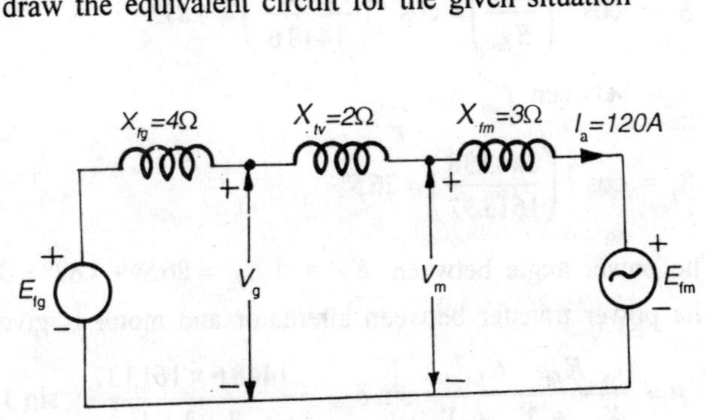

Fig. P. 5.40 (a)

Let us take terminal voltage of motor as reference phasor.

The phasor diagram can be drawn as

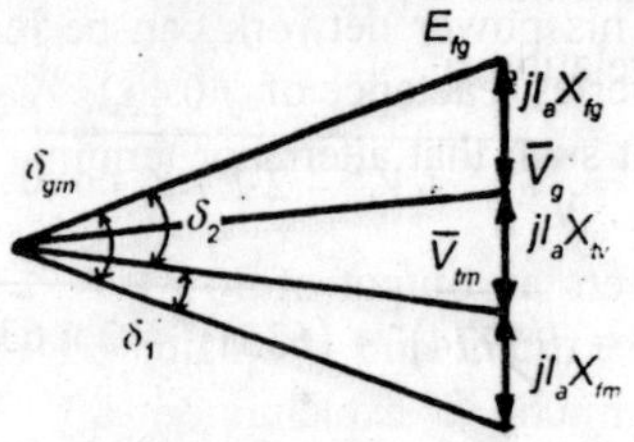

Fig. P. 5.40 (b)

$$\text{Rated motor terminal voltage/phase} = \frac{2500}{\sqrt{3}} = 1443.38 \text{ V}$$

From phasor diagram

$$\overline{E}_{fg} = \overline{V}_{tm} + j\, I_a \left(x_{tr} + x_{fg}\right) = 1443.38 + j\ 120\ (4 + 2)$$

$$= 1443.38 + j\ 720 = 1613.37 \angle 26.51$$

(i) Generator excitation voltage

$$= 1613.37 \text{ V/phase} = 2794.32 \text{ V (line to line)}$$ **Ans.**

$$\overline{E}_{fm} = \overline{V}_t - j\, I_a\, X_{fm} = 1443.38 - j\left(120 \times 3\right)$$

$$= 1443.38 - j\, 360 = 1487.6 \angle -14°$$

Motor excitation emf = 1487.6 V/phase

= 2576.6 V (line to line) **Ans.**

(ii) Angle between V_{tm} and E_g

$$\delta_2 = \cos^{-1}\left(\frac{V_{tm}}{E_{fm}}\right) = \cos^{-1}\left(\frac{1443.38}{1448.6}\right) = 4.87°$$

Angle between V_{tm} and E_{fg}

$$\delta_1 = \cos^{-1}\left(\frac{1443.38}{1613.37}\right) = 26.5°$$

The power angle between E_f and $E_{fg} = 26.5° + 4.87° = 31.37°$

The power transfer between alternator and motor is given by

$$P = \frac{E_{fg}\quad E_{fm}}{X_{fm} + X_{fg} + X_{tr}} \sin \delta_{gm} = \frac{1448.6 \times 1613.37}{2+3+4} \times \sin 31.37°$$

= 135180 = 135.18 kW/phase **Ans.**

= 405.54 kW (line) **Ans.**

Problem 5.41 : A 4-pole, star connected, 60 Hz, 10 kV, 30 MVA turbo generator with a synchronous reactance of 0.6 p.u. is connected to a power network. This power network can be represented by 10 kV infinite bus with a series reactance of $j\,0.4\,\Omega$. A voltage regulator adjusts the field current such that alternator terminal voltage remains constant at 10 kV.

The generator delivers an output of 30 MVA.

(a) Find the armature current and alternator power factor

(b) Find also the magnitude of excitation emf.

Solution :

The equivalent circuit diagram can be drawn as

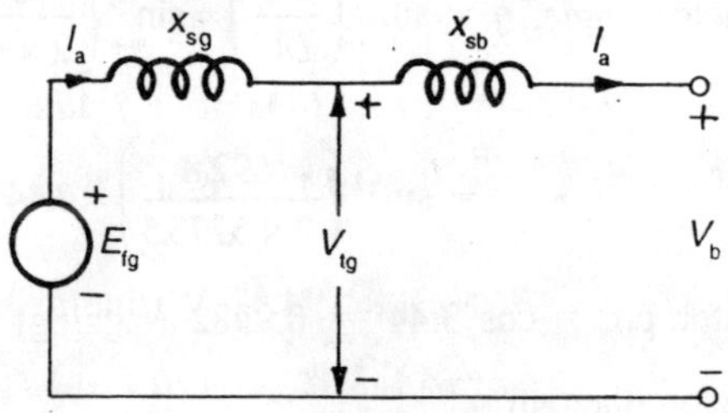

Fig. P. 5.41 (a)

Let us take generator terminal voltage as reference. The phasor diagram can be drawn as phasor.

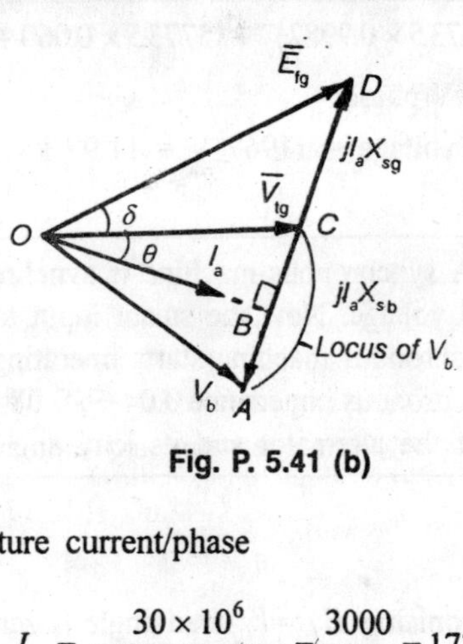

Fig. P. 5.41 (b)

(a) Rated armature current/phase

$$I_a = \frac{30 \times 10^6}{\sqrt{3} \times 10 \times 10^3} = \frac{3000}{\sqrt{3}} = 1732.1 \text{ A} \quad \textbf{Ans.}$$

$$X_s \text{ in ohms} = X_s \text{ in p.u.} \times \frac{V_t}{I_a}$$

$$= 0.6 \times \frac{10 \times 1000 \times \sqrt{3} \times 10 \times 10^3}{\sqrt{3} \times 30 \times 10^6} = 2\,\Omega$$

$$\text{Terminal voltage/phase} = \frac{10 \times 1000}{\sqrt{3}} = 5773.5 \text{ V}$$

$$\text{From phasor diagram, Voltage AC} = I_a \times X_{sh}$$

$$= 1732.1 \times 0.4 = 692.8 \text{ V}$$

$$\text{Power factor angle, } \theta = \sin^{-1}\left(\frac{BC}{DC}\right) = \sin^{-1}\left(\frac{AC}{2 \times OC}\right)$$

$$= \sin^{-1}\left(\frac{692.8}{2 \times 5773.5}\right) = 3.44°$$

$\therefore$ Alternator p.f. = cos 3.44° = 0.9982 (lagging) **Ans.**

(b) From phasor diagram

$$E_f^{\,2} = OB^2 + BD^2 = OB^2 + (BC + CD)^2$$

$$= (V_t \times \cos\theta)^2 + (V_t \sin\theta + I_a \times X_g)^2$$

$$\therefore E_f = \sqrt{(5773.5 \times 0.9982)^2 + (5773.5 \times 0.060 + 1732.1 \times 2)^2}$$

$$= 6909 \text{ V/phase}$$

line excitation voltage = 11967 V = 11.97 kV **Ans.**

Problem 5.42 : A synchronous machine is synchronized with an infinite bus at rated voltage. Now the steam input to prime mover is increased till synchronous machine starts operating at rated kVA the machine has synchronous impedance $0.04 + j\,0.08$ p.u. Determine the operating p.f. of the alternator and its load angle.

Solution :

At the time of synchronising $E_f = V_t$ load angle is zero. As the steam input is increased, E_f is pushed ahead V_t till rated kVA is delivered to the bus by the alternator. Let us take terminal voltage as reference phasor, the phasor diagram under loaded condition can be drawn as

$$\overline{V}_t = 1.00 \angle 0° \text{ p.u.}$$

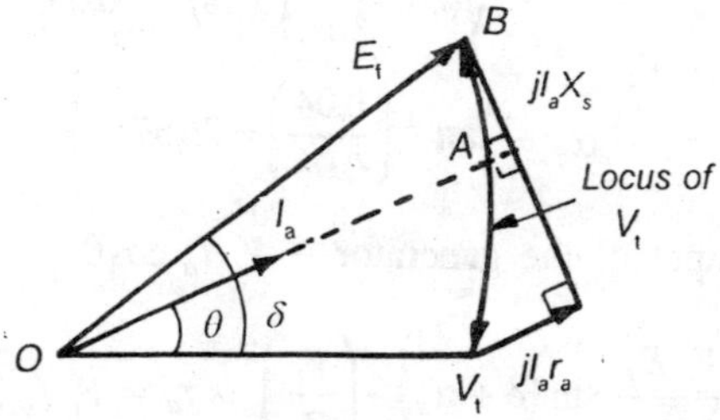

Fig. P. 5.42

From right angled triangle OAB

$$OB^2 = OA^2 + AB^2$$

or $$E_f^{\,2} = (V_t \cos\theta + I_a r_a)^2 + (-V_t \sin\theta + I_a X_s)^2$$

Since $$\overline{E}_f = \overline{V}_t = 1.00 \angle 0°$$

or $$1 = 1\left[\cos\theta + \frac{I_a r_a}{V_2}\right]^2 + \left(-\sin\theta + \frac{I_a X_s}{V_t}\right)^2$$

or $$1 = [\cos\theta + 0.04]^2 + (-1\sin\theta + 0.08)^2$$

or $$1 = \cos^2\theta + 0.08\cos\theta + 0.0016 + \sin^2\theta - 0.16\sin\theta + 0.0064$$

or $$0.08\cos\theta - 0.16\sin\theta + 0.0016 + .0064 = 0$$

or $$8\cos\theta - 0.16\cos\theta + 0.16 + .64 = 0$$

or $$16\sin\theta - 8\cos\theta = 0.80$$

or $$16\sin\theta - 8\cos\theta = 0.8$$

We know that, $A\sin\theta + B\cos\theta = \sqrt{A^2 + B^2}\,\sin\left[\theta + \tan^{-1}\frac{B}{A}\right]$

or $$\sqrt{(16)^2 + (8)^2}\,\sin\left(\theta - \tan^{-1}\frac{8}{16}\right) = 0.8$$

or $$\sin(\theta - 26.57)° = \frac{0.8}{\sqrt{16^2 + 8^2}} = \frac{0.8}{17.89}$$

or $\theta - 26.57° = 2.56°$

$\therefore$ $\theta = 26.57 + 2.56 = 29.13°$

Now, Power factor $= \cos 29.13 = 0.8735$ (leading) **Ans.**

Given, $\overline{Z}_s = 0.04 + j\,0.08$

$$Z_s = \sqrt{(0.04)^2 + (0.08)^2} = 0.089 \text{ p.u}$$

$$\alpha_z = \tan^{-1}\left(\frac{0.04}{0.08}\right) = 26.57°$$

Now, Power output of the generator $= V_t\, I_a \cos\theta$

$$\frac{V_t\, E_f}{Z_s} \sin(\delta + \alpha_z) - \left(\frac{V_t}{Z_s}\right)^2 \times r_a = V_t\, I_a \cos\theta$$

or $$\frac{1 \times 1}{0.089} \sin(\delta + 26.57°) - \left(\frac{1}{.089}\right)^2 \times 0.04 = 1 \times 1 \times .8735$$

or $$\frac{\sin(\delta + 26.57°)}{0.089} = 5.923$$

or $$\sin(\delta + 26.57°) = 0.527$$

or $$(\delta + 26.57) = \sin^{-1}(0.527) = 31.80°$$

$\therefore$ $$\delta = 31.80° - 26.57° = 5.23°$$

Thus, Load angle = 5.23° **Ans.**

Problem 5.43 : A 420 V, 50 Hz, 3-ϕ synchronous machine operates at rated voltage and at a leading p.f. of 0.8. Shaft power is 20 kW and excitation of emf is 390 V. If per phase resistance is 0.5 Ω, find the synchronous reactance. Neglect mechanical losses of the system

Solution :

Let us assume the synchronous machine being star connected.

Terminal voltage/phase $= \dfrac{420}{\sqrt{3}} = 242.5$ V

Excitation emf/phase $= \dfrac{390}{\sqrt{3}} = 225.2$ V

Given, p.f. = 0.8 (leading)

Power input $= \sqrt{3}\, V_L\, I_a \cos\theta$

$= \sqrt{3} \times 420 \times I_a \times 0.8 = 582\, I_a$

Shaft power output: P_{sh} = Power input – armature loss

$= 582\, I_a - 3 \times I_a^2 \times 0.5$

$= 582\, I_a - 1.5\, I_a^2$

Also Given, $P_{sh} = 20 \times 1000$

or $582\, I_a - 1.5\, I_a^2 = 20000$

or $I_a^2 - \dfrac{582}{1.5} I_a + \dfrac{20000}{1.5} = 0$

or $I_a^2 - 388\, I_a + 13333.3 = 0$

This yields, $I_a = \dfrac{388 \pm \sqrt{(388)^2 - 4 \times 13333.3}}{2} = \dfrac{388 \pm 311.8}{2}$

Neglecting higher value, $I_a = 38.11$ A

Now, from phasor diagram

$$E_f^2 = (V_t \cos\theta + I_a r_a)^2 + (V_t \sin\theta - I_a X_s)^2$$

or $(V_t \sin\theta - I_a X_s)^2 = E_f^2 - (V_t \cos\theta + I_a r_a)^2$

or $V_t \sin\theta - I_a X_s = \sqrt{E_f^2 - (V_t \cos\theta + I_a r_a)^2}$

or $I_a X_s = \left[V_t \sin\theta + \sqrt{E_f^2 - (V_t \cos\theta + I_a r_a)^2}\right]$

$\therefore$ $$X_s = \frac{1}{I_a}\left[V_t \sin\theta + \sqrt{E_f^2 - (V_t \cos\theta + I_a r_a)^2}\right]$$

$$= \frac{1}{38.11}\left[242.5 \times 0.6 + \sqrt{(225.2)^2 - (242.5 \times 0.8 + 38.11 \times 0.5)^2}\right]$$

$= 5.73\ \ \Omega$/phase

Thus, Synchronous reactance = 5.73 Ω/phase **Ans.**

Problem 5.44 : A 3-ϕ, 420 V, 8 pole, 60 Hz, star connected synchronous motor has an emf 540 V (L – L). The stator winding has an synchronous reactance of 4 Ω/phase, and the motor develops a torque of 240 Nm. The motor is operating off 420 V, 60 Hz bus. Calculate the current drawn from the supply and its power factor. Also draw the appropriate phasor.

Solution :

Given, $X_s = 4\ \Omega/\text{phase}$, $E_f = 540$ V,

$V_t = 420$ V, $P = 8$,

$f = 60$ Hz, Torque = 240 Nm

Synchronous speed, $W_s = \dfrac{120 \times f}{p} = \dfrac{120 \times 50}{8} = 750$ rpm

$$= \frac{750 \times 2\pi}{60} = 25\pi \text{ rad/s}$$

Excitation voltage/phase $= \dfrac{540}{\sqrt{3}}$ V, Terminal voltage/phase $= \dfrac{420}{\sqrt{3}}$ V

Since Torque =240 Nm or $\dfrac{3}{\omega_s} \cdot \dfrac{E_f V_t}{X_s} \sin\delta = 240$

or $$\frac{3}{25\pi} \times \frac{540}{\sqrt{3}} \times \frac{420}{\sqrt{3}} \times \frac{1}{4} \sin\delta = 240$$

or $$\sin\delta = \frac{4 \times 240 \times 25 \times \pi}{540 \times 420} = 0.3323$$

$\therefore$ Load angle $= \delta = \sin^{-1}(0.3323) = 19.41°$

Let us take terminal voltage as reference phasor. The phasor diagram can be drawn as

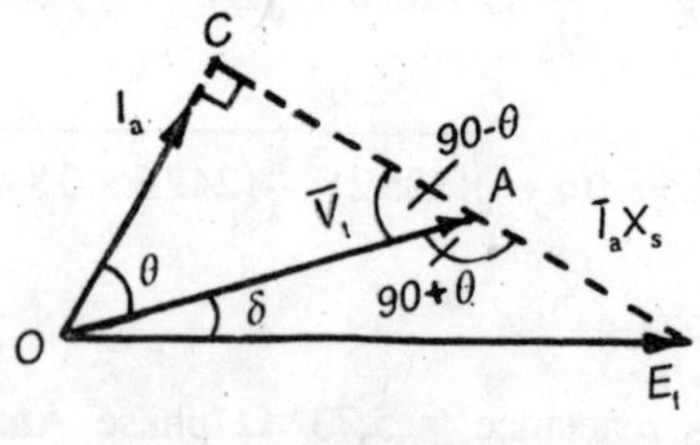

Fig. P 5.44

From phasor diagram using Caschie law,

$$(I_a X_s)^2 = E_f^2 + V_t^2 - 2 E_f V_t \cos\delta$$

or $$I_a X_s = \sqrt{E_f^2 + V_t^2 - 2 E_f V_t \cos\delta}$$

$\therefore$ $$I_a = \frac{\sqrt{E_f^2 + V_t^2 - 2 E_f V_t \cos\delta}}{X_s}$$

$$= \frac{\sqrt{(311.8)^2 + (242.5)^2 - 2 \times 242.5 \times 311.8 \times \cos 19.41°}}{4}$$

$$= 28.94 \text{ A}$$

Thus, Current drawn from supply = 28.94 A **Ans.**

Now, Power input $= \sqrt{3} V_t I_a \cos\theta = \frac{E_f V_t}{X_s} \sin' \delta$

or $$\cos\theta = \frac{E_f}{\sqrt{3} I_a X_s} \sin\delta$$

or $$\cos\theta = \frac{311.8 \times \sin 19.41°}{\sqrt{3} \times 28.94 \times 4} = 0.517 \text{ (leading)}$$

p.f. = 0.517 (leading) **Ans.**

> **Problem 5.45 :** A number of alternators are working in parallel with their terminal voltage equal to the rated value. One of the machine which has synchronous reactance of 40% and a resistance of 1% delivers a power input in kW equal to 60% of its rated kVA. If the emf of this unit equal to 1.1 times the terminal voltage. Find out the power factor at which machine is operating.

Solution :

Given, $$\frac{I_{rated}\, r_a}{V_t} = 1\% = 0.01 \text{ p.u}$$

$$\frac{I_{rated}\, X_s}{V_t} = 40\% = 0.40 \text{ p.u}$$

$$E_f / \text{phase} = 1.4 \times V_t / \text{phase}$$

Also given, $$\sqrt{3} V_t \times I_a = \sqrt{3} \times V_t \times I_{rated} \times 0.6$$

$\therefore$ $$I_a = 0.6 \times I_{rated}$$

Let θ be the operating p.f.

Now, Applying the relation

$$E_f^2 = (V_t \cos\theta + I_a r_a)^2 + (V_t \sin\theta + I_a X_s)^2$$

or
$$E_f^2 = V_t^2 \left(\cos\theta + \frac{I_a r_a}{V_t}\right)^2 + \left(\sin\theta + \frac{I_a X_s}{V_t}\right)^2$$

or
$$(1 - 1V_t)^2 = V_t^2 \left[\left(\cos\theta + \frac{0.6 \times I_{rated} \times X_s}{V_t}\right)^2 \right.$$
$$\left. \left(\sin\theta + \frac{0.6 \times I_{rated} \times X_s}{V_t}\right)^2\right]$$

or
$$1.21 = \left[(\cos\theta + 0.6 \times 0.1)^2 + (\sin\theta + 0.6 \times 0.40)^2\right]$$

or
$$\cos^2\theta + 0.012\cos\theta + 0.000036 + \sin^2\theta + 0.48\sin\theta + 0.0576 = 1.21$$

or
$$0.012\cos\theta + 0.000036 + 0.48\sin\theta + 0.0576 = 1.21 - 1$$

or
$$12\cos\theta + 0.036 + 57.6 + 480\sin\theta = 210$$

or
$$12\cos\theta + 480\sin\theta + 57.636 = 210$$

or
$$12\cos\theta + 480\sin\theta = 210 - 57.64 = 153.36$$

or
$$\cos\theta + 40\sin\theta = 12.7$$

Applying the formula

$$A\sin\theta + B\cos\theta = \sqrt{A^2 + B^2}\,\sin\left(\theta + \tan^{-1}(B/A)\right)$$

or
$$\sqrt{(1)^2 + (40)^2}\,\sin\left(\theta + \tan^{-1}\frac{1}{40}\right) = 12.7$$

or
$$\sin(\theta + 1.43°) = 0.3174$$

∴
$$\theta = \sin^{-1}(0.3174) - 1.43° = 17.08°$$

Power factor of the machine $= \cos 17.08°$

$= 0.9559$ (lagging) **Ans.**

> **Problem 5.46 :** A 4.4 kV star-connected 3-phase synchronous motor works at constant voltage and constant excitation. Its synchronous impedance is $1.2 + j\,10\,\Omega$ / phase. When the input power is 800 kW, the p.f. is 0.8 leading. Find the p.f. when input is increased to 900 kW.

Solution :

Given, $Z_s = 1.2 + j\,10\,\Omega/\text{phase}, \quad \alpha_z = 6.84°,$

$|Z_s| = 10.07\,\Omega$

$$\text{Armature current at 800 kW} = \frac{800 \times 1000}{\sqrt{3} \times 4400 \times .8} = 131\text{ A}$$

Terminal voltage $V_t = 4.4$ kV

$$\text{Terminal voltage per phase} = \frac{4400}{\sqrt{3}} = 2540\text{ V}, \quad \text{p.f.} = 0.8 \text{ (leading)}$$

For leading p.f. the excitation emf can be calculated as

$$E_f = \sqrt{(V_t \cos\theta - I_a r_a)^2 + (V_t \sin\theta + I_a X_s)^2}$$

$$E_f = \sqrt{(2540 \times 0.8 - 131 \times 1.2)^2 + (2540 \times 0.6 + 131 \times 10)^2} = 3398\text{ V}$$

When the input power is increased, load angle must be increased. The load angle at power input 900 kW can be calculated by using the formula

$$P_{in} = \frac{E_f V_t}{Z_s} \sin(\delta - \alpha_z) + \left(\frac{V_t}{Z_s}\right)^2 \times r_a$$

or
$$\frac{900 \times 10^3}{3} = \frac{3398 \times 2540}{10.07} \sin(\delta - \alpha_z) + \left(\frac{2540}{10.07}\right)^2 \times 1.2$$

or
$$300000 - 76346.6 = \frac{3398 \times 2540}{10.07} \sin(\delta - \alpha_z)$$

or $\sin(\delta - \alpha_z) = 0.261, \quad \delta - \alpha_z = \sin^{-1} 0.261$

$\therefore \quad \delta = 21.97°$

The current corresponding to this load angle can be calculated by

$$I_a = \frac{\sqrt{E_f^2 + V_t^2 - 2\,E_f\,V_t \cos\delta}}{Z_s}$$

$$= \frac{\sqrt{(3398)^2 + (2540)^2 - 2 \times 2540 \times 3398 \times \cos 21.97}}{10.07} = 140.1\text{ A}$$

$$\text{Now, Power input/phase} = \frac{900 \times 10^3}{3} = 300 \times 10^3$$

or $2540 \times 140.\text{I} \times \text{p.f.} = 300 \times 10^3$

$\therefore$ p.f. = 0.8430

Now, $E_f \cos\delta = 3398 \cos 21.97° = 3151 > V_t$

Hence, p.f. = 0.8430 (leading) **Ans.**

Problem 5.47 : A 4400 V, star-connected synchronous motor has synchronous impedance $0.5 + 9\,\Omega$ per phase. For an excitation emf of 5000 V and motor input power of 1320 kW at rated voltage, compute the line current and p.f.

Solution :

Given, $Z_s = 0.5 + 9\,\Omega$, $|Z_s| = 9.01\,\Omega$, $\alpha_z = 3.18°$

Terminal voltage = 4400 V

$$\text{Terminal voltage } (V_t)\text{/phase} = \frac{4400}{\sqrt{3}} = 2540 \text{ V}$$

$$\text{Excitation emf/phase} = \frac{5000}{\sqrt{3}} = 2887 \text{ V}$$

Power input = 1320 kW

$\therefore$ Power input per phase = 440 kW

or
$$\frac{E_f V_t}{Z_s}\sin(\delta - \alpha_z) + \left(\frac{V_t}{Z_s}\right)^2 \times r_a = 440 \times 1000$$

or
$$\frac{2887 \times 2540}{9.01}\sin(\delta - \alpha_z) + \left(\frac{2540}{9.01}\right)^2 \times 0.5 = 440000$$

or $\sin(\delta - \alpha_z) = 0.495$

$\therefore$ $\delta = \sin^{-1}(0.495) + 3.18°$

Load angle $\delta = 32.85°$.

To calculate armature current at load angle 32.85°. Applying,

$$I_a = \frac{\sqrt{E_f^2 + V_t^2 - 2\,E_f\,V_t \cos\delta}}{Z_s}$$

$$= \frac{\sqrt{(2540)^2 + (2887)^2 - 2 \times 2540 \times 2887 \times \cos 32.85°}}{9.01} = 174.3 \text{ A}$$

$\therefore$ Line current = 174.3 A **Ans.**

Now, Power input/phase = 440×1000

or $\quad V_t \times I_a \times p.f = 440 \times 1000$

$\therefore \quad \text{p.f.} = \dfrac{440 \times 1000}{174.3 \times 2540} = 0.9939$

Since, $\quad E_f \cos\delta = 2887 \cos 32.85 = 2425.4 < V_t$

Hence, $\quad$ p.f. = 0.9939 (lagging) **Ans.**

> **Problem 5.48 :** A 240 V, 6 pole, 60 Hz, star connected synchronous motor has synchronous impedance $0.4 + j\,2\,\Omega$/phase. Its field current is so adjusted that motor draws 10 A at a p.f. from rated voltage source. Now with the field current unchanged, the load on the motor is increased till it draws 30 A from the supply. Find the torque developed and new p.f.

Solution :

Given, $\quad P = 6, \qquad f = 60\text{ Hz},$

$$Z_s = 0.4 + j\,2\,\Omega/\text{phase}$$

$$\alpha_z = 11.31°. \qquad |Z_s| = 2.04\,\Omega/\text{phase}$$

$$I_{a1} = 10\text{ A}, \qquad I_{a2} = 30\text{ A}$$

Terminal voltage = 240 V

$$\text{Terminal voltage/phase} = \frac{240}{\sqrt{3}} = 138.6\text{ V}$$

The excitation at unity p.f. is given by

$$E_f = \sqrt{(V_t - I_{a1}\, r_a)^2 + (I_{a1}\, X_s)^2}$$

$$= \sqrt{(138.6 - 10 \times 0.4)^2 + (2 \times 10)^2} = 136.1\text{ V}$$

When load torque is increased keeping I_f constant i.e. E_f constant, load angle increases and the motor beings operating at a lagging p.f. with $I_{a2} = 30$ A

The load angle corresponding to $I_{e2} = 30$ A can be calculated by applying

$$I_{a2}\, Z_s = \sqrt{E_f^2 + V_t^2 - 2\, E_f\, V_t \cos\delta}$$

or $$\cos\delta = \frac{E_f^2 + V_t^2 - I_{a2}^2\, Z_s^2}{2\, E_f\, V_t}$$

$$= \frac{(136.1)^2 + (138.6)^2 - (30)^2 \times (2.04)^2}{2 \times 136.1 \times 138.6} = 0.901$$

$$\therefore \qquad \delta = \cos^{-1}(0.901) = 25.71°$$

Load angle $= 25.71°$

Per Phase Power input to the motor

$$= \frac{E_f V_t}{Z_s} \sin(\delta - \alpha_z) + \left(\frac{V_t}{Z_s}\right)^2 \times r_a$$

or, $$V_t \times I_{a2} \times \text{p.f.} = \frac{E_f V_t}{Z_s} \sin(\delta - \alpha_z) + \left(\frac{V_t}{Z_s}\right)^2 \times r_a$$

$$\therefore \ \text{p.f.} = \frac{1}{30 \times 138.6}\left[\frac{138.6 \times 136.1}{2.04} \sin(25.71 - 11.31°) + \left(\frac{138.6}{2.04}\right)^2 \times 0.4\right]$$

New p.f. = 0.9971 (lagging) **Ans.**

Power input per phase of the motor (P) = 4146 W

Gross power developed $= 3\left(P - I_{a2}^2 \, r_a\right)$

$$= 3\left(4146 - (30)^2 \times 0.4\right) = 11358 \text{ W}$$

Hence, Synchronous speed $= \dfrac{120 \times f}{P} = \dfrac{120 \times 60}{6}$

$$= \frac{1200 \times 2}{60} = 40\,\pi \text{ rad/sec}$$

Torque developed $= \dfrac{11358}{40 \times 3.14} = 90.42$ Nm **Ans.**

Problem 5.49 : A 440 V, 3-phase delta connected synchronous motor runs at rated voltage and with an excitation of 550 V. The synchronous impedance/phase is $0.5 + j\,6\,\Omega$ and friction windage and iron losses are 1 kW. Calculate the shaft power output, line current p.f. and efficiency for

(a) Maximum power output, and

(b) Maximum power input.

Solution :

Given, $P_{wf} + P_c = 1 \text{ kW} = 1000 \text{ W}$

$$V_t = 440 \text{ V}, \quad E_f = 550 \text{ V}$$

$$Z_s = 0.5 + j\,6\,\Omega, \quad \theta_z = 85.24°$$

$$|Z_s| = 6.02\,\Omega/\text{phase}$$

As we know that in Δ connection, phase voltage and line voltage are same.

(a) For maximum power output,

load angle = impedance angle = 85.24°

Maximum power output

$$P_{o\,(\max)} = \frac{E_f V_t}{Z_s} - \frac{E_f^2}{Z_s^2} \times r_a = \frac{550 \times 440}{6.02} - \left(\frac{550}{6.02}\right)^2 \times 0.5$$

$$= 36025.8 \text{ W}$$

$$\text{Shaft power output} = P_{o\max} - (P_{wf} + P_c) = 36025.8 \times 3 - 1000$$

$$= 107.1 \text{ kW} \quad \textbf{Ans.}$$

$$\therefore \quad \text{p.f.} = \cos 36.78° = 0.801 \text{ (lagging)} \quad \textbf{Ans.}$$

$$\text{Efficiency} = \left(\frac{\text{P}_{\text{output}}}{\text{P}_{\text{input}}}\right) \times 100\%$$

$$= \left(\frac{105415.6}{42870 \times 3}\right) \times 100\% = 81.97\% \quad \textbf{Ans.}$$

The armature current can be calculated by using the formula

$$I_a = \frac{\sqrt{E_f^2 + V_t^2 - 2\,E_g V_t \cos\theta_z}}{Z_s}$$

$$= \frac{\sqrt{(440)^2 + (550)^2 - 2 \times 550 \times 440 \times \cos 85.24°}}{6.82}$$

$$\text{Armature current} = 112.16 \text{ A}$$

$$\text{Line current} = 112.16 \times \sqrt{3} = 194.27 \text{ A} \quad \textbf{Ans.}$$

To calculate p.f., Applying the relation

$$\cos(\theta_z - \theta) = \frac{V_t Z_s - E_f r_a}{I_a Z_s^2}$$

$$= \frac{440 \times 6.02 - 550 \times 0.5}{112.16 \times 36.25} = 0.584$$

or $$\theta_z - \theta = \cos^{-1}(0.584)$$

$$\therefore \quad \theta = \theta_z - \cos^{-1}(0.584), \; \theta = 30.97°$$

$$\text{P.f.} = \cos 30.97 = 0.8574 \text{ (lagging)} \quad \textbf{Ans.}$$

$$\text{Power input} = V_t\, I_a \cos\varphi = 440 \times 112.16 \times 0.8574 = 42313 \text{ W}$$

$$\text{Efficiency} = \frac{\text{output}}{\text{input}} = \frac{107100}{42313 \times 3} = 84.35\% \quad \textbf{Ans.}$$

(b) For maximum power input to motor,

$$P_{in(\max)} = \frac{E_f V_t}{Z_s} + \left(\frac{V_t}{Z_s}\right)^2 \times r_a$$

$$= \frac{550 \times 440}{6.02} + \left(\frac{440}{6.02}\right)^2 \times 0.5 = 42870 \text{ W}$$

For maximum power input to motor, $\delta = 180 - \theta_z = 94.76°$

Armature current,

$$I_a = \frac{\sqrt{V_t^2 + E_f^2 - 2V_t E_f \cos\delta}}{Z_s}$$

$$= \frac{\sqrt{(440)^2 + (550)^2 - 2 \times 440 \times 550 \times \cos 94.76°}}{6.02} = 121.64 \text{ A}$$

Line current $= 121.64 \times \sqrt{3} = 210.7$ A **Ans.**

Shaft power output

$$= \text{Maximum input power} - 3 \times I_a^2\, r_a - (P_{wf} + P_c)$$

$$= 42870 \times 3 - 3 \times (121.64)^2 \times 0.5 - 1000$$

$$= 105415.6 \text{ W} \quad \textbf{Ans.}$$

To calculate p.f.,

$$\cos(\theta_z - \theta) = \frac{V_t + E_f \cos\theta_z}{I_a Z_s} = \frac{440 + 550 \cos 85.24}{121.64 \times 6.02}$$

$$= 0.6632$$

$$\therefore \theta = \theta_z - \cos^{-1} 0.6632 = 85.24° - \cos^{-1} 0.6632 = 36.78°$$

$$\therefore \quad \text{p.f.} = \cos 36.78° = 0.801 \text{ (lagging)}$$

$$\text{Efficiency} = \left(\frac{P_{output}}{P_{input}}\right) \times 100\% = 81.97\% \quad \textbf{Ans.}$$

Problem 5.50 : Two generators rated 500 MW and 600 MW are operating in parallel. The drop characteristics of their governers are 4% and 5% respectively from no load to full load. Assuming that the generators are operating at 60 Hz at no load, how would a load of 11000 MW be shared between them? What would be the system frequency at this load? Assuming linear governer operation.

Soluion :

The frequency-load characteristics of two generators can be drawn as below.

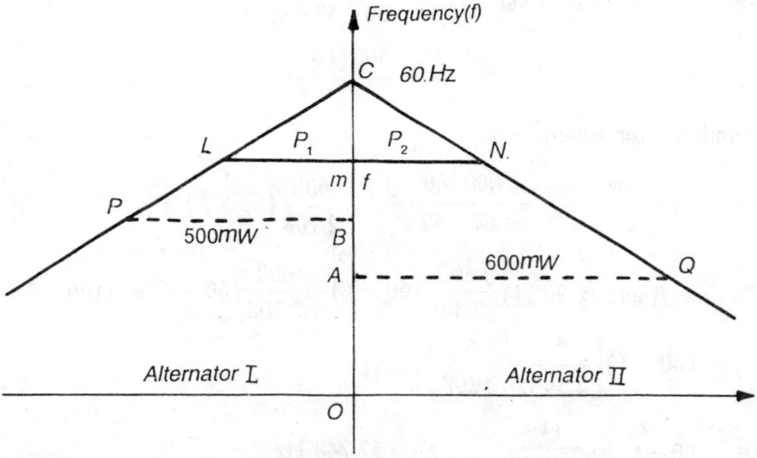

Fig. P. 5.50

Let f be the system frequency at load 1100 MW.

In figure, OA is the full-load of Alternator II.

OB is the full-load of Alternator I.

OC is the no-load of both alternators.

We know that, Speed regulation = $\dfrac{\text{No-load speed} - \text{full-load speed}}{\text{full-load speed}}$

For alternator I of 500 MW.

$$\frac{OC - OB}{OB} = 4\% \text{ or } \frac{60 - OB}{OB} = 0.04$$

or $$\frac{60}{OB} = 1.04$$

$$\therefore \quad OB = \frac{60}{1.04} = 57.692 \text{ Hz}$$

Similarly, for alternator II of 600 MW,

$$OA = \frac{60}{1.05} = 57.142 \text{ Hz}$$

Let P_1 be the load on machine I and P_2 be the load on machine II. So that, $P_1 + P_2 = 1100$ MW (1)

From similar triangles CLM and CPB,

Putting P_1 and P_2 in (1) $\frac{P_1}{500} = \frac{60-f}{60-57.692}$

$$\therefore \quad P_1 = \frac{500\,(60-f)}{2.308} \tag{2}$$

Similarly, for generator II,

$$P_2 = \frac{600\,(60-f)}{60-57.292} = \frac{600}{2.708}(60-f) \tag{3}$$

Putting P_1 and P_2 in (1) $\frac{500}{2.308}(60-f) + \frac{600}{2.208}(60-f) = 1100$

or $(60-f)\left[\frac{5}{2.308} + \frac{6}{2.208}\right] = 11$

or $60 - f = \frac{11}{4.884}$, $\therefore$ $f = 57.748$ Hz

System frequency = 57.748 Hz **Ans.**

Now, Load shared by alternator 500 MW

$$= \frac{500}{2.308}(60-57.748) = 487.87 \text{ MW } \textbf{Ans.}$$

Load shared by alternator 600 MW

$$= \frac{600}{2.208}(60-57.748) = 611.96 \text{ MW } \textbf{Ans.}$$

Problem 5.51 : Two identical synchronous machines of rating 200 MW, 50 Hz operating in parallel have the following characteristics.

Machine I speed drop, $R = 5\%$

Speed changer set to give 60% of rated load at rated speed.

Machine II speed drop, $R = 5\%$

Speed changer set to give 80% of rated load at rated speed.

(*a*) Determine the load taken by each machine for total load of 300 MW and frequency of operation.

(*b*) What adjustments should be made by the speed changers of the machines to share the loads as in *(a)* but with a frequency of 50 Hz.

Solution :

(*a*) Let f be the frequency of operation of the system and P_1 and P_2 be the load shared by machine I and II respectively.

$\therefore$ P_1 and P_2 = 300 MW

Also at 50 Hz (1)

$$\text{Speed changer output of machine I} = \frac{60 \times 200}{100} = 120 \text{ MW}$$

$$\text{Speed changer output of machine II} = \frac{80}{100} \times 200 = 160 \text{ MW}$$

No-load speed frequency = 100 + 5 = 105%

The frequency-load characteristics of machines can be drawn as

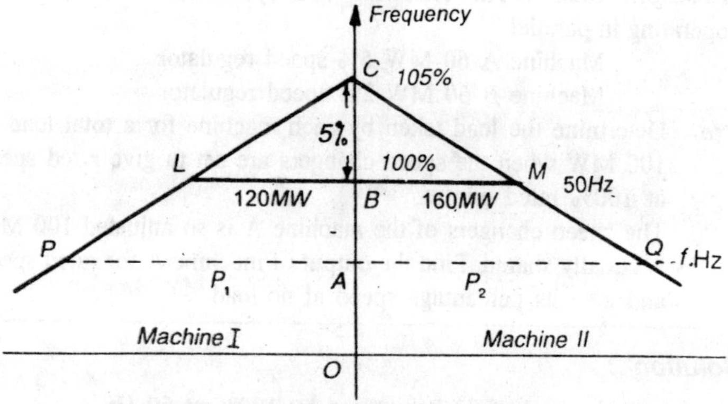

Fig. P. 5.51

For machine I, from similar triangles CLB and CPA,

$$\frac{120}{P_1} = \frac{5}{105 - f} \quad \therefore \; P_1 = 24(105 - f) \tag{3}$$

$$\text{Similarly, for machine II, } P_2 = \frac{160}{5}(105 - f)$$

$$P_2 = 32(105 - f) \tag{4}$$

Putting P_1 and P_2 in (1) $24(105-f)+32(105-f)=300$

or $\quad 24\times105-24f+32\times105-32f \;=\; 300$

or $\quad 56\times105-300 = 56\,f,\ f = 99.64\%$

Frequency of operation of system = $.9964\times50 = 49.82$ Hz **Ans.**

$\therefore$ Load shared by machine

$$\text{I} = 24(105-91.64) = 128.64\ \text{MW} \quad \textbf{Ans.}$$

$\therefore$ Load shared by machine

$$\text{II} = 32(105-99.64) = 171.52\ \text{MW} \quad \textbf{Ans.}$$

(b) For machine I, speed drop = 5%

Speed changer should be changed to give $\left(\dfrac{128.64}{2}\right)\%$

i.e. 64.32% rated load at rated speed

For machine II, speed drop = 5%

Speed changer should be changed to give $\left(\dfrac{171.52}{2}\right)\%$

i.e. 85.76% rated load at rated speed

> **Problem 5.52 :** The following two synchronous machines are operating in parallel.
>
> Machine A 60 MW 5% speed regulator
>
> Machine B 60 MW 2% speed regulator
>
> *(a)* Determine the load taken by each machine for a total load of 100 MW when the speed changers are set to give rated speed at 100% rated output.
>
> *(b)* The speed changers of the machine A is so adjusted 100 MW is equally shared. Find the output of machine A for rated speed and also its percentage speed at no-load.

Solution :

Let the rated speed of each machine by 100% at 50 Hz.

For machine A, speed regulation = 5%

$\therefore$ no-load speed = 105%

For machine B, speed regulation = 2%

$\therefore$ no-load speed = 102%

Let f be the new frequency and P_1 and P_2 be the load shared by machines A and B respectively such that

$$P_1 + P_2 = 100\ \text{MW} \tag{1}$$

The frequency-load characteristics can be drawn as

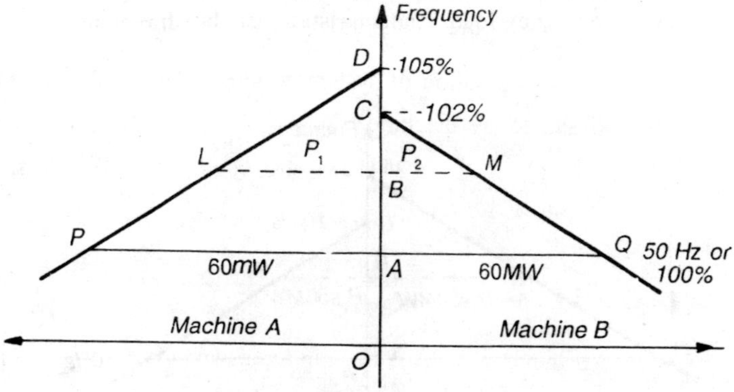

Fig. P. 5.52 (a)

(a) For machine A, from similar triangles DLB and DPA.

$$\frac{P_1}{60} = \frac{(105-f)}{50}$$

$$\therefore \qquad P_1 = \frac{60}{5}(105-f) = 12(105-f) \qquad (2)$$

Similarly, for machine B,

$$P_2 = \frac{60}{2}(102-f) = 30(102-f) \qquad (3)$$

Applying (2) and (3)

$$P_1 + P_2 = 12 \times 105 - 12f + 30 \times 102 - 30f$$

or $$100 = 1260 - 42f + 3060$$

or $$100 = 4320 - 42f$$

or $$42\,f = 4220$$

$$\therefore \qquad f = \frac{4220}{42} = 100.5\% = 100.5 \times \frac{50}{100} = 50.25 \text{ Hz}$$

New frequency = 50.25 Hz **Ans.**

Load shared by machine A

$$= 12(105 - 100.5) = 12 \times 4.5 = 54 \text{ MW} \quad \textbf{Ans.}$$

Load shared by machine B

$$= \frac{60}{2}(102 - 100.5) = 30 \times 1.5 = 45 \text{ MW} \quad \textbf{Ans.}$$

(b) In this case, the machines share 100 MW equally, i.e. each machine will share the load 50 MW.
Let rated speed of each machine be 100% at 50 Hz.
The frequency-load characteristics can be drawn as

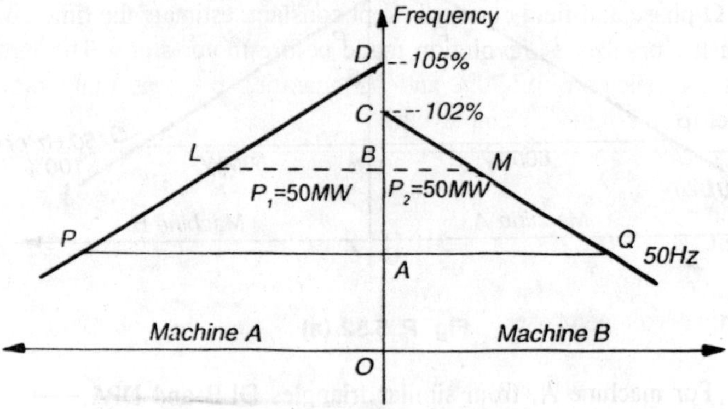

Fig. P. 5.52 (b)

For similar triangles CBM and CAQ

$$\frac{CB}{CA} = \frac{50}{60}, \text{CB} = \frac{50 \times 2}{60} = \frac{10}{6} = \frac{5}{3} = 1.67\%$$

Now $\quad$ OB = OC − CB = 102 − 1.67 = 100.33%

When $\quad P_1$ = 54 MW, BD = 105 − 100.5 = 4.5

$$\text{From fig, for } P_1 = 50 \text{ MW BD} = \frac{4.5 \times 50}{54} = 4.17$$

∴ Percentage speed at no-load = 100.33 + 4.17
= 104.50% **Ans.**

CD = 104.50 − 100 = 4.5%

From similar triangles, DLB and DPA

$$\frac{LB}{PA} = \frac{BD}{AD}$$

$$\therefore \quad \text{PA} = \frac{LB \times AD}{BD} = \frac{50}{4.17} \times 4.5 = 53.96 \text{ MW}$$

Output of machine A = 53.96 MW **Ans.**

Problem 5.53 : A 3000 V, 150 kW, 400 rpm, star-connected synchronous motor drives a fertilizer mill having a total moment of inertia of 500 kg-m^2, synchronous reactance of the motor is 10 Ω. When motor is running at full load, dynamic braking is employed by switching the armature to a bank of star connected resistors of 15 Ω/phase and field current is kept constant, estimate the time taken and the number of revolution made before motor stops. The motor has an efficiency of 90% and its operating p.f. was unity before electric braking was employed.

Solution :

Given, $X_s = 10\,\Omega/\text{phase} \quad R_b = 15\,\Omega/\text{phase}$

Terminal voltage/phase $= \dfrac{3000}{\sqrt{3}} = 1732.1\text{ V},\ J = 500\text{ kg m}^2,$

efficiency = 90% = 0.90, P.f. = 1

Motor full load current $= \dfrac{150 \times 1000}{\sqrt{3} \times 3000 \times 1 \times 0.9} = 32.1\text{ A}$

Motor full load torque $T_{fl} = \dfrac{150 \times 1000 \times 60}{2\pi \times 400} = 3582.8\text{ Nm}$

Now, let us take terminal voltage as reference phasor,
Excitation emf per phase

$$\bar{E}_f = \bar{V}_t - J I_a X_s$$

$$E_f = \sqrt{V_t^2 + (I_a X_s)^2} = \sqrt{(1732.1)^2 + (32.1 \times 10)^2} = 1761.6\text{ V}$$

Per phase armature current when dynamic braking is employed

$$I_{ab} = \frac{E_f}{\sqrt{R_b^2 + X_s^2}} = \frac{1761.6}{\sqrt{(15)^2 + (10)^2}} = 97.72\text{ A}$$

Per phase armature current 32.1 A gives a torque of 3382.8 Nm.

∴ per phase armature current of 97.72 A would produce a torque of

$$\frac{3582.8}{32.1} \times 97.72\text{ Nm} = 10906\text{ Nm}$$

Total initial braking torque = load torque + initial dynamic breaking torque

$$= (3582.8 + 10906)\text{ Nm} = 14488.8\text{ Nm}$$

Also, Initial dynamic breaking torque = 10906 Nm

or $K\omega_r = 10906$

$$\therefore \quad K = \frac{10906 \times 60}{2\pi \times 400} = 260.5 \text{ Nm-sec/rad}$$

Now, Time for stopping the mill

$$t = \frac{500}{260.5} ln\left[\frac{3582.8 + 10906}{3582.8 + 0}\right] = 2.68 \text{ sec} \quad \textbf{Ans.}$$

The number of revolution before the drive stops

$$= \frac{1}{2\pi \times 260.5}\left[\frac{500}{260.5}(2582.8 + 10906)\left(1 - e^{\frac{-260.5}{500} \times 2.68}\right) - 3582.8 \times 2.68\right] = 6.9 \text{ revolution} \quad \textbf{Ans.}$$

Problem 5.54 : A 3-phase 400 rpm synchronous motor and its driven equipment have a total moment of inertia 400 kgm^2 and a constant frictional torque 200 Nm. Calculate the time taken by the drive to come to rest and the number of revolution made during the retardation process when it is maked by the following methods.

(a) Braked by pugging with a constant breaking torque of 1400 Nm.

(b) Braked by rheostatic braking with field current and breaking resistances held constant at a value that gives an initial dynamic breaking torque of 5 kNm.

Solution :

Given, $J = 400$ kgm^2, $T_f = 200$ Nm

$$\omega_r = 400 \text{ rpm} = \frac{400 \times 2\pi}{60} = \frac{40\pi}{3} \text{ rad/sec} = 41.89 \text{ rad/sec.}$$

(a) Also, given $T_p = 1400$ Nm

During dynamic breaking,

$$T_b = T_p + T_f = 1400 + 200 = 1600 \text{ Nm}$$

Time taken for stopping the drive

$$t = \frac{J}{T_b}(\omega_s - \omega_r) = \frac{400}{1600}(41.89 - 0) = 10.47 \text{ sec} \quad \textbf{Ans.}$$

Total number of revolution in the time $\frac{1}{2\pi}\left[\omega_s t - \frac{T_b}{j}\frac{t^2}{2}\right]$

$$= \frac{1}{2\times 3.14}\left[41.89\times 10.47 - \frac{1600}{400}\times\frac{(10.47)^2}{2}\right]$$

= 34.93 revolution **Ans.**

(b) Given, initial dynamic breaking,

$$T_d = 5000 \text{ Nm}$$

or $\quad K\omega_s = 5000, \quad K = \dfrac{5000}{41.89} = 119.36$ Nm-sec/rad

Constant friction torque can be treated as $T_L = T_f = 200$ Nm

Time stopping for drive

$$t = \frac{J}{K}\ln\left(\frac{T_L + K\omega_s}{T_L + K\omega_r}\right) = \frac{400}{119.36}\ln\left(\frac{200+5000}{200}\right)$$

= 10.92 sec **Ans.**

Number of revolution made during this process

$$R = \frac{1}{2\pi K}\left[\frac{J}{K}(T_L + K\omega_s)\left(1 - e^{\frac{-k}{5}t}\right) - T_L t\right]$$

$$= \frac{1}{2\times\pi\times 119.36}\left[\frac{400}{119.36}(200+5000)\left(1 - e^{\frac{-119.36}{400}\times 10.92}\right) - 200\times 10.92\right]$$

= 19.43 revolution **Ans.**

Problem 5.55 : A star connected terbogenerator, with $Z_s = (0.5 + j\,8)\,\Omega$/phase, delivers 150 A at unity p.f. to an 10 kV infinite bus. If the alternator excitation is increased by 20%. Find the current and p.f. at which machine shall now work. Assume, a constant steam supply and unchanged efficiency.

Solution :

Given, $\quad I_a = 150$ A, $\quad \bar{Z}_s = (0.5 + j\,8)\,\Omega$/phase

$$|Z_s| = 8.02\,\Omega/\text{phase}, \quad \alpha_z = \tan^{-1}\frac{0.5}{8} = 3.58°$$

Terminal voltage/ phase $= \dfrac{10\times 1000}{\sqrt{3}} = 5773.5$ V

From phasor diagram. Excitation at unity p.f. is given by

$$E_f = \sqrt{(V_t + I_a r_a)^2 + (I_a X_s)^2}$$

$$= \sqrt{(5773.5 + 150 \times 0.5)^2 + (150 \times 8)^2}$$

$$= \sqrt{(5848.5)^2 + (1200)^2} = 5970 \text{ V}$$

$$\text{Load angle } (\delta) = \tan^{-1}\left(\frac{1200}{5848.5}\right) = 11.6^\circ$$

Now, excitation emf = $1.2 \times 5970 = 7164$ V

As the steam supply and alternator efficiency are unchanged,

Power input to alternator with E_f = Power input to alternator with 1.2 E_f

$$\frac{E_f V_t}{Z_s}\sin(\delta - \alpha_z) + \frac{E_f^2}{Z_s^2} \times r_a = \frac{1.2 E_f \times V_t}{Z_s}\sin(\delta' - \alpha_z) + \frac{1.44 E_f^2}{Z_s^2} \times r_a$$

$$V_t \sin(\delta - \alpha_z) + \frac{E_f}{Z_s} \times r_a = 1.2 V_t \sin(\delta' - \alpha_z) + \frac{1.44 E_f}{Z_s} \times r_a$$

or $$1.2 V_t \sin(\delta' - \alpha_z) = V_t \sin(\delta' - \alpha_z) + \frac{E_f}{Z_s} \times r_a (1 - 1.44)$$

or $$1.2 \times 5773.5 \sin(\delta' - \alpha_z) = 5773.5 \sin(11.6^\circ - 3.58^\circ) + \frac{5970}{8.02} \times 0.5\,(.44)$$

or $$\sin(\delta' - \alpha_z) = 0.093$$

$$\delta' - \alpha_z = 5.336^\circ$$

$$\therefore \quad \delta' = 5.34^\circ + 3.86^\circ, \quad \delta' = 9.2^\circ$$

Let I_{a1} be the armature current when excitation is increased.

$$\therefore I_{a1} = \frac{\sqrt{V_t^2 + E_f^2 - 2 E_f V_t \cos\delta}}{Z_s}$$

$$I_{a1} = \frac{\sqrt{(5773.5)^2 + (1.2 \times 5770)^2 - 2 \times 1.2 \times 5970 \times 5773.5 \times \cos 90^\circ}}{8.02}$$

I_{a1} = 332.3 A **Ans.**

For the same steam inputs $1.2\, E_f \times I_a \times \cos(\delta' + \theta) = E_f\, I_a \times \cos\delta$

$$\cos(\delta' + \theta) = \frac{150 \times \cos 11.6^\circ}{1.2 \times 332.3} = 0.368$$

$$\therefore \quad \delta' + \theta = \cos^{-1}(0.368)$$

$$\therefore \quad \theta = \cos^{-1}(0.368) - \delta' = 59.20^\circ$$

$\therefore$ New p.f. $= \cos 59.20^\circ = 0.5120$ lag **Ans.**

> **Problem 5.56 :** A 3-phase star-connected alternator with $Z_s = 0.5 + j\,6\ \Omega$ / phase, delivers 210 A at p.f. 0.8 lagging to 10 kV infinite bus
>
> (*a*) With steam supply unchanged, find the percentage increase or decrease in the excitation emf necessary to raise the power factor to unity. Assume rotational losses to remain constant.
>
> (*b*) In case alternator output is unchanged, determine % increase or decrease in the excitation emf needed to raise the p.f. to unity.

Solution :

Given $Z_s = 0.5 + j\,6\,\Omega/\text{phase} = 6.02\,\angle 85.24^\circ\,\Omega$

$$|Z_s| = 6.02\ \Omega/\text{phase}$$

$$\alpha_z = \tan^{-1}\left(\frac{0.5}{6}\right) = 4.96^\circ,\ \text{p.f.} = 0.8 \text{ lagging}$$

$$\text{Terminal voltage/phase} = \frac{10 \times 1000}{\sqrt{3}} = 5773.5\ \text{V}$$

$$\bar{I}_a = 210\ (0.8 - j\,0.6\ )$$

Let us take terminal voltage as reference phasor. The excitation emf can be calculated as

$$\bar{E}_f = \bar{V}_t + \bar{I}_a + \bar{Z}_s$$

$$= 5773.5\,\angle 0^\circ + 210\,(0.8 - j\,0.6) \times 6.02\,\angle 85 - 24^\circ$$

$$= 5773.5\,\angle 0^\circ + 210 \times 6.02\,\angle -36.9^\circ\,\angle 85.24^\circ$$

$$= 5773.5 + 840.32 + j\,944.49 = 6613.82 + j\,944.49$$

$$= 6680.9\,\angle 8.13^\circ$$

Alternator output power/phase = power delivered to infinite bus/phase

$$= V_t\, I_a \cos\theta = 5773.5 \times 210 \times 0.8 = 969948\ \text{W}$$

Power developed in alternator armature = electromagnetic power = I^2R armature loss + power delivered to bus bar.

$$= 969948 + (210)^2 \times 0.5 = 991998 \text{ W}$$

(*a*) As we know that, steam power input to alternator = rotational losses + electromagnetic power

As rotational losses are assumed constant, with steam supply unchanged, the electromagnetic power remains constant. When power factor is raised to unity let, I_e, be the armature current,

$\therefore$ Power delivered to infinite bus bar + I^2R loss = electromagnetic power

or, $$V_b\, I_{a1} + I_{a1}^{\,2}\, V_a = 991998$$

or, $$5773.5\, I_{a1} + 0.5\, I_a^{\,2} = 991998$$

or, $$I_a^{\,2} + 11547\, I_{a1} - 1983996 = 0$$

This yields, $$I_{a1} = \frac{-11547 \pm \sqrt{(11547)^2 + 4 \times 1983996}}{2}$$

$$= \frac{-11547 \pm 11885.7}{2}$$

Neglecting negative value,

$$I_{a1} = \frac{-11547 + 11885.7}{2} = 169.34 \text{ A.}$$

At unity p.f. $\bar{I}_{a1} = 169.34\ (1 + j\ 0)$

In this case the excitation emf.

$$\bar{E}_{f1} = \bar{V}_t + \bar{I}_{a1}\, \bar{Z}_s$$

$$= 5773.5 + 169.34 \angle 0° \times 6.02 \angle 85.24°$$

$$= 5858.1 + j\,1015.9 = 5945.54 \angle 9.84°$$

Now, percentage decrease in excitation

$$= \left(\frac{6680.9 - 5945.54}{6680 - 9}\right) \times 100\% = 11.01\% \quad \textbf{Ans.}$$

(*b*) For no change in alternator output,

$$V_t\, I_a \cos\theta = V_t \times I_{a2} \times \cos\theta_2$$

$$\Rightarrow \quad I_a \times 0.8 = I_{a2} \times 1$$

$$\therefore \quad I_{a2} = 0.8 \times 210 = 168 \text{ A}, \ \bar{I}_{a2} = 168 \angle 0°$$

Now, Excitation emf in this case,

$$\bar{E}_{f2} = \bar{V}_t + \bar{I}_{a2}\,\bar{Z}_s$$

$$= 5773.5\angle 0^\circ + 168\angle 0^\circ \times 6.02\angle 85.24$$

$$= 5857.4 + j\,1007.9$$

$$E_{f2} = 5943.5 \text{ V}$$

Percentage reduction in the excitation

$$= \left(\frac{6680.9 - 5943.5}{6680.9}\right) \times 100\% = 11.04\% = \textbf{Ans.}$$

> **Problem 5.57 :** A 3-phase, 440 V, 11 kVA, star-connected cylindrical rotor alternator with synchronous reactance 8 Ω/phase is driven by 240 V, 11 kW d.c. stunt motor. The alternator is synchronized with an infinite bus 440 V, 50 Hz. All the losses are to be ignored.
>
> (*a*) Determine the excitation emf and load angle of the alternator when it delivers rated kVA at unity p.f.
>
> (*b*) With the driving torque of dc shunt motor unaltered. The excitation voltage is increased by 20%. Find the, new load angle, current and p.f. Also obtain the maximum stable power output of the alternator and corresponding values of current and p.f.

Solution :

Given, $X_s = 8\ \Omega/\text{phase}, \quad \bar{X}_s = 8\angle 90^\circ\ \Omega$

Terminal voltage = 440 V

$$\text{Terminal voltage/phase} = \frac{440}{\sqrt{3}} = 254 \text{ V}$$

$$\text{Rated armature current } (I_a) = \frac{11 \times 1000}{\sqrt{3} \times 440} = \frac{25}{\sqrt{3}} = 14.43 \text{ A}$$

$$\therefore \quad \bar{I}_a = 14.33\angle 0^\circ$$

(*a*) Let us take terminal voltage as reference phasor.
Excitation emf/phase.

$$\bar{E}_f = \bar{V}_t + \bar{I}_a\,\bar{X}_s$$

$$= 254\angle 0 + 14.33\angle 0 \times 8\angle 90^\circ = 254 + j\,114.64$$

$$= 278.87\angle 24.29^\circ$$

Excitation emf = 278.8 V/phase **Ans.**

Load angle $\delta = 24.29^\circ$ **Ans.**

(b) When the excitation emf is increased by 20%

New excitation emf,

$$E_{f1} = 1.2 \times 278.87 = 334.64 \text{ V/phase.}$$

Since driving torque remains same, the alternator output with increased excitation voltage is unaltered.

$$\therefore \quad \frac{E_{f1} V_t \sin \delta_1}{X_s} = \frac{E_f V_t \sin \delta}{X_s}$$

$$\text{or,} \quad 1.2\, E_f V_t \sin \delta_1 = E_f V_t \sin \delta$$

$$\therefore \quad \sin \delta_1 = \frac{\sin \delta}{1.2} = \frac{\sin 24.29°}{1.2} = 0.343$$

$$\therefore \quad \delta_1 = \sin^{-1}(0.343) = 20.1°$$

New load angle $(\delta_1) = 20.1°$

The current can be calculated as

$$I_{a1} = \frac{\sqrt{E_f^2 + V_t^2 - 2 E_f V_t \cos \delta}}{X_s}$$

$$= \frac{\sqrt{(254)^2 + (334.64)^2 - 2 \times 254 \times 334.64 \cos 20.1°}}{8}$$

$$I_{a1} = 16.24 \text{ A}$$

New armature current = 16.24 A **Ans.**

Since alternator output is constant

$$V_t I_{a1} \cos \theta_1 = V_t \times I_a \times \cos \theta$$

$$\cos \theta_1 = \frac{14.33 \times 1}{16.24} = 0.8824 \text{ (lagging)} \textbf{ Ans.}$$

As we know that maximum output takes place when load angle is 90°

$$\therefore \quad P_{o(\max)} = \frac{E_{f1} \times V_t}{X_s} \sin 90° = \frac{334.64 \times 254}{8}$$

$$= 10624.82 \text{ W} = 10.625 \text{ kW} \textbf{ Ans.}$$

The corresponding armature current is given by

$$I_{a2} = \frac{\sqrt{E_{f1}^2 + V_t^2}}{X_s} = \frac{\sqrt{(334.64)^2 + (254)^2}}{8} = 52.51 \text{ A} \textbf{ Ans.}$$

The corresponding power factor is given by

$$V_t \times I_{aL} \times \cos \theta_2 = 10624.82$$

or, $\cos\theta_2 = \dfrac{10624.82}{254 \times 52.51} = \text{(lag)}$ **Ans.**

Problem 5.58 : A 3-phase star-connected turbogenerator feeds power to 10 kV infinite bus through two parallel transmission lines each having reactance 2 Ω and the generator synchronous reactance is 4 Ω.

The mechanical input to alternator and its excitation are adjusted so that the operating power factor at alternator terminals is unity and rated current of 400 A is delivered. Calculate the alternator terminal and excitation voltage. Also calculate the active bond reactive power delivered to infinite bus.

Solution :

Given $I_a = 400$ A $X_t = 2\,\Omega$, $X_s = 4\,\Omega$, p.f = 1, $V_b = 10$ kV

The single line diagram of this problem can be drawn as.

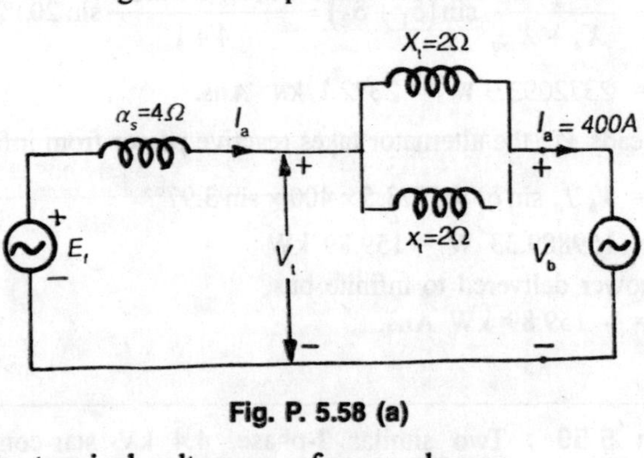

Fig. P. 5.58 (a)

Let us take terminal voltage as reference phasor.

Equivalent reactance of transmission lines = $\dfrac{2 \times 2}{4} = 1\,\omega$

Voltage drop across transmission lines $= I_a \times 1 = I_a$

The phasor diagram can be drawn as

From phasor diagram,

$$V_b^2 = V_t^2 + I_a^2$$

or, $$V_t = \sqrt{V_b^2 - I_a^2}$$

$$= \sqrt{(5773.5)^2 - (400)^2}$$

= 5759.6 V/phase **Ans.**

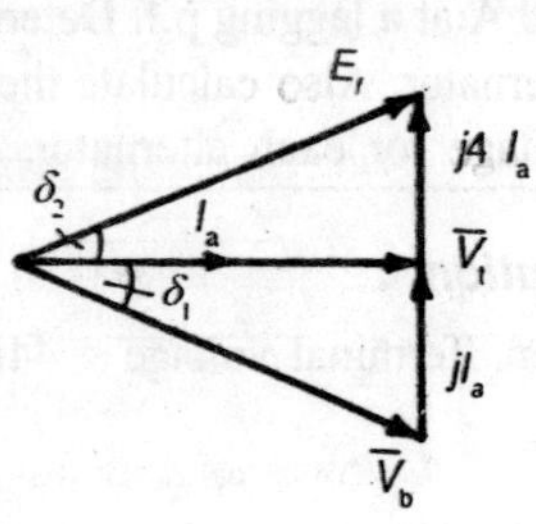

Fig. P. 5.58 (b)

Again, from phasor diagram,

$$E_f^2 = V_t^2 + (4\,I_a)^2$$

$$E_f = \sqrt{(5759.6)^2 + (4 \times 400)^2} = 5977.7 \text{ V / phase } \textbf{Ans.}$$

Power angle between V_b and V_t

$$\delta_1 = \sin^{-1}\left(\frac{I_a}{V_b}\right) = \sin^{-1}\left(\frac{400}{5773.5}\right) = 3.97°$$

Power angle between V_t and E_f

$$\delta_2 = \sin^{-1}\left(\frac{4\,I_a}{V_t}\right) = \sin^{-1}\left(\frac{4 \times 400}{5759.0}\right) = 16.13°$$

Active power delivered to infinite bus

$$= \frac{E_f\,V_b}{X_s + X_{teq}} \sin(\delta_1 + \delta_2) = \frac{5977.7 \times 5773.5}{4 + 1} \sin 20.1°$$

$$= 2372093.9 \text{ W} = 2372.1 \text{ kW } \textbf{Ans.}$$

Since I_a leads V_b, the alternator takes reactive power from infinite bus

$$= V_b\,I_a \sin\delta_1 = 5773.5 \times 400 \times \sin 3.97°$$

$$= 159889.33 \text{ W} = 159.89 \text{ kW}$$

Reactive power delivered to infinite bus.

$$= -\ 159.89 \text{ kW } \textbf{Ans.}$$

Problem 5.59 : Two similar 3-phase, 4.4 kV star-connected alternators running in parallel, supply a load 2000 kW at p.f 0.8 lagging. The synchronous impedance per phase for the first machine is $0.4 + j\,6\ \Omega$ and for the second machine is $0.5 + 8\ \Omega$. The total load is shared equally between them without changing. Their driving torques the excitation of first alternator is adjusted so that it delivers 150 A at a lagging p.f. Determine the current delivered by the second alternator, Also calculate the power factor, load angle and excitation voltage for each alternator.

Solution :

Given, Terminal voltage = 4400 V

$$\text{Terminal voltage/phase} = \frac{4400}{\sqrt{3}} = 7621\text{ V}$$

$$\bar{Z}_1 = 0.4 + j\,6 = 6.013\,\angle 86.19^\circ\ \Omega/\text{phase}$$
$$\bar{Z}_2 = 0.5 + j\,8 = 8.02\,\angle 86.42^\circ\ \Omega/\text{phase}$$
$$I_{L1} = 150\text{ A}$$

$$\text{Total load current } I_L = \frac{2000 \times 1000}{\sqrt{3} \times 0.8 \times 4.4 \times 1000} = 328\text{ A}$$

Let us take terminal voltage as reference phasor

$$\bar{I}_L = 328\ (0.8 - j\ 0.6)$$

Power factor of the first alternator.

$$\cos\theta_1 = \frac{1000}{\sqrt{3} \times 4.4 \times 150} = 0.875 \text{ lagging } \textbf{Ans.}$$

$$\therefore \quad \sin\theta_1 = \sqrt{1 - \cos^2\theta} = 0.484$$

$$\therefore \quad \bar{I}_{L1} = 150\ (0.875 - j\ 0.484)$$

Total load current = Sum of load current shared by each alternator

$$\bar{I}_L = \bar{I}_{L1} + \bar{I}_{L2}$$
$$\therefore \quad \bar{I}_{L2} = \bar{I}_L - I_{L1}$$
$$= 328\,(0.8 - j0.6) - 150\,(0.875 - j0.484)$$
$$= 131.15 - j124.2 = 180.63\,\angle -43.44$$

Current delivered by second alternator = 180.63 A **Ans.**

Power factor of the 2nd alternator.

$$\cos\theta_2 = \cos 43.44 = 0.726 \text{ (lagging) } \textbf{Ans.}$$

For Ist alternator,

$$\bar{E}_{f1} = \bar{V}_t + \bar{I}_{a1}\,\bar{Z}_s$$
$$= 7621 + 150\,\angle -28.95^\circ \times 6.013\,\angle 86.19^\circ = 8109.1 + j758.5$$
$$= 81470.1\,\angle 5.53^\circ = 141110.35\text{ V}\,(\text{L} - \text{L})$$

Excitation emf = 81470.1 V/phase

Load angle, $\delta_1 = 5.53^\circ$ **Ans.**

For second alternator

$$\bar{E}_{f2} = \bar{V}_t + \bar{I}_{a2}\,\bar{Z}_s$$
$$= 7621 + 180.63\,\angle -43.44 \times 8.02\,\angle 86.42^\circ$$
$$= 8680.8 + j\ 987.61 = 8736.8\,\angle 6.49^\circ$$

Excitation emf of 2nd alternator

Chapter 6

Induction Machines

EMF EQUATION

The magnitude of induced emf is stator winding in induction motor is given by

$$E_1 = \sqrt{2}\,\pi f\, k_{w1} N_{phs1}\, \phi \qquad (1)$$

Where f = Supply frequency,

N_{phs1} = Stator series turns per phase,

k_{w1} = Winding factor of stator winding,

ϕ = Air gap flux

The magnitude of induced emf in rotor winding in induction motor is given by:

$$E_2 \text{ (at standstill)} = \sqrt{2}\,\pi f\, K_{w2} N_{phs2}\, \phi \qquad (2)$$

Where f = Supply frequency,

N_{phs2} = Rotor series turns per phase.

K_{w2} = Winding factor of rotor winding

Now, $\therefore \quad \dfrac{E_1}{E_2} = \dfrac{K_{w1} N_{phs1}}{K_{w2} N_{phs2}} = \dfrac{N_s}{N_r} = a$

This ratio is know as transformation ratio at rotor standstill. Where N_s and N_r are effective no. of stator and rotor turns per phase respectively.

Slip and rotor frequency: Synchronous speed(N_s) $= \dfrac{120f}{p}$ (in rpm),

$$\text{Slip (s)} = \frac{N_s - N_r}{N_r}$$

Where N_r = rotor speed in rpm.

The frequency of rotor is given by.

$$f_2 = sf \quad \text{where,} \quad f = \text{supply frequency in Hz.}$$

When rotor is standstill, $N_r = 0$

$$s = \frac{N_s - 0}{N_s} = \frac{N_s}{N_s} \text{ and } s = 1$$

The equivalent circuit of rotor is given by

Rotor current is given by $I_2 = \dfrac{s\,E_2}{Z_2} = \dfrac{s\,E_2}{\sqrt{r_2^2 + (sx_2)^2}} = \dfrac{E_2}{\sqrt{\left(\dfrac{r_2}{s}\right)^2 + x_2^2}}$

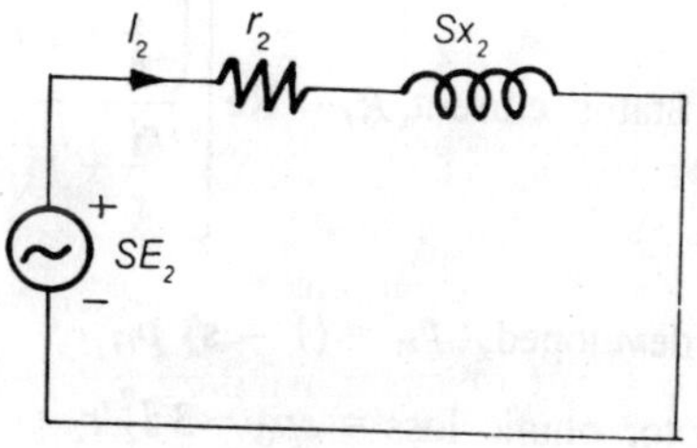

Fig. 6.1

EQUIVALENT CIRCUIT MODEL OF A 3-ϕ IM

The equivalent circuit model of a 3-ϕ induction motor can be drawn as.

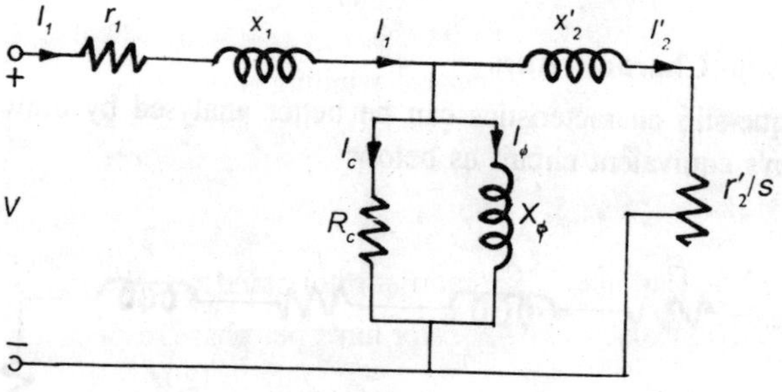

Fig. 6.2

In circuit diagram,

R_c = core loss-resistance,

X_ϕ = Magnetising reactance,

r_2' = Rotor resistance reffered to stator side = $a^2\, r_2$

x_2' = Rotor reactance reffered to stator side = $a^2\, x_2$

I_2' = Rotor current reffered to stator side = $\frac{I_2}{a}$

r_1 = Stator winding resistance

x_1 = Stator leakage reactance.

Power across air gap,

$$P_G = \frac{\text{rotor copper loss}}{\text{slip}} = \frac{3\, I_2^2\, r_2}{s} = 3\, I_1^2\, R_f$$

Where I_1 = Stator current, $R_f = \text{Re}\left[\frac{\left(\frac{r_2'}{s} + j\, x_2'\right)(j\, X_\varphi)}{\frac{r_2'}{s} + j\left(x_2' + X_\varphi\right)}\right]$

Mechanical power developed, $p_m = (1 - \text{s})\, p_G$

$$\text{Rotor ohmic loss} = sp_G = 3\, I_2^2\, r_2$$

$$\text{Shaft power output } (p_{sh}) = p_m - \text{fixed losses}$$

$$= p_m - \left(p_{wf} + p_c\right)$$

$$\text{Output torque } (T_{sh}) = \frac{p_{sh}}{\omega_r} = \frac{p_{sh}}{\omega_s(1-s)}$$

$$\text{Gross torque } (T_s) = \frac{p_G}{\omega_s}$$

Torque-slip Characteristics

The torque-slip characteristics can be better analysed by drawing the Thevenin's equivalent circuit as below:

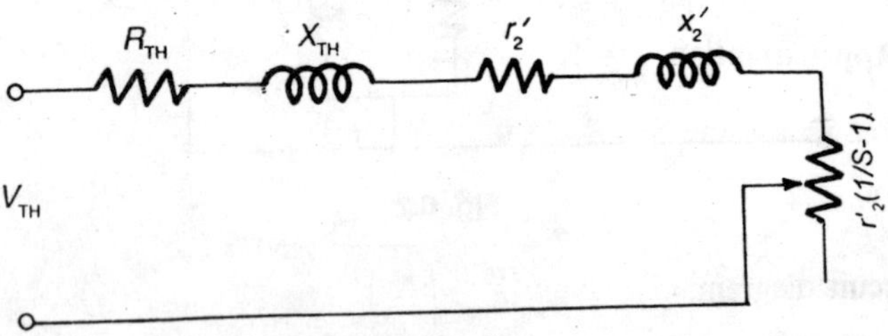

Fig. 6.3

Torque is given by

$$T_e = \frac{3}{\omega_s} \frac{V_{Th}^{\,2}}{\left(R_{Th} + \frac{r_2'}{s}\right)^2 + \left(x_{Th} + x_2'\right)^2}$$

Where $\quad V_{Th} = \dfrac{V_1 X_\varphi}{x_1 + X_\varphi}$, $R_{Th} = \dfrac{r_1 X_\varphi}{x_1 + X_\varphi}$ and $X_{Th} = \dfrac{x_1 X_\varphi}{x_1 + X_\varphi}$

The torque-slip characteristics can be drawn as

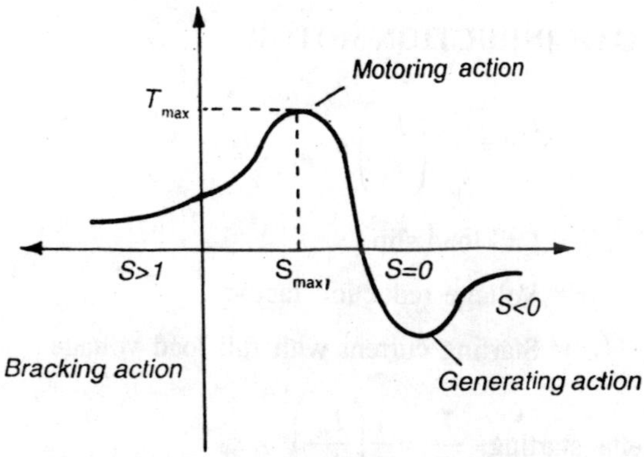

Fig. 6.4

Slip at maximum torque, $S_{\max, T} = \dfrac{r_2'}{\sqrt{R_{Th} + \left(X_{Th} + x_2'\right)^2}} = \dfrac{r_2'}{x_2'} = \dfrac{r_2}{x_2}$

$$= \frac{\text{Rotor resistance}}{\text{Standstill rotor reactance}}$$

Maximum torque, $T_{max} = \dfrac{3}{\omega_s} \cdot \dfrac{0.5\, V_{Th}^{\,2}}{R_{Th} + \sqrt{R_{Th}^{\,2} + \left(X_{Th} + x_2'\right)^2}}$

Approximation

Let us assume $\quad R_{Th} = 0, \quad X_{Th} = 0$ and $V_{Th} = V_1$

$$T = \frac{3}{\omega_s} \cdot \frac{V_1^2\,(r_2'/s)}{\left(\frac{r_2'}{s}\right)^2 + (x_2')^2}, \quad T_{max} = \frac{3}{\omega_s} \cdot \frac{0.5 V_1^2}{x_2'}$$

Starting Torque in Terms of Maximum torque

$$\frac{T_s}{T_{max}} = \frac{2}{\frac{1}{s_{max,T}} + s_{max,T}}$$

Full Load Torque is Terms of Maximum Torque

$$\frac{T_{fl}}{T_{max}} = \frac{2}{\frac{s_{fl}}{s_{max,T}} + \frac{s_{max,T}}{s_{fl}}}$$

STARTING OF INDUCTION MOTOR

$$\frac{T_s}{T_{fl}} = x^2 \left(\frac{I_{sc}}{I_{fl}}\right) \times s_{fl}$$

Where s_{fl} = Full load slip

x = Voltage reduction factor.

I_{sc} = Starting current with full load voltage

For star-delta starting, $\frac{T_s}{T_{fl}} = \frac{1}{3}\left(\frac{I_{sc}}{I_{fl}}\right)^2 \times s_{fl}$

Rotor Resistance Starting Method of Slipring Motors

$$T_s = \frac{3}{w_s} \cdot \frac{V_2 (r_2' + R_{exT})}{(r_2' + R_{exT})^2 + (x_2')^2}$$

Where R_{exT} = external resistance added to rotor circuit reffered to stator side.

Poly-phase Induction Motors Problems

Problem 6.1 : A 3-phase, 50 Hz induction motor has a full-load speed of 960 r.p.m. Calculate

(*i*) Number of the poles

(*ii*) Slip frequency

(*iii*) Speed of rotor field with respect to rotor structure; with respect to tator structure and with respect to stator field.

Solution :

Given $n_r = 960$ r.p.m., $f_1 = 50$ Hz

(*i*) Since the rotor and stator has same no. of poles Applying,

$$\frac{120 \times f}{p} = n, \text{ or } \frac{120 \times 50}{p} = 960$$

or, $$p = \frac{50}{8} \text{ or } 6\frac{1}{4}$$

Since the no. of poles can't be fractional it must be even in number.

no. of poles = 6 **Ans.**

(*ii*) Stator field speed $(n_s) = \frac{120 \times 50}{p} = \frac{120 \times 50}{6} = 1000$ rpm

Slip frequency $f_2 = s\, f_1$

$$= \left(\frac{n_s - n_r}{n_s}\right) \times 50 = \left(\frac{1000 - 960}{1000}\right) \times 50 = \frac{40 \times 50}{1000} = 2 \text{ Hz} \quad \textbf{Ans}$$

(*iii*) Speed of the rotor field with respect to the rotor structure
= speed of the rotor field – speed of the rotor structure
= 1000 – 960 = 40 r.p.m. **Ans.**
Speed of the rotor field with respect to stator = speed of the rotor field – speed of the stator = 1000 – 0 = 1000 r.p.m. **Ans.**
Speed of the rotor field with respect to stator field
= speed of the rotor field – speed of the stator field
= $n_s - n_s$ = 0 r.p.m. **Ans.**

Problem 6.2 : The stator of a 3-phase slip-ring induction motor is connected to 50 Hz supply. At the rotor terminals, a frequency of 30 Hz is required. Find the possible speeds at which the rotor must be driven. What is the ratio of slip-ring emfs at these speds at no load?

Solution :

Given, $f_1 = 50$ Hz, $f_2 = 30$ Hz
Let p be the no. of poles. Take $p = 4$

Stator field speed $$(n_s) = \frac{120 \times f_1}{p} = \frac{120 \times 50}{4} = 1500 \text{ rpm}$$

Rotor field speed $$(n_r) = \frac{120 \times f_2}{p} = \frac{120 \times 30}{4} = 900 \text{ rpm}$$

When the rotor is driven in the direction of stator field.

Speed = $n_s - n_r$, $n_1 = 1500 - 900 = 600$ rpm **Ans.**

$$s_1 = \left|\frac{1500 - 600}{1500}\right|$$

When the rotor is driven in the direction opposite to the stator field

Speed = $n_s - (-n_r)$, $N_2 = n_s + n_r = 1500 + 900 = 2400$ rpm **Ans.**

$$s_2 = \left|\frac{1500 - 2400}{1500}\right|, \quad \text{Here slip} = \frac{f_2}{f_1} = \frac{30}{50} = 0.6$$

Since the supply is constant, we know that
Slip rings emf at any speed = $s \times$ standstill emf = sE
The magnitude of the slips at these speeds is same.

The required ratio $= \dfrac{s_1 E}{s_2 E} = 1$ **Ans.**

Problem 6.3 : A 3-phase, delta-connected, 4-pole, 50 Hz induction motor has a stator resistance of 0.4 Ω per phase at the operating temperature. For a line current of 20 A, the total stator input is 4000 watts. For negligible stator core losses, find out the internal torque.

Solution :

Given, $f = 50$ Hz, $p = 4$

Power input $(p_{in}) = 4000$ W

Phase current $= \dfrac{I_L}{\sqrt{3}} = \dfrac{20}{\sqrt{3}}$ A

Total stator $i^2 r$ loss $= 3 \times I_s^2 \times r_s = 3 \times \left(\dfrac{20}{\sqrt{3}}\right)^2 \times 0.4 = 160$ W

Power across air gap, $P_G = p_{in} -$ stator $i^2 r$ loss

$= 4000 - 160 = 3840$ W

Synchronous speed, $n_s = \dfrac{120 f}{p} = \dfrac{120 \times 50}{4} = 1500$ rpm

$\therefore \quad \omega_s = \dfrac{2\pi \times 1500}{60}$ rad / sec.

Internal torque developed, $T_e = \dfrac{P_G}{\omega_s} = \dfrac{3840 \times 60}{2\pi \times 1500} = 24.45$ Nm **Ans.**

Problem 6.4 : A 3-phase, 400 V, 50 Hz induction motor takes a power input of 35 kW at its full-load speed of 890 r.p.m. The total stator losses are 1 kW and the friction and windage losses are 1.5 kW. Calculate (*a*) slip, (*b*) rotor ohmic losses, (*c*) shaft power, (*d*) shaft torque, and (*e*) efficiency.

Solution :

Given, $V_t = 400$ V, $f_1 = 50$ Hz, $p_{in} = 35$ kW

$n_r = 980$ r.p.m., stator loss = 1 kW

Let us take $n_s = 1000$ r.p.m.

(*i*) $$\text{Slip } (s) = \frac{n_s - n_r}{n_s} = \frac{1000 - 980}{1000} = \frac{20}{1000} = 0.02 \text{ **Ans.**}$$

(*ii*) Power across air gap,

p_G = power input – stator losses = 35 kW – 1 kW = 34 kW

Rotor ohmic losses = sp_G = 0.02 × 34 × 1000 = 680 W **Ans.**

(*iii*) Shaft power output = Gross power developed – rotor copper loss – windage and friction loss

$$= (34 - 0.68 - 1.5) \text{ kW} = 31.82 \text{ kW} \quad \textbf{Ans.}$$

(*iv*) $$\text{Shaft torque} = \frac{p_{sh}}{\omega_r} = \frac{31820 \times 60}{2\pi \times 980} = 310.06 \text{ Nm} \quad \textbf{Ans.}$$

(*v*) $$\text{Efficiency} = \frac{P_{\text{output}}}{P_{\text{input}}} = \frac{31.82}{35} = 0.9091 = 90.91\% \quad \textbf{Ans.}$$

Problem 6.5 : A 400 V, 3-phase, 6 pole, 50 Hz, induction motor draws a power of 2 kW at no load and at rated voltage and frequency. At a full-load slip of 3%, the power input to motor is 50 kW and the stator ohmic loss is 1.5 kW. Neglect I^2R loss at no load. If the stator core loss and mechanical losses are assumed equal, then at a slip of 3% calculate (*a*) rotor ohmic loss, (*b*) shaft (or output) power, (*c*) shaft torque, (*d*) internal torque, and (*e*) efficiency.

Solution :

Power input at no-load = 2 kW, Stator ohmic losses = 1.5 kW

Power input at slip 0.03 = 50 kW

Fixed loss = Power input at No-load = 2 kW

According to Question,

Mechanical losses = Stator core loss = 1 kW

Power across air gap, $p_G = p_{in}$ – stator core loss – stator $i^2 r$ loss

$$= 50 - 1.5 - 1 = 47.5 \text{ kW}$$

Given, Slip = 0.03

(*a*) Rotor ohmic loss $= s \times p_G \left(0.03 \times 47.5\right) \text{kW}$

$$= 1.425 \text{ kW} \textbf{ Ans.}$$

(*b*) Shaft power out put

$= p_G -$ Rotor ohmic loss – Mechanical losses

$= 47.50 - 1.425 - 1 = 45.075$ kW **Ans.**

(*c*) Synchronous speed $n_s = \dfrac{120 \times f}{p} = \dfrac{120 \times 50}{6} = 1000 \text{ rpm}$

$\therefore$ Rotor speed $\left(n_r\right) = \left(1-s\right) n_s = \left(1-0.03\right) \times 1000 = 970 \text{ rpm}$

$$\therefore \qquad \omega_r = \frac{2\,\pi \times 970}{60} \text{ rad / sec}$$

Shaft torque, $T_{sh} = \dfrac{p_{sh}}{w_r} = \dfrac{45.075 \times 1000 \times 60}{2\,\pi \times 970}$

$$= 443.75 \text{ Nm} \textbf{ Ans.}$$

(*d*) Internal torque developed,

$$T_e = \frac{p_G}{\omega_s} = \frac{47.5 \times 1000 \times 60}{2\,\pi \times 1000} = 453.59 \text{ Nm} \textbf{ Ans.}$$

$$\text{Efficiency} = \frac{\text{Power output}}{\text{Power input}} = \frac{45.075}{50} = 0.9015 = 90.15\% \textbf{ Ans.}$$

Problem 6.6 : A 20 kW, 6 pole, 400 V, 50 Hz, 3-phase induction motor has a full-load slip of 0.02. If the torque, lost in mechanical (friction and windage) losses is 20 Nm, find the rotor ohmic loss, motor input and efficiency. Stator losses total 900 watts.

Solution :

Given $p = 6$, $f_1 = 50$ Hz, $V_t = 400$ V

Slip $(s) = 0.02$, stator losses = 900 W

Power output at full load = 20 kW

Synchronous speed, $n_s = \dfrac{120 \times f_1}{p} = \dfrac{120 \times 50}{6} = 1000$ r.p.m.

Shaft torque $T_{sh} = \dfrac{\text{Shaft power}}{\text{Rotor speed}} = \dfrac{20 \times 1000 \times 60}{(1-5)\,\omega_s \times 2\pi}$

$$= \frac{20 \times 1000 \times 60}{2\pi \times (1-0.02) \times 1000} = 194.884 \text{ Nm}$$

Mechanical torque developed,

$$T_e = T_{sh} + w_f = 194.884 + 20 = 214.884 \text{ Nm.}$$

Power across air gap, $P_G = T_e \times \omega_s = 214.884 \times 2\pi \times \dfrac{1000}{60}$

$= 22.503$ kW

Rotor copper loss $= (0.02 \times 22.503)$ W $= 450.06$ kW **Ans.**

Power input to motor $= P_G$ + total stator losses = (22.503 + 0.9)

$= 23.403$ kW **Ans.**

Efficiency $= \dfrac{\text{output}}{\text{input}} = \dfrac{20}{23.403} = 0.85459 = 85.459\%$ **Ans.**

> **Problem 6.7 :** A 10 kW, 3-phase, 50 Hz, 4 pole induction motor has a full-load slip of 0.03. Mechanical and stray load losses at full-load are 3.5% of output power. Compute
> (*a*) Power delivered by stator to rotor.

Solution :

Full load power output = 10 kW, $f_1 = 50$ Hz, $p = 4$

Mechanical and stray load losses = 3.5 % of 10 × 1000 W

$$= \frac{3.5}{100} \times 10 \times 1000 = 350 \text{ W}$$

Full load slip (s) = 0.03

Synchronous speed $(n_s) = \dfrac{120 \times f_1}{p} = \dfrac{120 \times 50}{4} = 1500$ rpm

Mechanical power developed,

$$P_m = P_{out} + P_{wf} = (10 + 0.350) \text{ kW} = 10.350 \text{ kW}$$

Rotor ohmic losses $= \left(\dfrac{s}{1-s}\right) \times P_m = \dfrac{0.03}{0.97} \times 10.350$

$= 0.3201$ kW $= 320.1$ W

(*a*) Power delivered by stator to rotor,

$$p_G = \frac{\text{Rotor ohmic losses}}{\text{slip}} = \frac{0.3201}{0.03} = 10.67 \text{ kW} \textbf{ Ans.}$$

(*b*) Electromagnetic torque developed,

$$T_e = \frac{P_G}{\omega_s} = \frac{10.67 \times 1000 \times 60}{2\,\pi \times 1500} = 67.927 \text{ Nm} \textbf{ Ans.}$$

(*c*) Rotor ohmic loss = 0.3201 kW = 320.10 W **Ans.**

> **Problem 6.8 :** An induction motor has its stator copper loss equal to the sum of the mechanical and iron losses. Its rotor copper loss is equal to one-third of the stator copper loss. If its efficiency is 79 %, calculate its slip. Take mechanical loss equal to iron loss.
> (*b*) Electromagnetic (internal) torque at full load, and
> (*c*) Rotor ohmic losses at full load.

Solution :

Let the mechanical losses be A units. According to Question,

Iron loss = Mechanical loss = A units

Stator copper loss = Mechanical loss + iron loss = A + A = 2 A units

$$\text{Rotor copper loss} = \frac{1}{3} \times \text{stator copper loss} = \frac{1}{3} \times 2\text{A} = \frac{2\,\text{A}}{3}$$

Let the power input be 100 % Efficiency $= \dfrac{\text{output power}}{\text{input power}}$

$\therefore$ Power out put $= \eta \times p_{in} = 100 \times 0.79 = 79$ units

Now, Total losses in the motor = 100 − 79 = 21 units

or, Stator copper loss + stator core loss + rotor copper loss + Mechanical loss = 21 units

or, $2\text{A} + \text{A} + \frac{2}{3}\text{A} + \text{A} = 21$, or $\frac{14\,\text{A}}{3} = 21$

$$\therefore \qquad \text{A} = \frac{63}{14} = \frac{9}{2}$$

Now, Power across air gap

$$P_G = \text{Power input} - \text{stator copper loss} - \text{stator core loss}$$

$$= 100 - 2 \times \text{A} - \text{A} = 100 - 2 \times \frac{63}{14} - \frac{63}{14} = 86.5 \text{ units}$$

$$\text{Rotor copper loss} = \frac{2}{3}\text{A} = \frac{2}{3} \times \frac{63}{14} = 3 \text{ units}$$

$$\text{Now, Slip } (s) = \frac{\text{Rotor copper loss}}{\text{Power across air gap}} = \frac{3}{86.5} = 0.0347 \textbf{ Ans.}$$

> **Problem 6.9 :** The power supplied to a 3-phase induction motor is 40 kW and the corresponding stator losses are 1.5 kW. Calculate the net mechanical power developed and the rotor I^2R loss when the slip is 0.04 per unit. What will be the net power developed if the speed of the above motor is reduced to 40 % of the synchronous speed by means of external rotor resistors, assuming the torque and stator losses to remain unaltered? Friction and windage losses may be assumed to be 0.8 kW. *(I.A.S., 1980)*

Solution :

Let un choose 40 kW as base power and speed 1500 rpm as base synchronous speed

$$\therefore \quad \text{Power input, } p_i = \frac{40}{40} = 1.00 \text{ p.u.}$$

$$\text{Synchronous speed } (n_s) = \frac{1500}{1500} = 1 \text{ p.u., slip} = 0.04$$

$$\text{Stator losses in p.u.} = \frac{\text{Stator losses}}{\text{base power}} = \frac{1.5}{40} = 0.0375 \text{ p.u.}$$

$$\text{Friction and windage losses} = \frac{0.8}{40} = 0.02 \text{ p.u.}$$

$$\begin{aligned}\text{Now, Power across air gap, } p_G &= p_{in} - \text{Stator losses} \\ &= 1 - 0.0375 = 0.9625 \text{ p.u.}\end{aligned}$$

$$\text{Rotor copper loss} = s \times p_G = (0.04 \times 0.9625) \text{ p.u.} = 0.0385$$

$$\begin{aligned}\text{Net power output } P_{net} &= p_G' - \text{rotor loss} - \text{windage and friction loss} \\ &= 0.9625 - 0.0385 - 0.02 = 0.904 \text{ p.u.}\end{aligned}$$

$$\begin{aligned}\text{Actual power developed} &= \text{Power in p.u.} \times P_{base} \\ &= 40 \times 0.904 = 36.16 \text{ kW} \quad \textbf{Ans.}\end{aligned}$$

$$\text{Rotor speed } (n_r) = (1 - s)\, n_s = (1 - 0.04) \times 1 = 0.96 \text{ p.u.}$$

$$\begin{aligned}\text{Actual speed of the rotor} &= 0.96 \times \text{base speed} \\ &= 0.96 \times 1500 = 1440 \text{ rpm}\end{aligned}$$

When the speed is reduced to 40 % of synchronous speed

$$\text{New speed} = 1440 \times \frac{40}{100} = 576 \text{ rpm}$$

New rotor speed is p.u. $n_r' = n_r'$ p.u.

$$\text{New slip, } s' = \frac{n_s - n_r'}{n_s} = \frac{1 - 0.400}{1} = 0.6$$

Now, rotor copper loss = $s' \, P_G$

Net power out put in p.u. = P_G − rotor copper loss − windage and friction loss

= 0.9625 − 0.600 × 0.9625 − 0.02

= 0.365 p.u.

Actual power output = Base power × Power in p.u.

= (40 × 0.365) kW = 14.6 kW **Ans.**

Problem 6.10 : A 3-phase wound-rotor induction motor develops a maximum torque of 4 times the full-load torque at a slip of 0.20. The per phase rotor resistance is 0.04 Ω. The stator resistance and rotational losses are negligible.

(*i*) Calculate the slip at full-load torque. If the stator resistance were considered, would the value of slip at full-load torque be smaller, same or larger?

(*ii*) Calculate the value of external resistance that must be inserted in the rotor circuit in order to obtain maximum torque at starting.

Solution

Given $r_2 = 0.04 \, \Omega$ per phase, $s_{\max, T} = 0.20$

$$\therefore \quad x_2 = \frac{r_2}{s_{\max, T}} = \frac{0.04}{0.2} = 0.2 \, \Omega \text{ per phase}$$

$$T_{fl} = ?, \quad T_{\max} = 4 \times T_{fl}$$

(*i*) Let S_{fl} be the full load slip. Now, Applying the relation.

$$\frac{T_{fl}}{T_{\max}} = \frac{2}{\dfrac{s_{\max T}}{s_{fl}} + \dfrac{s_{fl}}{s_{\max T}}}$$

or,
$$\frac{1}{4}\left(\frac{0.2}{s_{fl}} + \frac{s_{fl}}{0.2}\right) = 2$$

or,
$$0.04 + s_{fl}^2 = 2 \times 4 \times 0.2 \times s_{fl}$$

or, $s_{fl}^2 - 1.6\, s_{fl} + 0.04 = 0$

This yields $$s_{fl} = \frac{1.6 \pm \sqrt{(1.6)^2 - 4 \times 0.04}}{2} = \frac{1.6 \pm 1.549}{2}$$

Neglecting higher value, $= \dfrac{1.6 - 1.549}{2} = s_{fl} = 0.0255$ **Ans.**

Since slip at full load is inversely proportional to stator resistance. If the stator resistances were considered the slip will be lowered.

(*ii*) Let r_2' be the total rotor resistance when the starting torque is maximum.

Now, $$\frac{T_{st,\,max}}{T_{max}} = \frac{2}{\dfrac{s_{max,\,T}}{1} + \dfrac{1}{s_{max,\,T}}} \qquad (1)$$

As we know that the maximum torque is very nearly equal to maximum starting torque.

$$\therefore \qquad \frac{T_{st\ max}}{T_{max}} = 1$$

Now, From (1) $$s_{max,\,T} + \frac{1}{s_{max,\,T}} = 2$$

or $$s^2_{max,\,T} - 2\, s_{max,\,T} + 1 = 0$$

This yields $s_{max,\,T} = 1$ or, $\dfrac{r_2'}{x_2} = 1$

or, $$r_2' = 1 \times x_2 = 1 \times 0.2 = 0.2\ \Omega.$$

Now, Additional resistance in the rotor,

$$= r_2' - r_2 = 0.2 - 0.04 = 0.16\ \Omega \quad \textbf{Ans.}$$

Problem 6.11 : For a 3-phase induction motor, the rotor ohmic loss at maximum torque is 16 times that at full-load torque. The slip at full-load torque is 0.03. If stator resistance and rotational losses are neglected, then calculate

(*i*) The slip at maximum torque,

(*ii*) The maximum torque in terms of full-load torque and

(*iii*) The starting torque in terms of full load torque.

Solution :

Given $s_{fl} = 0.03, \quad (I_{2m})^2 r_2 = 16 \times (I_{2fl})^2 \times r_2$

or, $$\left(\frac{I_{2m}}{I_{2fl}}\right)^2 = 16, \text{ or, } \frac{I_{2m}}{I_{2fl}} = 4$$

(*i*) Now, Applying, $$\frac{T_{fl}}{T_{max}} = \frac{2}{\dfrac{s_{max,T}}{s_{fl}} + \dfrac{s_{fl}}{s_{max\,T}}} \qquad (1)$$

Also $$\frac{T_{fl}}{T_{max}} = \frac{I_{2fl}^2}{I_{2m}^2} \times \frac{s_{max,T}}{s_{fl}} = \frac{1}{16} \times \frac{s_{max,T}}{s_{fl}} \qquad (2)$$

From (1) and (2)

$$\frac{1}{16} \times \frac{s_{max,T}}{s_{fl}} = \frac{2}{\dfrac{s_{max,T}}{s_{fl}} + \dfrac{s_{fl}}{s_{max\,T}}}$$

or, $$\left(\frac{s_{max,T}}{s_{fl}}\right)^2 + 1 = 32, \text{ or } \frac{s_{max,T}}{s_{fl}} = \sqrt{32-1}$$

$$\therefore \quad s_{max,T} = \sqrt{31} \times s_{fl} = \sqrt{31} \times 0.03 = 0.167 \quad \textbf{Ans.}$$

From relation (2) $$\frac{T_{max}}{T_{fl}} = 16 \times \frac{s_{fl}}{s_{max\,T}} = 16 \times \frac{0.03}{0.167}$$

$$\therefore \quad T_{max} = 2.874\, T_{fl} \quad \textbf{Ans.}$$

(*iii*) Applying, $$\frac{T_{st}}{T_{max}} = \frac{2}{\dfrac{s_{max,T}}{1} + \dfrac{1}{s_{max,T}}} = \frac{2}{.167 + \dfrac{1}{.167}} = 0.325$$

or, $$T_{st} = 0.325 \times T_{max} = 0.325 \times 2.874\, T_{fl}$$

$$T_{st} = 0.934\, T_{fl} \quad \textbf{Ans.}$$

Problem 6.12 : A 3-phase squirrel-cage induction motor has a rotor starting current of 6 times its full load value. The motor has a full load slip of 5 %. Determine

(*i*) The starting torque in terms of full-load torque;
(*ii*) The slip at which maximum torque occurs; and
(*iii*) Maximum torque in terms of full-load torque.

Solution :

Given $s_{fl} = 0.05, \; I_{st} = 6 \times I_{fl}$

(*i*) Applying the relation,

$$\frac{T_{st}}{T_{fl}} = \frac{I_{st}^2}{I_{fl}^2} \times s_{fl} = \left(\frac{I_{st}}{I_{fl}}\right)^2 \times s_{fl} = (6)^2 \times 0.05$$

$$\therefore \quad T_{st} = 1.8\, T_{fl} \quad \textbf{Ans.}$$

(*ii*) To calculate slip at maximum torque, Applying the relation,

$$\frac{I_{st}}{I_{fl}} = \sqrt{\frac{s_{fl}^2 + s_{\max,T}^2}{s_{fl}^2\left(1 + s_{\max,T}^2\right)}}$$

Squaring both side,

$$\text{or} \quad 36 = \frac{(0.05)^2 + s_{\max,T}^2}{(0.05)^2 \times \left(1 + s_{\max,T}^2\right)}$$

$$\text{or} \quad \left(1 + s_{\max,T}^2\right) = 0.0025 + s_{\max,T}^2$$

$$\text{or} \quad 0.09 + 0.09\, s_{\max,T}^2 = 0.0025 + s_{\max,T}^2$$

$$\text{or} \quad s_{\max,T}^2 \;(1 - 0.09) = 0.09 - 0.0025$$

$$\text{or} \quad s_{\max,T} = \sqrt{\left(\frac{0.09 - 0.0025}{1 - 0.09}\right)} = 0.31 \quad \textbf{Ans.}$$

(*iii*) To calculate maximum torque in terms of full load, Applying,

$$\frac{T_{fl}}{T_{\max}} = \frac{2}{\dfrac{s_{fl}}{s_{\max,T}} + \dfrac{s_{\max,T}}{s_{fl}}} = \frac{2}{\dfrac{0.05}{0.31} + \dfrac{0.31}{0.05}} = 0.3144$$

$$\therefore \quad T_{\max} = \frac{T_{fl}}{0.3144} = 3.181\, T_{fl} \quad \textbf{Ans.}$$

Problem 6.13 : The maximum torque of a 3-phase squirrel-cage induction motor is 4 times the full-load torque and the starting torque is 1.6 times the full-load torque. Neglect stator resistance. Calculate

(i) The slip at the maximum torque

(ii) Full-load slip and

(iii) The rotor current at starting in terms of full-load rotor current.

Solution :

Given, $T_{max} = 4 \times T_{fl}$

and $T_{st} = 1.6 \times T_{fl}$

(i) Applying the relation,

$$\frac{T_{st}}{T_{max}} = \frac{2}{\dfrac{s_{max,T}}{1} + \dfrac{1}{s_{max,T}}}$$

or $$\frac{1.6\,T_{fl}}{4\,T_{fl}} = \frac{2\,s_{max,T}}{s_{max,T}^2 + 1}$$

or $$0.4 = \frac{2\,s_{max,T}}{s_{max,T}^2 + 1}$$

or $$s_{max,T}^2 - \frac{2}{0.4}\,s_{max,T} + 1 = 0$$

or $$s_{max,T}^2 - 5\,s_{max,T} + 1 = 0$$

This yields $$s_{max,T} = \frac{5 \pm \sqrt{25-4}}{2}$$

$$= \frac{5 \pm 4.5826}{2}$$

Neglective higher value, $s_{max,T} = 0.21$ **Ans.**

(ii) To calculate full load slip, Applying the relation,

$$\frac{T_{fl}}{T_{max}} = \frac{2}{\dfrac{s_{fl}}{s_{max,T}} + \dfrac{s_{max,T}}{s_{fl}}}$$

or
$$\frac{T_{fl}}{4 \times T_{fl}} = \frac{2}{\dfrac{0.21}{s_{fl}} + \dfrac{s_{fl}}{0.21}}$$

or
$$\frac{0.21}{s_{fl}} + \frac{s_{fl}}{0.21} = 8$$

or
$$(0.21)^2 + s_{fl}^2 = 8 \times 0.21\, s_{fl}$$

or
$$s_{fl}^2 - 1.68\, s_{fl} + 0.0441 = 0$$

This yields
$$s_{fl} = \frac{1.68 \pm \sqrt{(1.68)^2 - 4 \times 0.0441}}{2}$$

Neglecting higher value,
$$s_{fl} = \frac{1.68 - 1.6267}{2} = 0.0267 \quad \textbf{Ans.}$$

(iii) To calculate rotor current at starting,

Applying the relation,
$$\frac{T_{st}}{T_{fl}} = \left(\frac{I_{st}}{I_{fl}}\right)^2 \times s_{fl}$$

or
$$1.6 = \left(\frac{I_{st}}{I_{fl}}\right)^2 \times 0.0267$$

or
$$\left(\frac{I_{st}}{I_{fl}}\right)^2 = \frac{1.6}{0.0267}$$

or
$$I_{st} = \sqrt{\frac{1.6}{0.0267}} \times I_{fl}$$

∴
$$I_{st} = 7.741\, I_{fl} \quad \textbf{Ans.}$$

Problem 6.14 : A 3-phase induction motor at rated voltage and frequency has a starting torque of 150% and a maximum torque of 200 per cent of full-load torque. Neglecting stator resistance and rotational losses, calculate the slip of full load and slip at maximum torque. *(I.A.S., 1984)*

Solution :

Given, $T_{st} = 1.5 \times T_{fl}, \quad T_{max} = 2 \times T_{fl}$

To Calculate slip at maximum torque,

Applying the relation, $$\frac{T_{st}}{T_{max}} = \frac{2}{s_{max,T} + \dfrac{1}{s_{max,T}}}$$

or $$s_{max,T}^2 + 1 = 2\, s_{max,T} \times \frac{T_{max}}{T_{st}}$$

or $$s_{max,T}^2 + 1 = 2\, s_{max,T} \times \frac{2}{1.5}$$

or $$s_{max,T}^2 + 1 = 2.667\, s_{max,T}$$

or $$s_{max,T}^2 - 2.667\, s_{max,T} + 1 = 0$$

This yields $$s_{max,T} = \frac{2.667 \pm \sqrt{(2.667)^2 - 4}}{2}$$

Neglecting higher value, $$s_{max,T} = \frac{2.667 - 1.764}{2} = 0.452$$

To calculate full load slip, $$\frac{T_{fl}}{T_{max}} = \frac{2}{\dfrac{s_{fl}}{s_{max,T}} + \dfrac{s_{max,T}}{s_{fl}}}$$

or $$\frac{1}{2} = \frac{2}{\dfrac{0.452}{s_{fl}} + \dfrac{s_{fl}}{0.452}}$$

or $$(0.452)^2 + s_{fl}^2 = 4 \times 0.452 \times s_{fl}$$

or $$s_{fl}^2 - 1.808\, s_{fl} + 0.204 = 0$$

This yields $$s_{fl} = \frac{1.808 \pm \sqrt{(1.808)^2 - 4 \times 0.204}}{2}$$

Neglecting higher value $$s_{fl} = \frac{1.808 - 1.5662}{2} = 0.121$$

Full load slip = 0.121 **Ans.**

Slip at maximum torque = 0.452 **Ans.**

Problem 6.15 : A 3-phase, 4-pole, 50 Hz, 400 V, delta-connected induction motor develops a maximum torque of 240 Nm at a speed of 1200 rpm. Stator impedance and rotational losses are assumed negligible. The rotor resistance is 0.2 Ω per phase. Calculate the additional rotor resistance required to give a starting torque of:

(a) 75% of maximum torque at rated voltage and

(b) 75% of maximum torque when the supply voltage is reduced to 360 volts.

Solution :

Given, $r_2 = 0.2\ \Omega/\text{phase}, \quad T_{max} = 240\ \text{N}$

$V_t = 400\ \text{V}, \quad f = 50\ \text{Hz}$

$P = 4, \quad n_r = 1200\ \text{r.p.m.}$

Synchronous speed $$(n_s) = \frac{120 \times f}{8} = \frac{120 \times 50}{4} = 1500\ \text{r.p.m.}$$

$\therefore$ $$\text{slip } (s) = \frac{n_s - n_r}{n_s} = \frac{1500 \times 1200}{1500}$$

$$s_{max,T} = 0.2 = \frac{r_2}{x_2}$$

$\therefore$ $$x_2 = 1\ \Omega/\text{phase}$$

Since r_1 and x_1 are negligible, Maximum torque $= \dfrac{1}{\omega_s} \dfrac{3 V_1^2}{2 x_2}$

Putting $\dfrac{3 V_1^2}{\omega_s} = K, \; T_{max} = \dfrac{K}{2 x_2}$, or $240 = \dfrac{K}{2 \times 1}$

$\therefore$ $K = 480$

(a) Let R be the sum of rotor resistance and external resistance,

$$T_{st} = \frac{3}{\omega_s} \frac{V_1^2}{\left(R^2 + x_2^{\,2}\right)} \times R$$

According to question, $T_{st} = 0.75 \times T_{max}$

$\therefore$ $$0.75 \times 240 = \frac{K}{R^2 + 1} \times R$$

or $$R^2 + 1 = \frac{480}{240 \times 0.75} \times R$$

or $$R^2 - 2.667\,R + 1 = 0$$

This yields $$R = \frac{2.667 \pm \sqrt{(2.667)^2 - 4}}{2}$$

Neglecting higher value,

$$R = 0.4513$$

or $$r_2 + R_{ext} = 0.4513$$

$$\therefore \quad R_{ext} = 0.4513 - 0.2 = 0.2513\ \Omega \quad \textbf{Ans.}$$

(b) When the supply voltage is reduced to 360 V.

We have, $$T_{st} = \left(\frac{360}{400}\right)^2 \times \frac{K}{R^2 + 1} \times R$$

Since $T_e \propto V^2$

or $$0.75 \times T_{max} = \left(\frac{360}{400}\right)^2 \times \frac{480}{R^2 + 1} \times R$$

or $$R^2 + 1 = \left(\frac{360}{400}\right)^2 \times \frac{480}{0.75 \times 240} \times R$$

or $$R^2 + 1 = 2.16\ R$$

or $$R^2 - 2.16\,R + 1 = 0$$

This yields $$R = \frac{2.16 \pm \sqrt{(2.16)^2 - 4}}{2}$$

Neglecting higher value,

$$R = 0.6721$$

or $$r_2 + R'_{ext} = 0.6721$$

$$\therefore \quad R_{ext} = 0.6721 - 0.2 = 0.4721\ \Omega \quad \textbf{Ans.}$$

Problem 6.16 : A 4-pole, 50 Hz, 3-phase induction motor develops a maximum torque at a speed of 1350 rpm, and per phase rotor resistance is 0.2 Ω Calculate the value of external resistance that must be inserted in series with each rotor phase to produce a starting torque equal to half the maximum torque. Neglect stator impedance.

Solution :

Given, $P = 4$, $f = 50$ Hz

$$r_2 = 0.2\ \Omega/\text{phase}, \quad n_{max} = 1350 \text{ rpm}$$

Synchronous speed, $n_s = \dfrac{120 \times f}{P} = \dfrac{120 \times 50}{4} = 1500$ r.p.m.

$$\therefore \quad s_{max,T} = \frac{n_s - n_{max}}{n_s} = \frac{1500 - 1350}{1500} = 0.1$$

$$\frac{r_2}{x_2} = 0.1, \quad \therefore\ x_2 = \frac{0.2}{0.1} = 2\ \Omega$$

Let S_{mT1} be the slip at the maximum torque when rotor resistance is varied.

According to question, $T_{st} = \dfrac{1}{2} \times T_{max}, \quad \therefore\ \dfrac{T_{st}}{T_{max}} = \dfrac{1}{2}$

Now, Appluing the relation, $\dfrac{T_{st}}{T_{max}} = \dfrac{2}{\dfrac{s_{max,T}}{1} + \dfrac{1}{s_{max,T}}}$

or $$\frac{1}{2} = \frac{2 \times s_{max,T}}{s_{max,T}^2 + 1}$$

or $$s_{max,T}^2 - 4\, s_{max,T} + 1 = 0$$

This yields $s_{max,T} = \dfrac{4 \pm \sqrt{16 - 4}}{2}$

Neglecting higher value, $s_{max,T} = 0.2679$

or $$\frac{r_2 + R_{ext}}{x_2} = 0.2679$$

$$\therefore \quad R_{ext} = 2 \times 0.2679 - 0.2 = 0.3358 \textbf{ Ans.}$$

Problem 6.17 : A 4-pole, 3-phase, star-connected slip-ring induction motor is to operate from 400 V, 3-phase, 50 Hz mains. Its rotor has a stand still leakage mpedance of 0.4 + $j\,2$ ohms per phase. For negligible stator impedance and rotational losses, compute:

(a) Maximum torque in newton-metres,

(b) The resistance to be included in the rotor circuit to develop 0.8 of maximum torque at start.

Solution :

Given, $f = 50$ Hz, $r_2 = 0.4\ \Omega/\text{phase}$

$x_2 = 2\ \Omega/\text{phase}$, $P = 4$

$$s_{\max, T} = \frac{0.4}{2} = 0.2$$

(a) Terminal voltage per phase

$$V_{ph} = \frac{V_t}{\sqrt{3}} = \frac{400}{\sqrt{3}}\ \text{V}$$

Now, rotor current per phase, $I_2 = \dfrac{V_{ph}}{Z}$

$$\therefore \quad I_2^2 = \frac{V_{ph}^2}{Z^2} = \frac{\left(\frac{400}{\sqrt{3}}\right)^2}{\left(\frac{r_2}{s_{\max,T}}\right)^2 + x_2^2}$$

$$= \frac{\left(\frac{400}{\sqrt{3}}\right)^2}{\left(\frac{0.4}{0.2}\right)^2 + (2)^2} = \frac{(400)^2}{8 \times 3}\ \text{amp}^2$$

$$= 6666.667\ \text{amp}^2$$

Synchronous speed, $n_s = \dfrac{120 \times f}{P}$

$$= \frac{120 \times 50}{4} = 1500\ \text{r.p.m.}$$

$$\omega_s = \frac{1500 \times 2\pi}{60}\ \text{rad/sec}$$

Maximum torque developed,

$$T_{\max} = \frac{3}{\omega_s} I_2^2 \times \frac{r_2}{s_{\max, T}}$$

$$= \frac{3 \times 60}{1500 \times 2\pi} \times \frac{0.4}{0.2} = 254.65\ \text{Nm}\ \textbf{Ans.}$$

(b) Let R be the sum of external resistance and rotor resistance. According to question,

$$T_{st} = 0.8 \times T_{\max}$$

or
$$\frac{3}{\omega_s}\frac{(400/\sqrt{3})}{R^2 + x_2^{\,2}} \times R = 0.8 \times \frac{3}{\omega_s} \times \frac{\left(\frac{400}{\sqrt{3}}\right)^2}{\left(\frac{r_2^{\,2}}{s_{\max,T}}\right) + x_2^{\,2}} \times \frac{r}{s_{\max,T}}$$

or
$$\frac{R}{R^2 + (2)^2} = 0.8 \times \frac{0.4}{0.2} \times \frac{1}{\left(\frac{0.4}{0.2}\right)^2 + (2)^2}$$

or
$$\frac{R}{R^2 + 4} = 0.8 \times \frac{0.4}{0.2} \times \frac{1}{8}$$

or
$$\frac{R}{R^2 + 4} = 0.2$$

or
$$R^2 + 4 = 5\text{ R}$$

or
$$R^2 - 5R + 4 = 0$$

This yields $R = \dfrac{5 \pm \sqrt{25 - 16}}{2}$

Neglecting higher value $R = 1$

or
$$r_2 + R_{ext} = 1$$

$\Rightarrow$
$$R_{ext} = 1 - 0.4 = 0.6\ \Omega \textbf{ Ans.}$$

Problem 6.18 : A 4-pole, 3-phase slip ring induction motor is coupled mechanically with a synchronous motor having 2 poles. The synchronous motor and stator of the induction motor are fed from 50 Hz voltage source. What will be the frequency of the emfs at the rotor terminals if the synchronous motor is driven:

(a) In a direction opposite to the induction-motor stator rotating field,

(b) In the direction of the induction-motor stator rotating field. If the frequency of the rotor voltage is required to be 300 Hz, then calculate

(c) The number of poles that the induction motor must have.

Solution :

For induction motor, $P = 4$, $f = 50$ Hz

$$N_{sI} = \frac{120 \times f}{P} = \frac{120 \times 50}{4} = 1500 \text{ r.p.m.}$$

For synchronous motor, $P = 2$, $f = 50$ Hz

$$N_{ss} = \frac{120 \times f}{P} = \frac{120 \times 50}{2} = 3000 \text{ r.p.m.}$$

(a) When the synchronous motor is driven in the direction opposite to the induction-motor stator rotating field,

$$\text{Relative speed} = N_{s1} - (-N_{ss})$$

$$N_R = 1500 + 3000$$

$$N_R = 4500 \text{ rpm}$$

Let f_1 be the frequency of the emfs at rotor terminals,

$$f_1 = \frac{N_R \times P}{120} = \frac{4500 \times 4}{120} = 150 \text{ Hz} \textbf{ Ans.}$$

(b) When the synchronous motor is driven in the direction of the induction-motor stator rotating field.

$$\text{Relative speed} = 3000 - 1500, \; N_R = 1500 \text{ rpm}$$

$$\text{Frequency at the rotor terminals} = \frac{N_R \times P}{120}$$

$$= \frac{1500 \times 4}{120} = 50 \text{ Hz} \textbf{ Ans.}$$

(c) Frequency of the rotor voltage = 300 Hz
Maximum relative speed between synchronous and induction motor stator field = 4500 rpm

$$\text{Number of poles} = \frac{120 \times f}{N_R} = \frac{120 \times 300}{4500} = 8$$

Total number of poles = 8 + synchronous poles
= 8 + 2 = 10 poles **Ans.**

Problem 6.19 : *(a)* A 3-phase, 6-pole, slip-ring induction machine is directly driven from the shaft by a 4-pole, 3-phase synchronous motor. If stators of both the machines are given a 50 Hz supply, what frequencies are available at the rotor slip-rings of the induction machine.

> *(b)* A 3-phase, 50 Hz, induction motor has a starting torque which is 1.25 times full-load torque and a maximum torque which is 2.50 times full-load torque. Neglecting stator resistance and rotational losses and assuming constant rotor resistance, find:

Solution :

(a) For induction motor $P = 6, \quad f = 50$ Hz

Synchronous speed $(N_{sI}) = \frac{120 \times f}{P} = \frac{120 \times 50}{6} = 1000$ r.p.m.

For synchronous motor, $P = 4, \quad f = 50$ Hz

Synchronous speed

$$(N_{ss}) = \frac{120 \times f}{P} = \frac{120 \times 50}{4} = 1500 \text{ r.p.m.}$$

When the stator fields of both the machines rotate in the same direction, Relative speed

$$(N_R) = N_{ss} - N_{sI} = 1500 - 1000 = 500 \text{ r.p.m}$$

Rotor frequency $= \frac{N_R \times P}{120} = \frac{500 \times 6}{120} = 25$ Hz **Ans.**

When the stator fields of both the machines rotate in opposite direction,Relative speed

$$(N_R) = N_{sI} + N_{ss} = 1500 + 1000 = 2500 \text{ r.p.m}$$

Rotor emf frequency $= \frac{N_R \times P}{120} = \frac{2500 \times 6}{120} = 125$ Hz **Ans.**

(b) Given, $T_{st} = 1.25\, T_{fl}$ (1)

$T_{max} = 2.50\, T_{fl}$ (2)

From (1) and (2) $T_{max} = 2 \times T_{st}$ (3)

To calculate slip at maximum torque,

$$\frac{T_{st}}{T_{max}} = \frac{2}{\frac{s_{max,T}}{1} + \frac{1}{s_{max,T}}} \quad \text{or} \quad \frac{1}{2} = \frac{2 \times s_{max,T}}{s^2_{max,T} + 1}$$

or $$s^2_{max,T} - 4\, s_{max,T} + 1 = 0$$

This yields $s_{max,T} = \frac{4 \pm \sqrt{16 - 4}}{2}$

Neglecting higher value, $s_{max,T} = 0.268$

(i) To calculate slip at full load, Again Applying the relation,

$$\frac{T_{fl}}{T_{\max}} = \frac{2}{\dfrac{s_{fl}}{s_{\max,T}} + \dfrac{s_{\max,T}}{s_{fl}}}$$

or
$$\frac{1}{2.50} = \frac{2}{\dfrac{s_{fl}}{0.268} + \dfrac{0.268}{s_{fl}}}$$

or
$$s_{fl}^2 + (0.268)^2 = 5 \times 0.268 \times s_{fl}$$

or
$$s_{fl}^2 - 1.34\, s_{fl} + 0.0718 = 0$$

This yields
$$s_{fl} = \frac{1.34 \pm \sqrt{(1.34)^2 - 4 \times 0.0718}}{2}$$

Neglecting higher value

$$s_{fl} = 0.056 \quad \textbf{Ans.}$$

(ii) From above, $s_{\max,T} = 0.268$ **Ans.**

(iii) to calculate rotor current per unit full-load rotor current, Applying the relation,

$$\frac{T_{st}}{T_{fl}} = \left(\frac{I_{st}}{I_{fl}}\right)^2 \times s_{fl}$$

or
$$\frac{I_{st}}{I_{fl}} = \sqrt{\frac{1}{s_{fl}} \times \frac{T_{st}}{T_{fl}}} = \sqrt{\frac{1}{0.056}} \times 1.25 = 4.725 \quad \textbf{Ans.}$$

Problem 6.20 : *(a)* A 3-phase induction motor has a slip of 4% at normal voltage. Find the slip of the induction motor when developing the same torque but at a voltage 10% less than the normal voltage.

(b) A 3-phase induction motor has 5% slip at full-load torque and at normal voltage. The rotor resistance and rotor standstill leakage reactance per phase are 0.1 Ω and 0.6 Ω respectively. What should be the percentage reduction in supply voltage if the motor has to develop full-load torque at 3/4th of normal full-load speed. Neglect stator impedance.

Solution :

Given, $s_{fl} = 0.04$

(a) Assuming linear-torque slip characteristics.

$$T = \frac{3}{\omega_s}\frac{s V^2}{r_2}$$

Let s_2 be the required slip. Since torque in constant,

$$\frac{3}{\omega_s}\frac{s_{fl} \times V_1^2}{r_2} = \frac{3}{\omega_s}\frac{s_2 \times (0.9 V_1)^2}{r}$$

or $$s_2 = \frac{s_{fl}}{0.81} = \frac{0.04}{0.81} = 4.938\% \quad \textbf{Ans.}$$

(b) Given, $s_{fl} = 0.05,\ r_2 = 0.1\ ,\ x_2 = 0.6$

Full load torque, $$T_{fl} = \frac{3}{\omega_s}\frac{V_1^2}{\left(\frac{r_2}{s_{fl}}\right)^2 + x_2^2} \times \frac{r_2}{s_{fl}} \qquad (1)$$

When speed is reduced to 3/4 of full load speed, slip also changes. Let s' be the new slip.

We have $$N_r = (1 - s) N_s \qquad (2)$$

When speed in reduced to 3/4th,

$$\frac{3}{4} N_r = (1 - s') N_s \qquad (3)$$

From (2) and (3) $$\frac{1 - s'}{1 - s} = \frac{3}{4}$$

or $$1 - s' = \frac{3}{4}(1 - s)$$

or $$s' = 1 - \frac{3}{4}(1 - 0.05) = 0.2875$$

Let x be the fraction of reduced supply voltage to keep full load torque constant.

$$\frac{3}{\omega_s}\frac{(x V_1)^2}{\left(\frac{r_2}{s'}\right)^2 + x_2^2} \times \frac{x_2}{s'} = \frac{3}{\omega_s}\frac{V_1^2}{\left(\frac{r_2}{s}\right)^2 + x_2^2} \times \frac{r_2}{s}$$

or $$x^2 = \frac{\left(\frac{r_2}{s}\right)^2 + x_2^2}{\left(\frac{r_2}{s}\right)^2 + x_2^2} \times \frac{s'}{s}$$

$$= \frac{\left(\frac{0.1}{0.2875}\right)^2 + (0.6)^2}{\left(\frac{0.1}{0.05}\right)^2 + (0.6)^2} \times \frac{0.2875}{0.05} = 0.6343 \therefore x = 0.7964$$

$$\% \text{ reduction} = \left(\frac{V_1 - xV_1}{V_1}\right) \times 100\% = \left(\frac{1 - 0.7964}{1}\right) \times 100\%$$

$$= 20.357\% \textbf{ Ans.}$$

Problem 6.21 : A 440-V, 3-phase, 4-pole, 50 Hz slip-ring star-connected induction motor has a voltage of 80 V between slip-rings when full-voltage is applied to the stator and the slip-rings are open-circuited with the rotor stationary. The stator current at no-load is 2A at a pf of 0.21 lagging. The rotor is star-connected with a per phase standstill leakage impedance of $0.05 + j\,0.25\,\Omega$ referred to rotor. For the motor running with the slip-rings short-circuited and at a slip of 5%, calculate

(a) The torque developed in Nm,

(b) The mechanical power developed,

(c) The rotor ohmic loss and

(d) The stator current and power factor.

Neglect stator leakage impedance and rotational losses.

Solution :

Given, $r_2 = 0.05\,\Omega/\text{phase}, \quad x_2 = 0.25\,\Omega/\text{phase}$

$$\text{Rotor induced emf per phase} = \frac{80}{\sqrt{3}} = 46.188 \text{ V}$$

$$\text{Rotor current at slip } s = 0.05 = \frac{s \times E_2}{\sqrt{r_2^2 + (s\,x_2)^2}}$$

$$= \frac{0.05 \times 46.188}{\sqrt{(0.05)^2 + (0.05 \times 0.25)^2}} = 44.809 \text{ A}$$

$$\text{Synchronous speed } (N_s) = \frac{120 \times f}{P} = \frac{120 \times 50}{4} = 1500 \text{ rpm}$$

$$\therefore \quad \omega_s = \frac{2\pi \times 1500}{60} \text{ rad/sec}$$

(a) Torque developed $T_e = \dfrac{3}{\omega_s} I_2^2 \times \dfrac{r_2}{s}$

$$= \frac{3 \times 60}{2\pi \times 1500} \times (44.809)^2 \times \frac{0.05}{0.05}$$

$$= 38.347 \text{ Nm} \quad \textbf{Ans.}$$

(b) Power across air gap,

$$P_G = \frac{3 I_2^2 r_2}{s}$$

$$= \frac{3 \times (44.809)^2 \times 0.05}{0.05} = 6023.539 \text{ W}$$

Mechanical power developed,

$$P_M = (1-s) P_G = (1-0.05) \times 6023.539$$

$$= 5722.4 \text{ W} \quad \textbf{Ans.}$$

(c) Rotor copper loss $= 3 \times I_2^2 \times r_2 = 3 \times (44.809)^2 \times 0.05$

$$= 301.18 \text{ W} \quad \textbf{Ans.}$$

(d) Rotor current at slip= 0.05

$$I_2 = 44.809 \text{ A, P.f.} = \frac{r_2}{\sqrt{r_2^2 + (s\,x_2)^2}}$$

$$= \frac{0.05}{\sqrt{(0.05)^2 + (0.05 \times .25)^2}} = 0.9701$$

$$\therefore \quad \theta = \cos^{-1}(0.9701) = 14.05°$$

In phasor form, $\bar{I}_2 = 44.809 \angle -14.05°$

$$= 43.469 - j\,10.878$$

When rotor is short circuited, a primary load current flow,

$$\bar{I}_1' N_1 = \bar{I}_2 N_2$$

$$\therefore \quad \bar{I}_1' = \frac{N_2}{N_1}(\bar{I}_2) = \frac{80}{440}(43.469 - j\,10.875)$$

$$= \frac{43.469}{5.5} - \frac{j\,10.875}{5.5} = 7.903 - j\,1.999$$

As we know that primary current is the sum of the exciting current and secondary load current,

$$\bar{I}_1 = \bar{I}_e + \bar{I}_1'$$

$$= 2(0.2 - j\,0.9798) + 7.903 - j\,1.999$$

$$= 0.4 - j\,1.9596 + 7.903 - j\,1.999$$

$$= 8.303 - j\,3.9586 = 9.198 \angle -25.49°$$

$$= 9.198 \text{ A at p.f. } 0.9027 \text{ lagging} \quad \textbf{Ans.}$$

Problem 6.22 : The starting current of a delta-connected 3-phase induction motor at rated voltage is 5 times the full-load current and the slip at full load is 5%. The no-load current is negligible.

(a) If an auto-transformer starter is used to limit the starting current from mains to 2 times the full-load current, estimate the starting torque then obtained as a percentage of the full-load torque.

(b) Neglecting stator impedance, determine the slip at which maximum torque occurs in the motor. (*I.A.S.,* 1991)

Solution :

Given, $s_{fl} = 0.05, \; I_{sc} = 5 \times I_{fl}$

This yields, $$\frac{I_{sc}}{I_{fl}} = 5 \tag{1}$$

When auto transformer is used,

$$\frac{I_{st}}{I_{fl}} = 2 \tag{2}$$

(a) To calculate starting torque in terms of full-load torque, applying the relation,

$$\frac{T_{st}}{T_{fl}} = \frac{I_{st} \times I_{sc}}{I_{fl}^2} \times s_{fl}$$

$$= \left(\frac{I_{st}}{I_{fl}}\right) \times \left(\frac{I_{sc}}{I_{fl}}\right) \times s_{fl}$$

$$= 2 \times 5 \times 0.05 = 0.5 = 50\%$$

$$T_{st} = 50\% \text{ of } T_{fl} \quad \textbf{Ans.}$$

(b) Since stator impedance is negligible. Applying the relation,

$$\frac{I_{sc}}{I_{fl}} = \sqrt{\frac{s_{fl}^2 + s_{\max T}^2}{s_{fl}^2\left(1 + s_{\max T}^2\right)}}$$

Squaring both sides

$$\left(\frac{I_{sc}}{I_{fl}}\right)^2 = \frac{s_{fl}^2 + s_{\max, T}^2}{s_{fl}^2\left(1 + s_{\max, T}^2\right)}$$

or $$(5)^2 = \frac{(0.05)^2 + s_{\max, T}^2}{(0.05)^2\left(1 + s_{\max, T}^2\right)}$$

or $$0.625\left(1 + s_{\max, T}^2\right) = 0.0025 + 0.625\, s_{\max, T}^2$$

or $$s_{\max T}^2 (1 - 0.0625) = 0.0625 - 0.0025$$

or $$s_{\max, T}^2 = \frac{0.0600}{1 - 0.0625}$$

$\therefore$ $$s_{\max, T} = \sqrt{\frac{0.0600}{1 - 0.0625}}$$

Slip at maximum torque, $s_{\max, T} = 0.253$ **Ans.**

Problem 6.23 : A 50 kVA, 440 V, 3-phase, 50 Hz induction motor is provided with a 3-phase step-down auto-transformer starter which steps down the voltage to 50% of the input. Given that the starting current of the motor on rated voltage is 6 times the rated full-load current, estimate the current drawn by the auto-transformer from the mains at starting. What is the starting kVA drawn the auto-transformer. (*I.A.S.*, 1987)

Solution :

Given, $I_{sc} = 6 \times I_f$, $x = 50\% = 0.5$

Full load current $$\left(I_{fl}\right) = \frac{\text{Power in kVA}}{\sqrt{3} \times V_L}$$

$$= \frac{50 \times 1000}{\sqrt{3} \times 440} = 65.61 \text{ A}$$

Current drawn by the autotransformer from the mains at starting

$$I_{st} = x^2 I_{sc} = x^2 \times 6 \times I_{fl}$$

$$= (0.5)^2 \times 6 \times 65.61 = 98.415 \text{ A} \quad \textbf{Ans.}$$

Starting current drawn by the autotransformer

$$= \sqrt{3} \times V_L \times I_{st} = \sqrt{3} \times 440 \times 98.415$$

$$= 75.002 \text{ kVA} \quad \textbf{Ans.}$$

Problem 6.24 : A 4-pole, 3-phase, 50Hz induction motor has a full-load slip of 3% and a maximum torque of twice the full-load torque. Clearly specifying any approximations you may need to make, calculate the starting torque of the motor as a percentage of full-load torque. (*I.A.S.,* 1995)

Solution :

Given, $s_{fl} = 0.03, \; T_{max} = 2 \times T_{fl}$

Now, applying,
$$\frac{T_{fl}}{T_{max}} = \frac{2}{\dfrac{s_{max,T}}{s_{fl}} + \dfrac{s_{fl}}{s_{max,T}}}$$

or
$$\frac{1}{2} = \frac{2}{\dfrac{s_{max,T}}{0.03} + \dfrac{0.03}{s_{max,T}}}$$

or
$$s_{max,T}^2 + 0.0009 = 0.12\, s_{max,T}$$

or
$$s_{max,T}^2 - 0.12\, s_{max,T} + 0.0009 = 0$$

This yields,
$$s_{max,T} = \frac{0.12 \pm \sqrt{(0.12)^2 - 4 \times 0.0009}}{2}$$

$$= 0.0081 \quad \text{or} \quad 0.112$$

Since the $s_{max,T}$ can be less than full-load slip,

$\therefore$
$$s_{max,T} = 0.112$$

Now, Applying
$$\frac{T_{st}}{T_{max}} = \frac{2}{s_{max,T} + \dfrac{1}{s_{max,T}}}$$

or
$$T_{st} = \left(\frac{2}{0.112 + \dfrac{1}{0.112}} \right) \times T_{max}$$

$$= 0.2212 \times T_{max} = 0.2212 \times 2 \times T_{fl}$$

$$= 0.4424\, T_{fl} = 44.24\% \text{ of } T_{fl}$$

This relation shows that the starting torque of the motor is 44.24% of the full load torque. **Ans.**

> **Problem 6.25 :** In a 3-phase induction motor, the stator reactance equals the rotor reactance at standstill while each resistance is one-fourth of this value. If the motor develops 220 Nm at 3% slip, what will be its
> *(a)* Starting and
> *(b)* Pull-out torques. Ignore no-load current.

Solution :

Given,

Stator reactance (x_1) = rotor reactance at standstill (x_2)

$$T_L = 220 \text{ Nm}, \; s_l = 0.03$$

rotor resistance (r_2) = Stator resistance

$$(r_1) = \frac{1}{4} \times \text{reactance}, \; x_1 = x_2 = 4\, r_1 = 4\, r_2$$

Neglecting shunt parameters,

(a) Now, Applying the relation,

$$T_e = \frac{3}{\omega_s} \frac{V^2}{\left(x_1 + \frac{r_2}{s}\right)^2 + (x_1 + x_2)^2} \times \frac{r_2}{s}$$

or
$$220 = \frac{3V^2}{\omega_s} \times \frac{1}{\left(r_2 + \frac{r_2}{0.03}\right)^2 + (4\, r_2 + 4\, r_2)^2} \times \frac{r_2}{0.03}$$

or
$$220 = \frac{3V^2}{\omega_s} \times \frac{1}{r_2^2\left[\left(1 + \frac{1}{0.03}\right)^2 + (4+4)^2\right]} \times \frac{r_2}{0.03}$$

or
$$220 = \frac{3V^2}{\omega_s \times r_2} \times \frac{1}{\left(1 + \frac{1}{0.03}\right)^2 + 64} \times \frac{1}{0.03}$$

$$\text{Putting } K = \frac{3V^2}{\omega_s \times r_2} \quad \text{or} \quad 220 = K \times 0.0268$$

$$\therefore \quad K = 8208.955$$

Now, At starting slip = 1

$$\therefore \quad T_{st} = \frac{3V^2}{\omega_s} \times \frac{1}{(r_1 + r_2)^2 + (x_1 + x_2)^2} \times r_2$$

or

$$T_{st} = \frac{3V^2}{\omega_s \times r_2} \times \frac{1}{(1+1)^2 + (4+4)^2}$$

$$= \frac{8208.955}{68} = 120.72 \text{ Nm} \quad \textbf{Ans.}$$

(b) For pull-out torques,

$$s_{max,T} = \frac{r_2}{\sqrt{r_1^2 + (x_1 + x_2)^2}} = \frac{r_2}{\sqrt{r_2^2 + (x_1 + x_2)^2}}$$

$$= \frac{1}{\sqrt{1 + (4+4)^2}} = \frac{1}{\sqrt{65}} = 0.124$$

Now, Applying the relation, Pull-out torque

$$= \frac{3V^2}{\omega_s} \frac{1}{\left(r_1 + \frac{r_2}{s_{max,T}}\right)^2 + (x_1 + x_2)^2} \times \frac{r_2}{s_{max,T}}$$

$$= K \frac{1}{\left(1 + \frac{1}{s_{max,T}}\right)^2 + (4+4)^2} \times \frac{1}{s_{max,T}}$$

$$= \frac{8208.955}{\left(1 + \frac{1}{0.124}\right)^2 + 64} \times \frac{1}{0.124} = 452.92 \quad \textbf{Ans.}$$

Problem 6.26 : A 6-pole, 50 Hz, 3-phase induction motor has a rotor resistance of 0.2 Ω per phase and a maximum torque of 160 Nm at 875 rpm. Calculate:

(*a*) The torque for a full-load slip of 4% and

(*b*) The resistance to be added to the rotor circuit to obtain 80% of full-load torque at starting.

Rotational losses and stator impedance are neglected.

Solution :

Given, $P = 6$, $f = 50$ Hz, $N_{max} = 875$ rpm,

$T_{max} = 160$ Nm, $s_{fl} = 0.04$, $r_2 = 0.2\ \Omega/\text{Phase}$

Synchronous speed $$(N_s) = \frac{120 \times f}{P} = \frac{120 \times 50}{6} = 1000 \text{ rpm}$$

$$\therefore \quad s_{max, T} = \frac{N_s - N_{max}}{N_s} = \frac{1000 - 875}{1000}$$

$$= \frac{125}{1000} = 0.125$$

or $$\frac{r_2}{x_2} = 0.125 \therefore x_2 = \frac{r_2}{0.125} = \frac{0.2}{0.125} = 1.6\ \Omega$$

(*a*) Now, Applying the relation,

$$\frac{T_{fl}}{T_{max}} = \frac{2}{\dfrac{s_{fl}}{s_{max, T}} + \dfrac{s_{max, T}}{s_{fl}}}$$

$$\therefore \quad T_{fl} = \left(\frac{2}{\dfrac{s_{fl}}{s_{max, T}} + \dfrac{s_{max, T}}{s_{fl}}} \right) \times T_{max}$$

$$= \left(\frac{2}{\dfrac{0.04}{0.125} + \dfrac{0.125}{0.04}} \right) \times 160 = 92.89 \text{ Nm} \quad \textbf{Ans.}$$

(*b*) Given, $T_{st} = 80\% \text{ of } T_{fl} = 0.8 \times 92.89$

$= 74.312$ Nm

Let $s'_{max, T}$ be the maximum slip when additional resistance is added to rotor circuit.

Now, Applying the relation,

$$\frac{T_{st}}{T_{max}} = \frac{2}{\dfrac{s'_{max,T}}{1} + \dfrac{1}{s'_{max,T}}}$$

or
$$\frac{74.312}{160} = \frac{2\, s'_{max,T}}{s'^2_{max,T} + 1}$$

or
$$s'^2_{max,T} - 4.306\, s'_{max,T} + 1 = 0$$

This yields
$$s'_{max,T} = \frac{4.306 \pm \sqrt{(4.306)^2 - 4}}{2}$$

Neglecting the higher value, $s'_{max,T} = 0.2463$

or
$$\frac{r'_2}{x_2} = 0.2463$$

$\therefore$
$$r'_2 = 1.6 \times .2463 = 0.3941\,\Omega$$

Additional resistance to the rotor $= r'_2 - r_2 = 0.3941 - 0.2$

$= 0.1941\,\Omega$ **Ans.**

> **Problem 6.27 :** A 150 kW, 2500 V, 6 pole, 50 Hz, star connected slip ring induction motor has a star connected rotor. Stator to rotor turn ratio is 4 : 2. Rotor resistance and leakage reactance are 0.1 Ω and 1 Ω respectively. Neglect stator impedances. Find *(a)* current and torque at starting on rated voltage and with slip rings short circuited and *(b)* the external resistance required to reduce the starting current to 40 A with across the line starting. Compute also the starting torque under these conditions.

Solution :

Given, $f = 50$ Hz, $P = 6$

$N_s : N_r = 4 : 2$, $r_2 = 0.1\,\Omega$, $x_2 = 1\,\Omega$

Synchronous speed
$$(W_s) = \frac{120 \times f}{P} = \frac{120 \times 50}{6}$$

$$= \frac{1000 \times 2\pi}{60} \text{ rad/s} = \frac{100\pi}{3} \text{ rad/s.}$$

Terminal voltage/phase $= \dfrac{2500}{\sqrt{3}} = 1443.4$ V

Rotor resistance on stator side,

$$r_2' = \left(\frac{4}{2}\right)^2 \times 0.1 = 4 \times 0.1 = 0.4\,\Omega, \quad x_2' = 4 \times 1 = 4\,\Omega$$

(a) At starting, slip = 1

Starting current, $I_{st} = \dfrac{1443.4}{\sqrt{r_2'^2 + x_2'^2}} = \dfrac{1443.4}{\sqrt{16.16}} = 359.1\text{ A}$ **Ans.**

Starting torque, $T_{st} = \dfrac{3}{\omega_s} I_2^2\, r_2' = \dfrac{3 \times 3}{100\pi} \times (359.1)^2 \times 0.4$

$= 1478.1\text{Nm}$ **Ans.**

(b) Let R be the external resistance connected in each phase of the rotor circuit. $I_{st} = \dfrac{V_t}{\sqrt{(r_2' + R)^2 + x_2'^2}}$

or $$40 = \frac{1443.4}{\sqrt{(0.4 + R)^2 + 16}}$$

or $$(0.4 + R)^2 = \left(\frac{1443.4}{40}\right)^2 - 16$$

or $$0.4 + R = \sqrt{\left(\frac{1443.4}{40}\right)^2 - 16} = 35.86$$

$\therefore$ $$R = 35.46\,\Omega$$

When this resistance is reffered to rotor side.

$$R' = \left(\frac{1}{2}\right)^2 \times 35.46 = 8.97\,\Omega \quad \textbf{Ans.}$$

Starting torque in this situation,

$$T_{est} = \frac{3}{W_s}(I_{st})^2 \times (r_2' + R)$$

$$= \frac{3 \times 3}{100\pi} \times 1600 \times 35.86 = 1644.5 \text{ Nm} \quad \textbf{Ans.}$$

Problem 6.28 : Design the 4 section of a 5 stud-starter for a 3 phase slip-ring induction motor. The full load slip is 3% and the maximum starting current is limited by 3 times the full load current. Resistance per phase is $0.02\,\Omega$.

Solution :

Given, $r_2 = 0.02\,\Omega$

Number of section $(n) = 4,\ S_{fl} = 0.03,\ I_{sm} = 3 \times I_{fl}$

Rotor current, $$I_2 = \frac{E_2}{\sqrt{\left(\frac{r_2}{S}\right)^2 + x_2}}$$

For very small slip, $\frac{r_2}{s} >> x_2$, hence x_2 can be neglected in comparison

to $\frac{r_2}{s}$ $\therefore\ I_2 = \frac{s\,E_2}{r_2}$ (1)

Hence, $\therefore$ $$I_{2fl} = \frac{s_{fl} \times E_2}{r_2} = \frac{0.03\,E_2}{r_2}$$

From (1), Maximum starting torque in given by,

$$I_{sm} = \frac{E_2}{r_2} \times s_{\max} \text{ or } 3 \times I_{fl} = \frac{E_2}{r_2} \times s_{\max}$$

or $$3 \times 0.03\frac{E_2}{r_2} = \frac{E_2}{r_2} \times s_{\max}$$

$\therefore$ $$s_{\max} = 0.09$$

We know that $$\alpha = (s_{\max})^{\frac{1}{n}} = \left(\frac{1}{0.09}\right)^{\frac{1}{4}} = 0.548,\ [\because n = 4]$$

The total resistance in the rotor circuit on first step is given by

$$R_1' = \frac{r_2}{s_{\max}} = \frac{0.02}{0.09} = 0.222\,\Omega$$

The resistance of various elements are

$$R_1 = R_1'(1-\alpha) = 0.222(1-0.548) = 0.1004\,\Omega \textbf{ Ans.}$$

$$R_2 = \alpha\,R_1 = 0.548 \times 0.1004 = 0.055\,\Omega \textbf{ Ans.}$$

$$R_3 = \alpha\,R_2 = 0.055 \times 0.548 = 0.0302\,\Omega \textbf{ Ans.}$$

$$R_4 = \alpha\,R_3 = 0.548 \times 0.0302 = 0.0165\,\Omega \textbf{ Ans.}$$

Problem 6.29 : Calculate the values of resistance elements of 5 step starter for a 3-phase 440 V, wound rotor induction motor. The full load slip is 2% and the maximum starting current is limited to twice its full load value. Rotor resistance per phase is 0.02.

Solution

Given, number of steps $(n) = 5$

$$s_{fl} = 0.02, \quad r_2 = 0.02\ \Omega$$

The rotor current is given by $I_2 = \dfrac{E_2}{\sqrt{\left(\dfrac{r_2}{s}\right)^2 + x_2^2}}$

For very small value of slip, $\dfrac{r_2}{s} >> x_2$

$$\therefore \quad I_2 = \frac{s\,E_2}{r_2} \qquad (1)$$

$$\text{Now, Hence } I_{fl} = s_{fl} \times \frac{E_2}{r_2} \qquad (2)$$

$$\text{and } I_{sm} = s_{max} \times \frac{E_2}{r_2} \qquad (3)$$

From (2) and (3) $\dfrac{I_{sm}}{I_{fl}} = \dfrac{s_{max}}{s_{fl}}$

or $\dfrac{2 \times I_{fl}}{I_{fl}} = \dfrac{s_{max}}{s_{fl}}$

or $s_{max} = 2 \times s_{fl} = 2 \times 0.02 = 0.04$

$$s_{max} = 0.04$$

$$\text{We have, } \alpha = (s_m)^{\frac{1}{n}} = (0.04)^{\frac{1}{5}} = 0.525$$

$$\text{Total resistance, } R_1' = \frac{r_2}{s_m} = \frac{0.02}{0.04} = 0.500\ \Omega$$

The resistance of various elements are

$$R_1 = R_1'(1-\alpha) = 0.5(1-0.525) = 0.2375\ \Omega$$

$$R_2 = \alpha\ R_1 = 0.525 \times 0.2375 = 0.1247\ \Omega$$

$$R_3 = \alpha\ R_2 = 0.525 \times 0.1247 = 0.0655\ \Omega$$

$$R_4 = \alpha\ R_3 = 0.525 \times 0.0655 = 0.0344\ \Omega$$

$$R_5 = \alpha\ R_4 = 0.525 \times 0.0344 = 0.0180\ \Omega \quad \textbf{Ans.}$$

Note: This can be checked by

$$R_1' = R_1 + R_2 + R_3 + R_4 + R_5 + r_2 = 0.5\ \Omega$$

Problem 6.30 : A 50 kW, 3-ϕ induction motor has standstill impedance $1 + j\,4.0\ \Omega$ per phase. The motor is fed from 500 V, 60 Hz source through a feeder 600 m long. Determine the minimum allowable cross-section of each conductor of feeder in order that starting torque shall not be reduced by more than 40% due to voltage drop in the feeder the machine is started through a star-delta starter. Take the resistivity (ρ) of feeder material is $0.03/\text{m}/\text{mm}^2$.

Solution :

Given, $$Z_s = 1 + j\,4.0\ \Omega$$

$$\text{Terminal voltage per phase} = \frac{500}{\sqrt{3}} = 288.7\ \text{V}$$

Length of the feeder (l) = 600 m.

The starting torque without feeder is given by

$$T_{sw} = \frac{3V^2}{\omega_s} \frac{1}{(1)^2 + (4)^2} \times 1.0$$

Let R be the resistance of each feeder. The starting torque in this case,

$$T_{sf} = \frac{3V^2}{\omega_s} \times \frac{1}{(R+1)^2 + (4)^2} \times 1.0$$

According to question, the starting torque when fed through feeder should at least be equal to 0.6 times the starting torque without feeder

$$\frac{3V^2}{\omega_s} \times \frac{1}{(R+1)^2 + 16} \times 1 = 0.6 \times \frac{3V^2}{\omega_s} \frac{1}{1^2 + 16} \times 1$$

or $$\frac{1}{(R+1)^2 + 16} = 0.6 \text{ or } \frac{17}{0.6} = (R+1)^2 + 16$$

or $$(R+1)^2 = 28.33 - 16 = 12.33$$

$$\therefore \quad R+1 = \sqrt{12.33} = 3.511$$

$$\therefore \quad R = 2.511\,\Omega$$

Now we know that $R = \rho \dfrac{l}{A}$

$$\therefore \quad A = \frac{\rho l}{R} = \frac{0.03 \times 600}{2.511} = 7.168 \text{ mm}^2$$

Thus, Minimum allowable cross-section of the feeder conductor

$$= 7.168 \text{ mm}^2 \quad \textbf{Ans.}$$

Problem 6.31 : A 50 kW induction motor with star-delta starter is supplied through a feeder from 500 V, 50 Hz main, owing to the line drop, the starting torque is found to be the same with star as with delta connections. Determine the resistance of the feeder. From a short circuit test on the motor, when delta connected, the following data were obtained:

V = 250 V, I = 100 A, Power factor = (0.3)

If a second feeder of the same cross-section is run in parallel with original one. Find the percentage increase in starting torque in star connection.

Solution :

Given, V_{sc} = 250 V, I_{sc} = 100 A, p.f = 0.3

From above data, the per phase Quantity can be calculated as

$$Z_{sc} = \frac{250 \times \sqrt{3}}{100} = 4.333\,\Omega/\text{phase}$$

$$R_{sc} = Z_{sc} \cos\theta_{sc} = 4.333 \times 0.3 = 1.2999\,\Omega$$

$$\therefore \quad X_{sc} = \sqrt{Z_{sc}^{\,2} - R_{sc}^{\,2}}$$

$$= \sqrt{(4.338)^2 - (1.2999)^2} = 4.133\,\Omega$$

Let R be the resistance of each feeder line. The starting torque with star connection is given by

$$= \frac{3}{\omega_s}\left(\frac{500}{\sqrt{3}}\right)^2 \frac{1}{(R+1.299)^2 + (4.133)^2} \times 1.2999$$

When the stator winding is connected in delta,

$$T_{st} = \frac{3}{\omega_s}\left(\frac{500}{\sqrt{3}}\right)^2 \frac{1}{\left(R+\dfrac{1.299}{3}\right)^2+\left(\dfrac{4.133}{3}\right)^2} \times \left(\frac{1.2999}{3}\right)$$

Since starting torque is same

$$= \frac{3}{\omega_s}\left(\frac{500}{\sqrt{3}}\right)^2 \times \frac{1}{(R+4.33)^2+(1.378)^2} \times \frac{1.2999}{3}$$

$$= \frac{3}{\omega_s}\left(\frac{500}{\sqrt{3}}\right)^2 \times \frac{1}{(R+1.299)^2+(4.133)^2} \times 1.2999$$

or $$\frac{1}{(R+0.433)^2+(1.378)^2} = \frac{3}{(R+1.299)^2+(4.133)^2}$$

or $$(R+1.2999)^2+(4.133)^2 = 3\times\left[(R+0.433)^2+(1.378)^2\right]$$

or $$(R+1.2999)^2 - 3(R+0.433)^2 = (1.378)^2\times 3-(4.133)^2$$

or $$R^2+2.6\,R+1.69-3\left(R^2+0.866\,R+0.187\right) = -\ 11.31$$

or $$R^2+2.6\,R+1.69-3\,R^2-2.598\,R-0.561 = -\ 11.31$$

or $$2\,R^2-0.002\,R-11.31-1.129 = 0$$

or $$2\,R^2-0.002\,R-12.439 = 0$$

This yields $$R = \frac{0.002\pm\sqrt{(0.002)^2+4\times 12.439}}{4} = \frac{0.002\pm 7.054}{4}$$

Neglecting negative value,

$$R = \frac{7.056}{4} = 1.764\ \Omega \textbf{ Ans.}$$

The starting torque with star connection and with one feeder is given by

$$T_{st1} = \frac{3}{\omega_s}\frac{(288.7)^2}{(1.764+1.299)^2+(4.133)^2}\times 1.2999$$

Substituting $$K = \frac{3}{\omega_s}(288.7)^2 = K\times\frac{1.2999}{(3.063)^2+(4.133)^2} = 0.049\,\text{K}$$

The starting torque with star connection and with two feeder

$$T_{st2} = K \times \frac{1}{\left(\frac{1.764}{2}+1.2999\right)^2+(4.133)^2} \times 1.299$$

$$= K \times \frac{1.2999}{(0.882+1.2999)^2+(4.133)^2} = 0.0595\,\text{K}$$

% increase in starting torque when star connected.

$$= \left(\frac{0.0595\,K - 0.049\,K}{0.049\,K}\right) \times 100\% = 21.45\% \textbf{ Ans.}$$

Problem 6.32 : A 15 kW, 500 V, 3-phase induction motor has full load efficiency of 90% and power factor of 0.8. At standstill and at rated voltage, the motor draws 4 times its full load current and develops a starting torque of 2 times its full load torque. An autotransformer is installed to reduce the starting current and to give full load torque at starting. Neglect exciting current of auto transformer, determine at the time of starting

(i) Voltage applied to motor terminals
(ii) Current drawn by the motor
(iii) Line-current drawn from the supply mains.

Solution :

Given, Efficiency = 90% = 0.9 , p.f = 0.8

$$I_{st} = 4 \times I_{fl} \tag{1}$$

$$T_{st1} = 2 \times T_{fl} \tag{2}$$

$$T_{st2} = T_{fl} \tag{3}$$

(i) We know that electromagnetic torque is proportional to the square of the voltage $T_{st1} \propto (500)^2$ and $T_{st2} \propto V_1^2$

Now, $$\frac{T_{st1}}{T_{st2}} = \left(\frac{500}{V_1}\right)^2$$

$$\frac{2\,T_{fl}}{T_{fl}} = \left(\frac{500}{V_1}\right)^2 \quad \text{or} \quad \frac{500}{V_1} = \sqrt{2}$$

$$\therefore \qquad V_1 = \frac{500}{\sqrt{2}} = 353.55 \text{ V } \textbf{Ans.}$$

(ii) For 500 V, the motor starting current $= 4\, I_{fl}$

$\therefore$ For 353.55 V, the motor starting current,

$$= 4 \times I_{fl} \times \frac{353.55}{500} = 2.83\, I_{fl}$$

Motor full load current,

$$I_{fl} = \frac{15 \times 1000}{\sqrt{3} \times 0.9 \times 500 \times .8} = 24.06 \text{ A}$$

$\therefore$ Current drawn by the motor $= 2.83 \times 24.06 = 68.08$ A **Ans.**

(iii) On the auto-transformer secondary, the motor current is 68.08 A and the secondary voltage is 353.55 V. Therefore, the current on the auto-transformer primary side of 500 V in given by

$$= \frac{353.55}{500} \times 68.08 = 48.12 \text{ A}$$

$\therefore$ Current drawn from the supply mains = 48.12 A **Ans.**

Problem 6.33 : A three phase 400 V distribution circuit is designed to supply not more than 1000 amperes. Assuming that a three phase squirel cage induction motor has a full load efficiency of 90% and a full load power factor is 0.8 and that the starting current at rated voltage is 4 times the rated full load current, what is the maximum permissable kW rating of the motor.

(a) If it is to be started at full load voltage

(b) If it is to be started using an auto-transformer stepping down the voltage to 85%.

Solution

Given, $\qquad I_{max} = 1000$ A, full load p.f = 0.8

Full load efficiency = 0.9, $\qquad I_{sc} = 4 \times I_{fl} \qquad (1)$

(a) Maximum permissible line current that the 3-ϕ induction motor can take from distribution circuit is 1000 A at the starting. It is given that the starting torque at rated voltage is 4 times the full load current. Therefore, the rated line current of 3-phase induction motor will full voltage starting is $\frac{1000}{4} = 250$ A. Thus the maximum permissible kW rating of the induction motor

$$= \sqrt{3} \times 400 \times 250 \times 0.8 \times 0.9 = 124.708 \text{ kW } \textbf{Ans.}$$

(b) The maximum permissible starting current from main supply

$= 1000$ A $\qquad I_{st} = 1000 = x^2 I_{sc}$ (2)

Since $\qquad I_{sc} = 4 \times I_{fl}$

Here $\qquad x = 8.5$

From (2), $\quad (0.85)^2 \times 4 \times I_{fl} = 1000$

$$\therefore \qquad I_{fl} = \frac{1000}{4 \times (0.85)^2} = 346.02 \text{ A}$$

$\therefore$ Maximum permissible kW rating of the induction motor,

$$= \sqrt{3} \times 900 \times 346.02 \times 0.8 \times 0.9$$

$$= 172.605 \text{ kW } \textbf{Ans.}$$

Problem 6.34 : A three phase squirrel cage induction motor, with an applied voltage of 50%, gives a block rotor current of 300% and internal starting torque of 40% of their corresponding, rated values. If an auto transformer limits the starting line current to 200% of the motor full load current, calculate the percentage starting torque.

Solution

Let V_1 be the rated voltage.

A/Q, 50% of applied voltage, i.e. $0.5\,V_1$ gives a blocked rotor current of 3 times the full load current.

$\therefore$ Rated voltage V_1 would give a block rotor current or short circuit current,

$$I_{sc} = \frac{V_1}{0.5\,V_1} \times 3I_{fl}, \quad I_{sc} = 6\,I_{fl}$$

Now, According to Question, the starting line current I_{st} is limited to 200%, i.e. $2\,I_{fl}$.

$\therefore \qquad I_{st} = 2\,I_{fl}$, Since $\quad I_{st} = x^2\,I_{sc}$

or $\qquad x^2\,I_{sc} = 2\,I_{fl}$ or $\quad x^2 \times 6\,I_{fl} = 2\,I_{fl}$

or $$x^2 = \frac{1}{3}$$

For an induction motor, starting torque $\propto (\text{voltage})^2$, $T_{st1} \propto (0.5\,V_1)^2$

or $$0.4\,T_{fl} \propto (0.5\,V_1)^2 \qquad (1)$$

Also $$T_{st2} \propto (x\,V_1)^2 \qquad (2)$$

From (1) and (2) $$\frac{T_{st2}}{T_{st1}} = \frac{(x\,V_1)^2}{(0.5\,V_1)^2}$$

or $$T_{st2} = \frac{x^2}{(0.5)^2} \times T_{st1}$$

or $$T_{st2} = \frac{1}{3 \times (0.25)} \times 0.4\,T_{fl}$$

or $$T_{st2} = \frac{1}{0.75} \times 0.4 \times T_{fl} = 0.5333\,T_{fl}$$

This relation shows that the starting torque is 53.33% of full load torque T_{st2} = 53.33% of full load torque **Ans.**

Problem 6.35 : A 3-phase, 500 kW, 2.5 kV, 50 Hz, star connected induction motor works at a full load power factor of 0.8 lagging. A delta connected capacitor bank is used to raise the full-load power factor to 0.9 lag. Calculate the capacitance of the bank. If each capacitor is rated 400 V, 50 Hz, compute the capacitance of each unit.

The motor efficiency is 85%.

Solution :

Given, Terminal Voltage = 2500 V

$$\text{Terminal Voltage/phase} = \frac{2500}{\sqrt{3}} = 1443.4 \text{ V}$$

$$\text{Efficiency} = 85\% = 0.85$$

Initial operating power factor, $\cos\theta_1 = 0.8$ (lag)

Final operating p.f. $(\cos\theta_2) = 0.9$ (lag)

$$\text{Motor input rated current/phase} = \frac{\text{rated power}}{\sqrt{3} \times V_L \times \text{p.f} \times \eta_m}$$

$$= \frac{500 \times 1000}{\sqrt{3} \times 2500 \times .85 \times 0.8} = 169.81 \text{ A}$$

Let us take terminal voltage as reference phasor.

$\therefore \qquad \bar{I}_a = 169.81(0.8 - j0.6) = 135.85 - j101.87$

The magnitude of per phase motor current at improved p.f. 0.9 is given

by $$I'_a = \frac{\text{Real component of } I_a}{0.9} = \frac{135.85}{0.9} = 150.94\text{A}$$

In phasor form, $\bar{I}'_a = 150.94(0.9 - j0.436) = 135.85 - j65.79$

Now, Reactive component of current per phase that capacitor must supply is equal to difference between reactive component of I_a and I'_a.

$\therefore$ Current through capacitor = 101.87 – 65.79 = 36.08 A

As induction motor is star connected and the capacitor bank is delta connected, the delta connected capacitor bank is firstly converted equivalentally into star.

We have $$X_Y = \frac{X_\Delta}{3} = \frac{1}{2\pi f 3c}$$

Per phase voltage = 1443.4 V

Per phase current = 36.08 A

$\therefore$ Per phase reactance, $$X_Y = \frac{1443.4}{36.08}$$

or $$\frac{1}{2\pi f 3C} = \frac{1443.4}{36.08}$$

or $$C = \frac{36.08}{1443.4 \times 2 \times \pi \times 50}$$

Capacitance of bank = 79.61 μF **Ans.**

It is given that each capacitor is rated at 400 V,

$\therefore$ Number of series capacitors $= \frac{2500}{400} = 6.25 \cong 7$

$\therefore$ Capacitance of each capacitor $= 79.61 \times 7 = 557.27\ \mu\text{F}$ **Ans.**

Problem 6.36 : An induction motor operating from a balanced three phase 440 V, 50 Hz supply, takes a line current of 12 A at 0.8 lagging. Calculate the per phase value of capacitance and the total kVA rating of the 3-ϕ capacitor bank, which when connected to motor terminals would improve the line power factor to 0.9 lag. Assume that stator is star connected and capacitor is also star connected.

Solution

Given, Terminal Voltage = 440 V

$$I_a = 12 \text{ A}, \quad \text{p.f} = 0.8 \text{ (lag)}$$

$$\overline{I}_a \, \overline{I}_a = 12(0.8 - j0.6)$$

Improved p.f., $\cos\theta' = 0.9$ (lagging)

Let us take phase terminal voltage as reference phasor.

$$\therefore \quad \overline{I}_a = 9.6 - j\,7.2$$

Magnitude of motor current at p.f. 0.9 is

$$I'_a = \frac{\text{real part of } I_a}{\text{New p.f.}} = \frac{9.6}{0.9} = 10.67 \text{ A}$$

In phasor form, $\overline{I}'_a = 10.67(0.9 - j\,0.436) = 9.6 - j\,4.65$

The reactive component of current per phase that capacitor must supply is equal to the difference between reactive component of I_a and I'_a

$$\therefore \quad I_c = 7.2 - 4.65 = 2.55 \text{ A}$$

Now,
$$X_c = \frac{V}{I_c} = \frac{440}{\sqrt{3} \times 2.55} = 99.62\ \Omega$$

Since capacitor bank is star connected,

$$X_c = \frac{1}{2\pi f C}$$

$$\therefore \quad C = \frac{1}{2 \times \pi \times 50 \times 99.62} = 31.97\ \mu\text{F} \quad \textbf{Ans.}$$

Total kVA rating of the 3-phase capacitor bank

$$= 3\,I_c \times V = 3 \times 2.55 \times \frac{440}{\sqrt{3}} = 1943.4 \text{ VA}$$

$$= 1.943 \text{ kVA} \quad \textbf{Ans.}$$

Problem 6.37 : A 100 kW, 400 V, 6 pole, 60 Hz, star connected induction motor has the following impedance parameters in ohms per phase reffered to stator side.

$$r_1 = 0.01\ \Omega,\ r_2 = 0.02\ \Omega$$

$$x_1 = x_2 = 0.2\ \Omega,\ X_d = 9\ \Omega$$

At rated shaft output, it runs with a slip of 5% and with an efficiency of 90%

Calculate stator current stator and rotor I^2R loss, power across air gap and total losses.

Solution

Given, $r_1 = 0.01\,\Omega, r_2 = 0.02\,\Omega, \quad x_1 = x_2 = 0.2\,\Omega, X_\phi = 9\,\Omega$

Terminal Voltage = 400 V, $P = 6$, $f = 60$ Hz
Efficiency = 0.9, slip = 0.05
The equivalent circuit on stator side can be drawn as

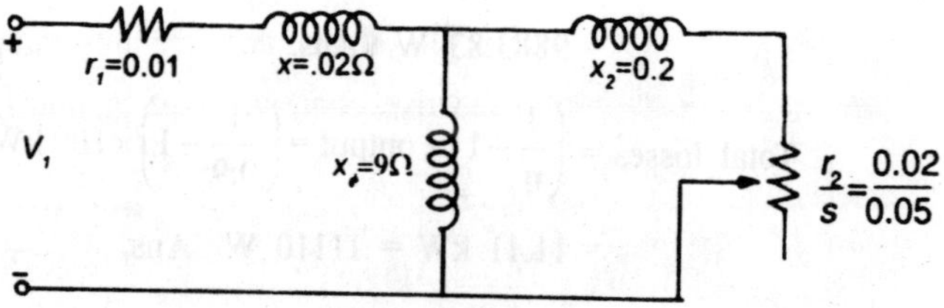

Fig. P. 6.37

Applying the relation,

$$Z_f = R_f + i\,X_f = \frac{\left(\frac{r_2}{s} + j\,x_2\right) \times \left(j\,X_\phi\right)}{\frac{r_2}{s} + j\left(x_2 + X_\phi\right)}$$

$$= \frac{\left(\frac{0.02}{0.05} + j\,0.2\right) \times j\,9}{\frac{0.02}{0.05} + j\,(9 + 0.2)} = \frac{j\,3.6 - 1.8}{0.4 + j\,9.2}$$

$$= \frac{-1.8 + j\,3.6}{0.4 + j\,9.2} \times \frac{0.4 - j\,9.2}{0.4 - j\,9.2}$$

$$= \frac{32.4 + j\,18}{84.8} = 0.382 + j\,0.212$$

We have, $\bar{Z}_s = r_1 + j\,x_1 = (0.01 + j\,0.2)\,\Omega$

Now, Total impedance, $\bar{Z} = \bar{Z}_f + \bar{Z}_s$

$$= 0.01 + j\,0.2 + 0.382 + j\,0.212$$
$$= 0.392 + j\,0.412 = 0.569\ \angle 46.42°\ \Omega$$

Stator input current $(I_1) = \dfrac{V_{Ph}}{Z} = \dfrac{400}{\sqrt{3} \times 0.569} = 405.87$ A **Ans.**

Air gap power $(P_G) = 3\,I_1^2\,R_f = 3 \times (405.87)^2 \times 0.382$

$= 188781$ W **Ans.**

Total stator I^2R loss $= 3\,I_1^2 \times r_1 = 3 \times (405.87)^2 \times 0.01$

$= 4941.91$ W **Ans.**

Total rotor I^2R loss $= 3 \times (I_1)^2 \times r_2 = 3 \times (405.87)^2 \times 0.02$

$= 9883.83$ W **Ans.**

$$\text{Total losses} = \left(\frac{1}{\eta} - 1\right) \times \text{output} = \left(\frac{1}{0.9} - 1\right) \times 100\ \text{kW}$$

$= 11.11$ kW $= 11110$ W **Ans.**

Problem 6.38 : A 15 kW, 440 V, 3-phase, 6 pole, 50 Hz, delta connected squirrel cage induction motor gave the following data for block rotor test.

220 V, 25 A, 6 kW

Stator core loss at rated voltage and frequency is 400 watts. The d.c. resistance measured between any two stator terminals is $0.8\ \Omega$. Determine the starting torque at rated voltage and frequency.

Solution

Given $R_{eq} = 0.8\ \Omega$

Let R be the resistance per phase. The resistance per phase R for the stator winding in delta can be obtained by

$$\frac{R \times 2R}{3R} = 0.8$$

or $$R = \frac{0.8 \times 3}{2} = 1.2\ \Omega$$

Effective stator resistance per phase $= 1.2 \times 1.2 = 1.44\ \Omega$

Power input at rated voltage during block rotor test

$$= 6 \times \left(\frac{440}{220}\right)^2 = 24 \text{ kW}$$

Stator current at rated voltage during block rotor test

$$= 25 \times \left(\frac{440}{220}\right) = 50 \text{ A.}$$

$\therefore$ Air gap power at rated voltage and frequency is

$$p_g = 24000 - 3 \times \left(\frac{50}{\sqrt{3}}\right)^2 \times 1.44 - 400 = 20000 \text{ W}$$

Synchronous speed, $\omega_s = \dfrac{120 \times f}{p} = \dfrac{120 \times 50}{6}$

$$= \frac{1000 \times 2\pi}{60} \text{ rad / sec} = \frac{100\pi}{3} \text{ rad / sec}$$

$\therefore$ Starting torque $= \dfrac{p_g}{\omega_s} = \dfrac{18800 \times 3}{100 \times \pi}$ Nm = 191.08 Nm **Ans.**

> **Problem 6.39 :** A 15 kW, 440 V, 6 pole, delta-connected squirrel cage induction motor gave the following test results:
>
> | | |
> |---|---|
> | No-load test: | 440 V, 10 A, 280 Watts |
> | Block rotor test: | 100 V, 40 A, 1600 Watts |
>
> The d.c resistance of the stator winding per phase measured immediately after block rotor test is 0.8 Ω. Calculate the rotational losses and the equivalent circuit parameters.

Solution

Given, V_{nl} = 400 V, I_{nl} = 10 A, P_{nl} = 280 Watts

V_{pr} = 100 V, I_{br} = 40 A, P_{br} = 1600 Watts

r_{dc} = 0.8 Ω

Effective stator resistance per phase

$$r_{eff} = 1.2 \times 0.8 = 0.96 \ \Omega$$

$$\text{No-load rotational losses} = p_{nl} - 3 \times \left(\frac{I_{nl}}{\sqrt{3}}\right)^2 \times r_{eff}$$

$$= 280 - 3 \times \left(\frac{10}{\sqrt{3}}\right)^2 \times 0.96 = 280 - 96 = 184 \text{ W}$$

For no load test, $Z_{nl} = \frac{V_{nl}}{I_{nl}} = \frac{440}{10} = 44\ \Omega$

$$R_{nl} = \frac{280}{3 \times \left(\frac{10}{\sqrt{3}}\right)^2} = 2.8\ \Omega$$

$$\therefore \quad X_{nl} = \sqrt{Z_{nl}^2 - R_{nl}^2} = \sqrt{(44)^2 - (2.8)^2} = 43.91\ \Omega$$

From block rotor test,

$$Z_{br} = \frac{100}{\left(\frac{40}{\sqrt{3}}\right)} = \frac{100 \times \sqrt{3}}{40} = 4.33\ \Omega = R_{br} = \frac{1600}{\left(\frac{40}{\sqrt{3}}\right)^2 \times 3} = 1\ \Omega$$

$$\therefore \quad X_{br} = \sqrt{(4.33)^2 - 1} = 4.213\ \Omega$$

Now, $x_1 = x_2 = \frac{1}{2} X_{br} = 2.107\ \Omega$

and $X_\varphi = X_{nl} - x_1 = 43.91 - 2.107 = 41.803\ \Omega$

$$x_2 = X_\varphi + x_2 = 41.803 + 2.107 = 43.91\ \Omega$$

Per phase rotor resistance is given by

$$r_2 = (R_{br} - r_1)\left(\frac{x_2}{X_\varphi}\right)^2 = (1 - 0.96)\left(\frac{43.91}{41.803}\right)^2 = 0.044\ \Omega$$

Thus, the parameters of the equivalent circuit are

$r_1 = 0.96\ \Omega, r_2 = 0.044\ \Omega, x_1 = x_2 = 2.107\ \Omega$ and $X_\varphi = 41.803$ **Ans.**

Problem 6.40 : A Three phase induction motor is designed to operate at 60 Hz and V_1 Volts. Its breakdown torque occurs at 0.3 and its stator impedance is negligible. In case the motor is made to operate from V_1 Volts, 50 Hz, find the ratio of starting current, starting torque and maximum torque at 50 Hz operation to their corresponding values at 60 Hz operation.

Solution

Given $s_{\max, T} = 0.3$, or $\frac{r_2}{x_2} = 0.3 \quad \therefore \quad r_2 = 0.3\, x_2$

Let, x_2 be the leakage reactance of the rotor at 60 Hz.

$\therefore$ At 50 Hz, leakage reactance of the rotor

$$= \left(\frac{50}{60}\right) x_2 = \frac{5}{6} x_2 = 0.83\, x_2$$

At $\quad f_1 = 60$ Hz, $I_{st1} = ?$

At $\quad f_2 = 50$ Hz, $I_{st2} = ?$

Starting currents, $\quad I_{st1} = \dfrac{V_1}{\sqrt{r_2^2 + x_2^2}}, \quad I_{st2} = \dfrac{V_1}{\sqrt{r_2^2 + (.83\, x_2)^2}}$

$\therefore$ A/O

$$\frac{I_{st2}}{I_{st1}} = \frac{\sqrt{r_2^2 + x_2^2}}{\sqrt{r_2^2 + (0.83\, x_2)^2}} = \sqrt{\frac{s_{\max, T}^2 + 1}{s_{\max, T} + (0.83)^2}}$$

$$= \sqrt{\frac{0.09 + 1}{0.09 + (.83)^2}} = 1.183 \quad \textbf{Ans.}$$

Starting torque, $\quad T_{st1} = \dfrac{r_2 \times V_1^2}{r_2^2 + x_2^2} \times \dfrac{3}{\omega_s}$

$$T_{st2} = \frac{r_2^2 + V_1^2}{r_2^2 + (0.83\, x_2)^2} \times \frac{3}{\omega_s}$$

$\therefore$

$$\frac{T_{st2}}{T_{st1}} = \frac{r_2^2 + x_2^2}{r_2^2 + (0.83\, x_2)^2} = \frac{s_{\max, T}^2 + 1}{s_{\max, T}^2 + (0.83)^2} = \frac{0.09 + 1}{0.09 + (0.83)^2}$$

$\therefore$

$$\frac{T_{st2}}{T_{st1}} = 1.399 \quad \textbf{Ans.}$$

Maximum torques $\quad T_{m1} = \dfrac{3}{\omega_s} \cdot \dfrac{V_1^2}{2\, x_2}$ and $T_{m2} = \dfrac{3}{\omega_s} \cdot \dfrac{V_2^2}{2 \times (0.83\, x_2)^2}$

Now,

$$\frac{T_{m2}}{T_{m1}} = \frac{1}{(0.83)^2} = 1.452$$

$\therefore$

$$\frac{T_{m2}}{T_{m1}} = 1.452 \quad \textbf{Ans.}$$

Problem 6.41 : A 3-phase, 6 pole, 80 kW, 50 Hz induction motor connected to rated supply voltage and running without load consumes 4 kW. When prevented from rotating it draws rated current at 40% rated supply and takes power input 5 kW. Assuming that under rated conditions, the stator and rotor copper losses are equal and that the mechanical losses are 40% of no-load losses. Determine

(*a*) Slip at rated load and

(*b*) Starting torque of the motor with rated applied voltage.

Solution

Given, No-load power input = 4 kW

Mechanical loss = 40% of no-load losses

$$= \frac{40}{100} \times 4 \times 1000 = 1600 \text{ Watts.}$$

Stator core loss = Power input at no-load – Mechanical loss – Stator I^2R loss at no-load

Neglecting stator I^2R loss at no-load

Stator core loss = 4000 – 1600 = 2400 W

Power input during block rotor test = Stator I^2R loss + rotor I^2R loss = 5000 W.

It is given that stator copper loss is equal to rotor copper loss.

∴ Stator I^2R loss = rotor I^2R loss = 2500 W

(*a*) At rated load, Air gap power,

P_g = output + mechanical loss + rotor I^2R loss = 80000 + 2500 + 1600 = 84100 W

We have, rotor I^2R loss = sP_G

or $2500 = s \times P_G$

$$\therefore \quad s = \frac{2500}{84100} = 0.03$$

Slip at rated load, $s = 0.03$ **Ans.**

(*b*) At rated voltage, power input to motor during block rotor test

$$= 5 \times \left(\frac{100}{40}\right)^2 = 31.25 \text{ kW}$$

Air gap power = Power input – Stator core loss – Stator I^2R loss

= 31250 – 2400 – 2500 = 26350 W

$$\text{Synchronous speed,}\quad \omega_s = \frac{120\,f}{p} = \frac{120 \times 50}{6} = 1000 \text{ rpm}$$

$$= \frac{1000 \times 2}{6} \text{ rad / sec} = \frac{100\,\pi}{3} \text{ rad / sec}$$

$$\therefore \quad \text{Starting torque} = \frac{p_G}{\omega_s} = \frac{26350 \times 3}{100 \times \pi} = 251.75 \text{ Nm} \quad \textbf{Ans.}$$

> **Problem 6.42 :** A three phase induction motor is designed to operate on 60 Hz and V_1 volts. In case the motor is fed from 50 Hz, 0.8 V_1 voltage source, find the ratio of 60 Hz to 50 Hz operation for the starting currents, starting torques and maximum torques. Also find the ratio of V_2 at 50 Hz to V_1 at 60 Hz for the same starting and maximum torques.

Solution

Subscripts 1 and 2 are used for the quantities at 60 Hz and 50 Hz respectively.

Let x_2 be the rotor reactance at 60 Hz.

Rotor reactance at 50 Hz = 0.83 x_2

$$I_{st1} = \frac{V_1}{\sqrt{r_2^2 + x_2^2}}, \quad I_{st2} = \frac{0.8\,V_1}{\sqrt{r_2^2 + (0.83\,x_2)^2}}$$

Neglecting x_2 in comparison to x_2,

$$\frac{I_{st1}}{I_{st1}} = \frac{V_1 / x_2}{0.8\,V_1 / 0.83\,x_2} = \frac{0.83}{0.8} = 1.0375$$

$$\frac{I_{st1}}{I_{st2}} = 1.0375 \quad \textbf{Ans.}$$

Starting torque of the motor is given by

$$T = \frac{3}{\omega_s} \cdot \frac{V_1^2}{x_2^2 + r_2^2} \times r_2$$

Neglecting x_2 in comparison to x_2, $T = \dfrac{3}{\omega_s}, \dfrac{V_1^2}{x_2^2}\, r_2$

$$\therefore T_{st1} = \frac{3}{2\,\pi\,f_1} \times \frac{V_1^2}{x_2^2} \times r_2 \quad \text{and} \quad T_{st2} = \frac{3}{2\,\pi\,f_2} \times \frac{(0.8\,V_1)^2}{(0.83\,x_2)^2} \times r_2$$

Now $$\frac{T_{st1}}{T_{st2}} = \frac{f_2}{f_1} \times \left(\frac{0.83}{0.8}\right)^2 = 0.893$$

$$\therefore \quad \frac{T_{st1}}{T_{st2}} = 0.893 \quad \textbf{Ans.}$$

Maximum torque is given by $T_{max} = \frac{3}{\omega_s} \times \frac{V_1^2}{2\,x_2}$

$$T_{m1} = \frac{3}{2\,\pi\, f_1} \times \frac{V_1^2}{2\,x_2}, \quad T_{m2} = \frac{3}{2\,\pi\, f_2} \times \frac{(0.8\,V_1)}{2 \times 60.83\,x_2}$$

$$\therefore \quad \frac{T_{m1}}{T_{m2}} = \frac{f_2}{f_1} \times \frac{0.83}{(0.8)^2} = \frac{50}{60} \times \frac{0.83}{(0.8)^2} = \frac{(0.83)^2}{(0.8)^2}$$

$$\frac{T_{m1}}{T_{m2}} = 1.076 \quad \textbf{Ans.}$$

For the same starting torqu

$$\frac{3}{\omega_{s1}} \times \frac{V_2^2}{(0.83\,x_2)^2} = \frac{3}{\omega_{s2}} \times \frac{V_1^2}{x_2^2} \quad \text{or,} \quad \frac{1}{f_1} \times \frac{V_2^2}{(0.83)^2} = \frac{1}{f_2} \times V_1^2$$

or $$\left(\frac{V_2}{V_1}\right)^2 = \frac{f_1}{f_2} \times \frac{1}{(0.83)^2} = \left(\frac{60}{50}\right) \times \frac{1}{(0.83)^2} = \frac{1}{(0.83)^3} = 1.749$$

$$\therefore \quad \frac{V_2}{V_1} = 1.322 \quad \textbf{Ans.}$$

For the same maximum torque,

$$\frac{3}{2\,\pi\, f_1} \times \frac{V_2^2}{2 \times (0.83\,x_2)} = \frac{3}{2\,\pi\, f_2} \times \frac{V_1^2}{2\,x_2}$$

or, $$\left(\frac{V_2}{V_1}\right)^2 = \frac{f_1}{f_2} \times 0.83 = 1 \quad \therefore\ V_2 = V_1 \quad \textbf{Ans.}$$

Problem 6.43 : A 8 kW, 440 V, 3-phase, 50 Hz, 10 pole squirrel cage induction motor operates at 5% slip at full load when rated voltage and supply is applied. Assuming a linear relationship between torque and slip in the operating region. Calculate no-load speed of the motor when the supply voltage is reduced to one-third its rated value. The no-load torque is 8 Nm.

Solution :

Given, $s_{fl} = 0.05,$ $p = 10,$
$f = 50$, Rated voltage = 440 V

The linear torque-slip characteristic is given by $T = \frac{3}{\omega_s}\frac{sV_2}{r_2}$

$$N_s = \frac{120 \times f}{p} = \frac{120 \times 50}{10} = 600 \text{ rpm}$$

Full load torque, $T_{fl} = \frac{8000 \times 60}{2\,\pi \times 600 \times (1-0.05)} = \frac{800}{2 \times \pi \times .95}$

$$= 134.1 \text{ Nm}$$

Now, At rated voltage and supply

$$134.1 = \frac{3}{\omega_s} \cdot \frac{(440)^2}{r_2} \times s_{fl} \qquad (1)$$

At one-third of the rated voltage. T_e = 8 Nm

$$\therefore \qquad 8 = \frac{3}{\omega_s} \cdot \frac{\left(\frac{440}{3}\right)^2}{r_2} \times s \qquad (2)$$

From (1) and (2) $\frac{s \times \left(\frac{440}{3}\right)^2}{s_{fl} \times (440)^2} = \frac{8}{134.1}$

or, $s = \frac{8 \times 9 \times s_{fl}}{134.1} = \frac{8 \times 9 \times 0.05}{134.1} = 0.027$

No-load speed of the motor, $N_o = N_s(1-s) = 600(1-0.027)$

$= 583.9$ rpm **Ans.**

Problem 6.44 : A 4 pole, 50 Hz, 3-phase wound rotor induction motor has maximum torque of 200% of full-load torque at 15% slip. The rotor resistance per phase is 0.5 at 50 Hz. The input supply voltage is subject to large fluctuations. Determine the minimum voltage that could be impressed so that the motor can still supply its rated torque. Also determine the resistance to be inserted in the rotor circuit so that the motor will develop full-load torque at starting at this reduced voltage.

Solution

Given, r_a 0.5 Ω, $p = 4$, $f = 50$ Hz, $s_{mt} = 0.15$, $T_m = 2 \times T_{efl}$ (1)

Since we have, $s_{mt} = \dfrac{r_2}{x_2}$

$$\therefore \quad x_2 = \frac{0.5}{0.15} = 3.33\ \Omega$$

Let S_{fl} be the full load slip. Applying the relation,

$$\frac{T_{efl}}{T_{em}} = \frac{2}{\frac{s_{mt}}{s_{fl}} + \frac{s_{fl}}{s_{mt}}} \text{ or, } \frac{1}{2} = \frac{2}{\frac{0.15}{s_{fl}} + \frac{s_{fl}}{0.15}}$$

$$\text{or} \quad \frac{0.15}{s_{fl}} + \frac{s_{fl}}{0.15} = 4, \text{ or } (0.15)^2 + s_{fl}^2 = 4 \times 0.15 \times s_{fl}$$

$$\text{or} \quad s_{fl}^2 - 0.6\, s_{fl} + 0.023 = 0$$

This yields, $$s_{fl} = \frac{0.6 \pm \sqrt{(0.6)^2 - 4 \times 0.023}}{2} = \frac{0.6 \pm 0.158}{2}$$

Neglecting higher value, $$s_{fl} = \frac{0.6 - 0.518}{2} = \frac{0.082}{2} = 0.042$$

Applying the relation, $$T_e = \frac{3}{\omega_s} \cdot \frac{V^2}{\left(\frac{r_2}{s}\right)^2 + x_2^2} \times \frac{r_2}{s}$$

The full load torque is given by

$$T_{efl} = \frac{3}{\omega_s} \cdot \frac{V^2}{\left(\frac{0.5}{0.042}\right)^2 + (3.33)^2} \times \frac{0.5}{0.042} \qquad (1)$$

Let the minimum voltage be xV, where $x < 1$,

$$T_{er} = \frac{3}{\omega_s} \cdot \frac{(xV)^2}{2 \times x} = \frac{3}{\omega_s} \cdot \frac{(xV)^2}{2 \times 3.33}$$

A/Q, $$T_{er} = T_{efl} \Rightarrow \frac{3}{\omega_s} \cdot \frac{x^2 V^2}{2 \times 3.33} = \frac{3}{\omega_s} \cdot \frac{V^2}{152.8} \times 11.90$$

$$\text{or} \quad x^2 = \frac{11.90 \times 2 \times 3.33}{152.8} = 0.519$$

$$\therefore \quad x = 0.7204 = 72.04\ \% \quad \textbf{Ans.}$$

Let R be the external resistance added to rotor circuit

A/Q $$\frac{3}{\omega_s}\cdot\frac{0.7204^2\times(R+0.5)}{(R+0.5)^2+(3.33)^2} = \frac{3}{\omega_s}\cdot\frac{V^2}{152.8}$$

or $$\frac{(0.7204)^2\times(R+0.5)}{(R+0.5)^2+(3.33)^2} = \frac{1}{152.8}$$

Put $$(R + 0.5) = A$$

or $$(0.7204)^2\times 152.8\times A = A^2+(3.33)^2$$

or $$79.30\ A = A^2+11.09 \text{ or } A^2-79.30\,A+11.09=0$$

This yields $$A = \frac{79.30\pm\sqrt{(79.30)^2-44.36}}{2} = \frac{79.30\pm 79.02}{2}$$

Either $$A = \frac{0.28}{2} = 0.14 \textit{ or } A = \frac{158.32}{2} = 79.16\,\Omega$$

Neglecting $A = 0.14$, $A = 79.16$ or $R + 0.5 = 79.16$

$\therefore$ $R = 79.66\ \Omega$ **Ans.**

> **Problem 6.45 :** A 4-pole, 8 kW, 50 Hz, 3-phase induction motor develops rated torque at 1440 rpm. In case load torque is reduced to one-fourth, find the motor speed and the power output. Assume linear-torque-slip characteristics near the operating range.

Solution

Given $p = 4$, $f = 50$ Hz, $N_r = 1440$ rpm.

Synchronous speed $$(N_s) = \frac{120\times 50}{4} = 1500 \text{ rpm.}$$

$\therefore$ Slip at rated torque, $$s = \frac{1500-1440}{1500} = 0.04$$

The linear torque-slip characteristics is given by $T_e = \dfrac{3}{\omega_s}\cdot\dfrac{V^2}{r_2}\times s$

Since, w_s, V and r_2 is constant, $T_e \propto s$

When slip is 0.04, $T_{e1} \propto 0.04$ (1)

and at reduced load torque, $\frac{1}{4}T_{e1} \propto s_1$ (2)

From (1) and (2) $\frac{s_1}{0.04} = \frac{1}{4}$

$\therefore$ $s_1 = 0.01$

New motor speed, $N_{new} = 1500\ (1 - 0.01) = 1485$ rpm **Ans.**

Rated torque $= \frac{8 \times 1000 \times 60}{2\,\pi \times 1440} = 53.08$ Nm

Power output at one-fourth of rated torque

$$= \frac{1}{4} \times \text{Rated torque} \times \text{New speed}$$

$$= \frac{1}{4} \times 53.08 \times \frac{1485 \times 2\pi}{60} = 2.063 \text{ kW } \textbf{Ans.}$$

Problem 6.46 : A 3-phase, 1440 rpm induction motor drives a load requiring constant torque. In case supply drops to 80% find what happens to motor speed and losses. Take linear torque-slip characteristics over the working torque.

Solution

Given, Synchronous speed,

$N_s = 1500$ rpm (it not mentioned)

Load speed, $N = 1440$ rpm.

$\therefore$ Load slip, $s = \frac{1500 - 1440}{1500} = 0.04$

The linear torque-slip characteristics is given by $T_e = \frac{3}{\omega_s} \cdot \frac{V^2}{r_2} \times s$

At slip $S = 0.04$ and voltage $= V$, $T_{e1} = \frac{3}{\omega_s} \cdot \frac{V^2}{r_2} \times 0.04$

At reduced supply, $T_{e2} = \frac{3}{\omega_s} \cdot \frac{V^2}{r_2} \times 0.04 = \frac{3}{\omega_s} \cdot \frac{(0.8\,V)^2}{r_2} \times S$

A/O, $T_{e1} = T_{e2}$ [for constant torque]

$$\frac{3}{\omega_s} \cdot \frac{(0.8)^2 \times V^2}{r_2} \times s_2 = \frac{3}{\omega_s} \cdot \frac{V^2}{r_2} \times 0.04$$

or $$s_2 = \frac{0.04}{(0.8)^2} = 0.063$$

$$\text{New speed} = 1500(1 - 0.063) = 1406.3 \text{ rpm.}$$

At reduced supply, the speed gets reduced,

$$\% \text{ reduced speed} = \left(\frac{1440 - 1406.3}{1440}\right) \times 100\% = 2.34\% \quad \textbf{Ans.}$$

At small values of slip, $I_2 = \dfrac{sV}{r_2}$

At slip $$s = 0.04, \; I_{21} = \frac{0.04\,V}{r_2} \quad (1)$$

At slip $$s = 0.063, \; I_{22} = \frac{0.63\,V}{r_2} \quad (2)$$

From (1) and (2) $$\frac{I_{22}}{I_{21}} = \frac{0.063}{0.04} = 1.575, \; \therefore \left(\frac{I_{22}}{I_{21}}\right)^2 = 2.481$$

$$\left(\frac{I_{22}}{I_{21}}\right)^2 \times \frac{r_2}{r_2} = 2.481$$

$$\frac{\text{Ohmic loss at rated voltage}}{\text{Ohmic loss at normal voltage}} = 2.481$$

At reduced supply, ohmic loss increases.

$$\% \text{ increase in loss} = \left(\frac{2.481 - 1}{1}\right) \times 100\% = 148.1\% \quad \textbf{Ans.}$$

Problem 6.47 : A 400 V, 50 Hz, three phase, 6 pole induction motor develops half the rated torque at 980 rpm with the applied voltage magnitude remaining at the rated value. What should be its frequency if the motor has to develop same torque at 1150 rpm? Neglect stator and rotor winding resistance, leakage resistance and iron losses.

Solution

Given $$p = 6, \quad f = 50 \text{ Hz}$$

Synchronous speed, $$N_s = \frac{120 \times 50}{6} = 1000 \text{ rpm.}$$

Rated slip $$s = \frac{1000 - 980}{1000} = 0.02$$

The electromagnetic torque is given by

$$T = \frac{3}{\omega_s}, \frac{V_1^2}{\left(\frac{r_2}{s} + r_1\right)^2 + (x_1 + x_2)^2} \times \frac{r_2}{s}$$

Neglecting r_1, x_1, x_2, ω_i $T = \frac{3}{\omega_s} \times \frac{V_1^2}{r_2'} \times S$

$$T_{e1} = \frac{3}{2\pi f_1} \times \frac{V_1^2}{r_2} s_1$$

$$T_{e1} = \frac{3}{2\pi \times 50} \cdot \frac{(400)^2}{r_2} \times 0.02 \qquad (1)$$

Let f_2 be the new frequency at speed 1150 rpm

$$T_{e2} = \frac{3}{2\pi f_2} \cdot \frac{(400)^2}{r_2} \times s_2 \qquad (2)$$

From (1) and (2) $\quad 1 = \frac{f_2}{50} \times \frac{s_2}{0.02}$

or $\quad = s_2 f_2 = 50 \times 0.02, \; s_2 f_2 = 1$

Now $\quad N_2 = \frac{120 \times f_2}{6} = 20 f_2$

The slip s_2 is given by $\quad s_2 = \frac{20 f_1 - 20 f_2}{20 f_1}$

or $\quad \frac{1}{f_2} = \frac{1000 - 20 f_2}{1000}$ or $\frac{1}{f_2} = \frac{50 - f_2}{50}$

or $\quad 50 f_2 - f_2^2 = 50$ or $f_2^2 - 50 f_2 + 50 = 0$

This yields, $f_2 = \frac{50 \pm \sqrt{2500 - 200}}{2} = \frac{50 \pm 47.95}{2}$

Neglecting lower value $f_2 = \frac{97.95}{2} = 48.98 \text{ Hz}$ **Ans.**

Problem 6.48 : The following data refers to a 10 pole, 440 V, 50 Hz, 3-phase mesh connected induction motor $r_1 = 2.5\,\Omega$, $x_1 = 6.5\,\Omega$, $r_2' = 1.8\,\Omega, x_2' = 4.0/\text{phase}$. On no-load, the line value of the magnetising current is 6 A and total core loss is 200 W. Determine the power factor, input current, equivalent rotor current and torque developed by the motor at a slip of 25% using exact equivalent circuit. Determine the maximum torque developed and the corresponding speed.

Solution

Given, Core loss = 240 W

no-load magnetising current,

$$I_\phi = 6 \text{ A}, \qquad r_1 = 2.5\,\Omega,$$

$$r_2' = 1.8\,\Omega, x_1 = 6.5\,\Omega, \qquad x_2' = 4.0\,\Omega/\text{phase}$$

$$p = 10, f = 50 \text{ Hz}$$

$$\text{Synchronous speed, } n_s = \frac{120 \times 50}{10} = 600 \text{ r.p.m.}$$

$$\text{Magnetising current per phase} = \frac{6}{\sqrt{3}} = 3.46 \text{ A}$$

The equivalent circuit can be drawn as

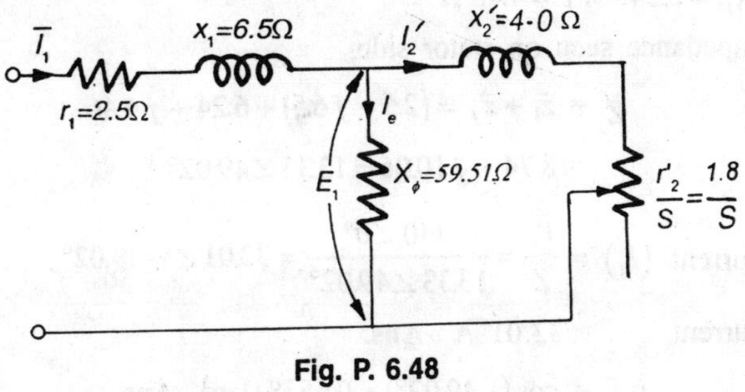

Fig. P. 6.48

Let us take E_1 as reference phasor,

$$\bar{V}_1 = \bar{E}_1 + \bar{I}_\phi (r_1 + j\,x_1)$$

or $$\bar{V}_1 = \bar{E}_1 - j\,3.46(2.5 + j\,6.5\,\Omega) \text{ or } \bar{V}_1 = \bar{E}_1 - j8.66 + 234$$

or $$V_1 = (E_1 + 234) - j\,8.66$$

$$\therefore \quad V_1^2 = (E_1 + 234)^2 + (8.66)^2$$

or, $$(440)^2 = E_1^2 + 468\,E_1 + 54830.996$$

or $$E_1^2 + 468\,E_1 - 138770.04 = 0$$

This yields, $$E_1 = \frac{-468 \pm \sqrt{(468)^2 + 4 \times 138770.4}}{2}$$

$$= \frac{-468 \pm 879.8}{2}$$

Neglecting negative values $E_1 = 206$ V

$\therefore$ Magnetising reactance, $$X_\phi = \frac{E}{I_\phi} = \frac{206}{3.46} = 59.51\,\Omega$$

From equivalent circuit,

$$\bar{Z}_f = \frac{\left(\frac{1.8}{0.25} + j\,4\right) \times j\,59.51}{\frac{1.8}{0.25} + j\,(59.51 + 4)} = \frac{(7.2 + j\,4) \times 59.51\,\angle 90^\circ}{7.2 + j\,63.51}$$

$$= \frac{8.24\,\angle 29.05^\circ \times 59.51\,\angle 90^\circ}{63.92\,\angle 83.53^\circ} = 7.67\angle 35.52^\circ$$

$$= 6.24 + j\,4.46\,\Omega\,/\,\text{phase} = R_f + j\,X'_f$$

Here, $R_f = 6.24$, $X_f = 4.46\,\Omega$

Total impedance seen on stator side,

$$\bar{Z} = \bar{z}_1 + \bar{z}_f = (2.5 + j\,6.5) + 6.24 + j\,4.46$$

$$= 8.74 + j\,10.06 = 13.33\,\angle 49.02^\circ$$

Input current $(I_1) = \dfrac{\bar{V}_t}{\bar{Z}} = \dfrac{440\,\angle 0^\circ}{13.33\,\angle 49.02^\circ} = 33.01\,\angle -49.02^\circ$

Input current $= 33.01$ A **Ans.**

$$\text{p.f.} = \cos(-49.02^\circ) = 0.6558\,(\text{lag}) \quad \textbf{Ans.}$$

Equivalent rotor current,

$$\bar{I}_2 = I_1 \cdot \frac{j \times X_\phi}{\frac{r_2}{s} + j\,(x_2 + X_\phi)} = \frac{33.01\,\angle -49.02^\circ \times 59.51\,\angle +90^\circ}{63.92\,\angle 85.53^\circ}$$

$$= 30.73\,\angle -44.55^\circ \quad \textbf{Ans.}$$

$$\text{Now, Torque} = \frac{3 I_1^2 R_f}{\omega_s} = \frac{3 \times (33.01)^2 \times 6.20 \times 60}{2 \pi \times 600}$$

$$= 324.82 \text{ Nm} \quad \textbf{Ans.}$$

The maximum torque can be calculated by using. Thevenin's Theorem

$$X_1 = x_1 + X_\varphi = 59.51 + 6.5 = 66.01\,\Omega$$

$$R_{Th} = \frac{2.5 \times 59.51}{X_1} = \frac{2.5 \times 59.51}{66.01} = 2.25\,\Omega$$

$$X_{Th} = \frac{x_1 \cdot X_\varphi}{X_1} = \frac{6.5 \times 59.51}{66.01} = 5.86\,\Omega$$

Slip at which maximum torque occurs,

$$S_{\max, T} = \frac{r_2}{\sqrt{R_{Th}{}^2 + (x_{Th} + x_2)^2}} = \frac{4}{\sqrt{(2.25)^2 + (5.86 + 1.8)^2}}$$

$$= \frac{4}{\sqrt{(2.25)^2 + (6.66)^2}} = 0.57$$

$$V_{Th} = \frac{V_1 X_\phi}{X_1} = \frac{440 \times 59.51}{66.01} = 396.67 \text{ V}$$

Speed at maximum torque = $600(1 - 0.57) = 600 \times 0.43 = 158$ rpm **Ans.**

∴ Maximum torque,

$$T_{\max} = \frac{3 V_{Th}^2}{2 \pi \eta} \left[\frac{1}{2\left[R_{Th} + \sqrt{R_{Th}^2 + (X_{Th} + x_2)^2} \right]} \right]$$

$$= \frac{3 \times 60}{2 \times \pi \times 600} \times \frac{396.67^2}{\left[2\left[2.25 + \sqrt{2.25^2 + (5.86 + 1.8)^2} \right] \right]}$$

$$= 367.26 \text{ Nm} \quad \textbf{Ans.}$$

Problem 6.49 : An induction motor runs at a slip frequency of 4 Hz when supplied from a 3-phase, 440 V, 50 Hz supply. For the same developed torque, find the slip frequency at which it will run when supplied from a 3-phase, 400 V, 40 Hz system. The slip at which the machine develops maximum torque using 50 Hz supply is 0.2. Neglect the stator impedance and assume linear torque slip characteristics between zero torque and maximum torque in the working region.

Solution

Given, $f_{s1} = 4$ Hz, $f_{s2} = ?$

$f_1 = 50$ Hz, $f_2 = 40$ Hz

$V_1 = 440$ V, $V_2 = 400$ V

$s_{max, T} = 0.2$

The maximum torque in 3-ϕ induction motor is given by

$$T_{max} = \frac{3}{\omega_s} \cdot \frac{V_t}{2x_2}$$

For 440 supply, $s_{max, T} = 0.2 = \frac{r_2}{x_2}$

When $f_1 = 50$ Hz, $s_{mT1} = \frac{r_2}{x_2} = 0.2 \propto \frac{r_2}{50}$ (1)

For $f_2 = 40$ Hz, $s_{mT2} \frac{r_2}{x_2}$ (2)

From (1) and (2) $\frac{s_{mT2}}{s_{mT1}} = \frac{50}{40}$

or $s_{sT2} = \frac{50}{40} \times 0.2 = 0.25$

We have $T_{max} = \frac{3}{w_s} \cdot \frac{V^2}{2x^2} = \frac{3}{4\pi f} \times \frac{V^2}{2\pi f l_2}$

$$= \left(\frac{3}{4\pi f \times 4\pi f \times l_2}\right) \times V^2 = \frac{3}{16\pi^2 l_2} \times \frac{V^2}{f^2}$$

Since $\frac{3}{16\pi^2 l_2}$ is constant, $T_{max} \propto \frac{V^2}{f^2}$

For linear torque characteristics.

Developed torque at a slip $= \frac{4}{50} = 0.08$ is

$$T_{ed1} = \left(\frac{440}{50}\right)^2 \times \frac{0.08}{0.2}$$

For 400 V, 40 Hz, $$T_{ed2} = \left(\frac{400}{40}\right)^2 \times \frac{0.5}{0.25}$$

According to question $T_{ed1} = T_{ed2}$

or $$\left(\frac{440}{80}\right)^2 \times 0.4 = \left(\frac{440}{40}\right)^2 \times 0.\frac{5}{0.25} = 0.08$$

Slip frequency at 40 Hz = $40 \times 0.08 = 3.2$ Hz **Ans.**

> **Problem 6.50 :** A 3-phase, 60 Hz, 6 pole induction motor has rated output of 12 kW at 960 rpm and maximum torque is developed at 1000 rpm. Calculate the starting torque. Neglect stator resistance and rotational losses.

Solution

Given, $p = 6$, $f = 60$ Hz

Synchronous speed $$(n_s) = \frac{120 \times 60}{6} = 1200 \text{ rpm}$$

$$\text{Slip} = \frac{1200 - 960}{1200} = 0.2$$

Since rotational losses is neglected

Power developed = power output, $P_m = 12$ kW

$\therefore$ Electromagnetic torque $$T_{efl} = \frac{P_m}{\omega_s(1-s)}$$

$$= \frac{12 \times 1000 \times 60}{2 \times \pi \times 1200\,(1 - .2)} = 119.43 \text{ Nm}$$

Slip at which maximum torque occurs

$$s_{\max, T} = \frac{1200 - 1000}{1200} = \frac{200}{1200} = 0.167$$

Taking $s = 0.2$ as full load slip

Applying the realtion

$$\frac{T_{efl}}{T_{em}} = \frac{2}{\frac{s_{fl}}{s_{\max, T}} + \frac{s_{\max, T}}{s_{fl}}}$$

or $$\frac{119.43}{T_{em}} = \frac{2}{\frac{0.2}{0.167}+\frac{0.167}{0.2}} = 0.984$$

$$\therefore \quad T_{em} = 121.37 \text{ Nm}$$

To obtain starting torque. Apply

$$\frac{T_{est}}{T_{em}} = \frac{2}{\frac{1}{s_{max,T}}+\frac{s_{max,T}}{1}} = \frac{2}{\frac{1}{0.2}+\frac{0.2}{1}} = \frac{2}{5.2}$$

$$\text{T}_{\text{est}} = 46.06 \text{ Nm}$$

$\therefore$ Starting torque = 46.68 Nm **Ans.**

> **Problem 6.51 :** A 3-phase, 60 Hz, 6 pole induction motor has a shaft output of 12 kW at 960 rpm. Friction and windage loss to 2% of output. Total stator losses are 500 W
> *(a)* Determine the stator input and rotor input
> *(b)* If the maximum torque is developed at 900 rpm
> Compute the starting torque with rated voltage starting.

Solution

Given, $p = 6, \ f = 60 \text{ Hz}$

$P_{sh} = 12 \times 1000 \text{ W}$, stator losses = 500 W

$$\text{Synchronous speed, } N_s = \frac{120 \times f}{p} = \frac{120 \times 60}{6} = 1200 \text{ rpm}$$

$$\text{Slip at speed 960 rpm } s = \frac{1200-960}{1200} = \frac{240}{1200} = 0.2$$

$$\text{Also, Friction and windage } (W_{wf}) = 12000 \times \frac{2}{100} = 240 \text{ W}$$

Mechanical power developed = 1200 + 240 = 12,240 W

(a) Rotor input $(P_G) = \frac{P_m}{(1-s)} = \frac{12{,}240}{(1-0.2)} = 15300 = 15.3 \text{ kW}$ **Ans.**

Stator input = 15300 + 500 = 15800 = 15.8 kW **Ans.**

(b) Slip at maximum torque, $s_{max,T} = \frac{1200-900}{1200} = \frac{300}{1200} = 0.25$

Full load torque, $T_{fl} = \frac{P_G}{\omega_s} = \frac{15300 \times 60}{2\,\pi \times 1200} = 121.82$ Nm

Applying the relation

$$\frac{T_{fl}}{T_{max}} = \frac{2}{\frac{s_{fl}}{s_{max,\,T}} + \frac{s_{max,\,T}}{s_{fl}}} = \frac{2}{\frac{0.2}{0.25} + \frac{0.25}{0.2}} = 0.976$$

$$\therefore \qquad T_{max} = \frac{T_{fl}}{0.976} = \frac{121.82}{0.976} = 124.8 \text{ Nm}$$

To compute stating torque. Applying the relation

$$\frac{T_{st}}{T_{max}} = \frac{2}{\frac{s_{max.T}}{1} + \frac{1}{s_{max.T}}}$$

$$\text{or} \qquad T_{st} = \left(\frac{2}{\frac{0.25}{1} + \frac{1}{0.25}}\right) \times T_{max} = \frac{2}{4.25} \times 124.82 = 58.74 \text{ Nm}$$

∴ Starting torque = 58.74 Nm **Ans.**

Problem 6.52 : A 3-phase induction motor has a starting of 200% and a maximum torque of 250% of full load torque. Neglect stator resistance and assume rotor resistance. Compute *(a)* slip at which maximum torque *(b)* full load slip and *(c)* rotor current at starting in terms of full load rotor current.

Solution

A/O $\qquad T_{st} = 2\,T_{fl}$ (1)

and $\qquad T_{max} = 2.5\,T_{fl}$ (2)

Let S_{fl} be the full load slip and $s_{max,\,T}$ be the full load slip

(a) Applying $\qquad \frac{T_{st}}{T_{max}} = \frac{2}{s_{max,\,T} + \frac{1}{s_{max,\,T}}}$

(c) To calculate rotor current at the starting, Applying the relation,

$$\frac{T_{st}}{T_{fl}} = \frac{\frac{1}{\omega_s} I_{2st}^{\ 2} \times r_2}{\frac{1}{\omega_s}\left(I_{fl}\right)^2 \times \frac{r_2}{s_{fl}}}$$

or $$\frac{2 \times T_{fl}}{T_{fl}} = \left(\frac{I_{2st}}{I_{fl}}\right)^2 \times s_{fl}$$

$\therefore$ $$\frac{I_{2st}}{I_{fl}} = \sqrt{\frac{2}{s_{fl}}} = \sqrt{\frac{2}{0.21}} = 3.09$$

$\therefore$ $$I_{2st} = 3.09 \times I_{fl} \quad \textbf{Ans.}$$

Problem 6.53 : In a 3-phase induction motor, leakage reactance is 2 times. The resistance for both stator and rotor circuits. But stator leakage reactance is equal to rotor standstill leakage reactance reffered to stator. The motor develops a torque of 150 Nm at a slip of 0.02. Compute the starting and maximum torques. Neglect exciting current.

Solution

According to Question, $x_1 = 2 \times r_2$ and $x_2 = x_1 = 2r_1 = 2r_2$

Given, Torque at slip $0.02 = 150$

or $$\frac{3V^2}{\omega_s} \times \frac{1}{\left[\left(r_1 + \frac{r_2}{0.02}\right)^2 + \left(x_2 + x_1\right)^2\right]} \times \frac{r_2}{0.02} = 150$$

or $$\frac{3V^2}{\omega_s} \times \frac{1}{\left[\left(1 + \frac{1}{0.02}\right)^2 + (2+2)^2\right] r_2} \times \frac{r_2}{0.02} = 150$$

or $$\frac{3\,V^2}{\omega_s} \times \frac{1}{(5)^2 + (4^2)} = 150 \times 0.02 \text{ or } \frac{3\,V^2}{\omega_s} = 7851$$

Now, starting torque is given by

$$T_{st} = \frac{3V^2}{\omega_s} \times \frac{1}{(r_1 + r_2)^2 + (x_1 + x_2)^2} \times r_2$$

$$= 7851 \times \frac{1}{(1+1)^2 + (2+2)^2} \times 1$$

$$= 7851 \times \frac{1}{20} = 392.55 \text{ Nm} \quad \textbf{Ans.}$$

Slip at maximum torque is given by

$$S_{mt} = \frac{r_2}{\sqrt{r_1^2 + (x_1 + x_2)^2}} = \frac{r_2}{\sqrt{r_2^2 + (2r_2 + 2r_2)^2}}$$

$$= \frac{1}{\sqrt{1+16}} = 0.243$$

Maximum torque,

$$T_{max} = \frac{3V^2}{W_s} \times \frac{1}{\left(1 + \frac{1}{0.243}\right)^2 + (2+2)^2} \times \frac{r_2}{0.243}$$

$$= \frac{7851}{10.25} = 765.95 \text{ Nm} \quad \textbf{Ans.}$$

Problem 6.54 : A 80 kW, 3 phase, 400 V, 6 pole, 50 Hz wound rotor induction motor, with its rotor winding short circuited, has a full load slip 0.02 and the slip at which maximum torque occurs is 0.2 when operating at rated voltage and frequency. Neglect stator resistance and rotational losses. Find *(a)* maximum torque *(b)* starting torque and *(c)* full load ohmic loss.

Solution

Given, $s_{fl} = 0.02, \quad s_{max, T} = 0.2, \; P = 6, \; f = 50$ Hz

$$N_s = \frac{120 \times 50}{6} = 1000 \text{ rpm}$$

(a) Since rotational losses are negligible

$\therefore$ Mechanical power developed (P_m) = Power output

$$P_m = 80 \times 1000 \text{ W}$$

$\therefore$ Power in air-gap

$$(P_G) = \frac{P_m}{1 - s_{fl}} = \frac{80 \times 1000}{1 - 0.02} = 81632.7 \text{ W}$$

$\therefore$ Full load torque,

$$T_{fl} = \frac{P_G}{\omega_s} = \frac{181632.7 \times 60}{2 \times \pi \times 1000} = 779.93 \text{ Nm}$$

(a) To calculate maximum torque, Applying the relation,

$$\frac{T_{fl}}{T_{max}} = \frac{2}{\dfrac{s_{fl}}{s_{max,\,T}} + \dfrac{s_{max,\,T}}{s_{fl}}} = \frac{2}{\dfrac{0.02}{0.2} + \dfrac{0.2}{0.02}} = 0.198$$

$$\therefore \quad T_{max} = \frac{T_{fl}}{0.198} = \frac{779.93}{0.198} = 3939 \text{ Nm} \quad \textbf{Ans.}$$

(b) To calculate starting torque, applying the relation,

$$\frac{T_{st}}{T_m} = \frac{2}{s_{max,\,T} + \dfrac{2}{s_{max,\,T}}} = \frac{2}{0.2 + \dfrac{1}{0.2}} = 0.385$$

$\therefore$ Starting torque $= 0.385 \times 3939 = 1515$ Nm **Ans.**

(c) Rotor ohmic loss $= sP_G$

$= 0.02 \times 81632.7 = 1632.7$ W **Ans.**

Problem 6.55 : A 8 kW, 440 V, 3 phase, star connected, 50 Hz, 6 pole squirel cage induction motor develops full load torque at a slip of 5% under rated voltage and frequency conditions. Friction windage and core losses are to be neglected. The machine has following impedance data:

$$r_1 = 1\,\Omega,\ x_2 = x_1 = 1.2\,\Omega,\ x_\phi = 36\,\Omega$$

Determine the maximum torque at rated voltage and frequency and slip at which it occurs.

Solution

Given, $r_1 = 1\,\Omega, X_\phi = 36\,\Omega, x_1 = x_2 = 1.2\,\Omega$

$s_{fl} = 0.05\ ,\ P = 6\ ,\ f = 50$

Synchronous speed $(n_s) = \dfrac{120 \times 50}{6} = 1000$ rpm

Let us calculate the equivalent parameters

$$V_{Th} = \frac{V_1 X_\phi}{x_1 + X_\phi} = \frac{440 \times 36}{36 + 1.2} = 425.81 \text{ V}$$

$$R_{Th} = \frac{r_1 X_\phi}{x_1 + X_\phi} = \frac{1 \times 36}{1.2 + 36} = 0.968\,\Omega$$

$$X_{Th} = \frac{x_1\, X_\phi}{x_1 + X_\phi} = \frac{1.2 \times 36}{37.2} = 1.16\,\Omega$$

Since windage and friction losses are negligible.

$\therefore$ Mechanical power developed (P_m) = Power output = 8000 W

We have $3(1-s)P_g = 8000$

The slip at which maximum torque,

or $$3(1-s) \times \frac{I_2^2 r_2}{s} = 8000$$

or $$3(1-s) \times \frac{s^2 V_{Th}^2}{r_2^2} \times \frac{r_2}{s} = 8000$$

or $$r_2 = \frac{3(1-s) \times s \times V_{Th}^2}{8000}$$

$$= \frac{3(1-0.05) \times 0.05 \times (425.81)^2}{8000} = 3.23\,\Omega$$

The slip at which maximum torque is given by

$$s_{\max, T} = \frac{r_2}{\sqrt{R_{Th}^2 + (X_{Th} + x_2)^2}}$$

$$= \frac{3.23}{\sqrt{(0.968)^2 + (1.16 + 1.2)^2}}$$

$$= \frac{3.23}{\sqrt{(0.968)^2 + (2.36)^2}} = 1.27 \textbf{ Ans.}$$

Slip at maximum torque = 1.27 (Generating mode) **Ans.**
The maximum torque is given by

$$T_m = \frac{3}{W_s} \frac{V_{Th}^{\ 2}}{\left[R_{Th} + \sqrt{R_{Th}^{\ 2} + (X_{Th} + x_2)^2} \right]}$$

$$= \frac{3 \times 60}{2\pi \times 1000} \times \frac{(425.81)^2}{\left[0.968 + \sqrt{(0.968)^2 + (1.16 + 1.2)^2} \right]}$$

$= 1745.9$ Nm **Ans.**

> **Problem 6.56 :** A three phase induction motor with maximum slip = 0.4, has a starting torque of 40 Nm. For negligible stator impedance and no-load current, determine the starting torque when the rotor resistance per phase is doubled.

Solution

We know that $s_{max, T} = \frac{r_2}{x_2} = 0.4$

$\therefore$ $r_2 = 0.4\, x_2$ or $x_2 = 2.5\, r_2$

Given, $T_{st} = 40$ Nm, at no-load,

or $$\frac{1}{\omega_s} \frac{3 V_1^2}{r_2^{\ 2} + x_2^{\ 2}} \times r_2 = 40$$

or $$\frac{3 V_1^2}{\omega_s} \times \frac{r_2}{r_2^{\ 2} + (2.5\, r_2)^2} = 40$$

or $$\frac{3 V_1^2}{\omega_s \times r_2} \times \frac{1}{1 + (2.5)^2} = 40$$

or $$\frac{3 V_1^2}{\omega_s \times r_2} = 40 \times 7.25 = 290$$

When rotor resistance is doubled, $x_2 = 5\, r_2$,

Starting torque in this case

$$T_s = \frac{3V_1^2}{\omega_s} \times \frac{2r_2}{2r_2 + x_2^2}$$

$$= \frac{3V_1^2}{\omega_s} \times \frac{2r_2}{(2r_2)^2 + (5r_2)^2}$$

$$= \frac{3V_1}{\omega_s \times r_2} \cdot \frac{2}{4+25} = 290 \times \frac{2}{29} = 20 \text{ Nm} \quad \textbf{Ans.}$$

Problem 6.57 : For a three phase induction motor, maximum torque is 200% of full load torque and starting torque is 150% of the full load torque. In order to get a full load slip of 4%, calculate the percentage reduction in rotor circuit resistance. Neglect stator impedance.

Solution

Given, $s_{fl} = 0.04$

According to the Question,

$$T_{m1} = 2 \times T_{fl} \quad (1)$$

and

$$T_s = 1.5 \times T_{fl} \quad (2)$$

Applying the relation, $$\frac{T_s}{T_{m1}} = \frac{2}{s_{\max, T1} + \dfrac{1}{s_{\max, T1}}}$$

or $$\frac{1.5 \times T_{fl}}{2 \times T_{fl}} = \frac{2}{s_{\max, T1} + \dfrac{1}{s_{\max, T1}}}$$

or $$\frac{s_{\max, T1}^2 + 1}{s_{\max, T1}} = \frac{2 \times 2}{1.5}$$

or $$s_{\max, T1}^2 + 1 = 2.67\, s_{\max, T1}$$

or $$s_{\max, T1}^2 - 2.67\, s_{\max, T1} + 1 = 0$$

This yields $$s_{\max, T_1} = \frac{2.67 \pm \sqrt{(2.67)^2 - 4}}{2} = \frac{2.67 \pm 1.77}{2}$$

Neglecting higher value,

$$s_{\max, T_1} = \frac{0.90}{2} = 0.45 = \frac{r_2}{x_2}$$

$$\therefore \quad r_2 = 0.45\, x_2$$

For a full load slip,

$$s_{fl} = 0.04$$

$$\frac{T_{efl}}{T_m} = \frac{2}{\dfrac{s_{fl}}{s'_{\max T}} + \dfrac{s'_{\max T}}{s_{fl}}}$$

or $$\frac{1}{2} = \frac{2}{\dfrac{0.04}{s'_{\max T}} + \dfrac{s'_{\max T}}{0.04}}$$

or $$\frac{(0.04)^2 + (s'_{\max, T})^2}{0.04 \times s'_{\max, T}} = 4$$

or $$0.0016 + s^2_{\max, T} = 0.16\, s'_{\max, T}$$

or $$s'^2_{\max, T} - 0.16\, s'_{\max, T} + 0.0016 = 0$$

This yields $$s'_{\max, T} = \frac{0.16 \pm \sqrt{(0.16)^2 - 0.0064}}{2}$$

$$= \frac{0.16 \pm 0.1386}{2}$$

Neglecting higher value,

$$s'_{\max, T} = 0.021 = \frac{r'_2}{x_2}$$

$$\therefore \quad r'_2 = 0.021\, x_2$$

Now, percentage reduction in rotor circuit resistance,

$$= \left(\frac{0.45 - 0.021}{0.45}\right) \times 100\% = 95.33\% \ \textbf{Ans.}$$

Problem 6.58 : The rotor of a 3-phase induction motor has 0.01 per phase and 0.1 Ω standstill reactance per phase. What external resistance is required in the rotor circuit in order to get one-fourth of the maximum torque at starting?

Solution

Given, $r_2 = 0.01\,\Omega, \ r_2 = 0.1\,\Omega$

According to the Question,

$$T_s = \frac{1}{4} \times T_m$$

or
$$\frac{T_s}{T_m} = \frac{2}{\dfrac{1}{s_{max,T}} + s_{max,T}}$$

or
$$\frac{1}{4} = \frac{2}{s_{max,T} + \dfrac{1}{s_{max,T}}}$$

or
$$s_{max,T}^2 + 1 = 8\, s_{max,T}$$

or
$$s_{max,T}^2 - 8\, s_{max,T} + 1 = 0$$

This yields
$$s_{max,T} = \frac{8 \pm \sqrt{8^2 - 4}}{2} = \frac{8 \pm 7.746}{2}$$

Neglecting higher value,

$$s_{max,T} = \frac{8 - 7.746}{2} = \frac{0.254}{2} = 0.127$$

or
$$\frac{r_2'}{x_2} = 0.127$$

$\therefore$
$$r_2' = 0.127 \times 0.1 = 0.0127\,\Omega$$

External resistance that must be inserted in the rotor

$$= r_2' - r_2 = 0.0127 - 0.01$$

$$= 0.0027\,\Omega \quad \textbf{Ans.}$$

Problem 6.59 : A 12 kW, 440 V, 3 phase, 4 pole, 50 Hz slip-ring induction rotor develops rated output at rated voltage and frequency and with its slip ring short circuited. The maximum torque equal to twice the full load torque occurs at a slip of 20% with zero external resistance in the rotor circuit, Stator resistance and rotational losses are negligible. Determine

(i) slip and rotor speed at full-load torque
(ii) rotor ohmic loss at full load and
(iii) starting current in terms of full load current.

Solution

Given, $s_{mT} = 0.2, \; P = 4, \; f = 50$ Hz

$$T_m = 2 \times T_{fl} \tag{1}$$

(i) To calculate full load slip, Applying

$$\frac{T_{fl}}{T_m} = \frac{2}{\dfrac{s_{fl}}{s_{\max,T}} + \dfrac{s_{\max,T}}{s_{fl}}}$$

or
$$\frac{1}{2} = \frac{2}{\dfrac{0.2}{s_{fl}} + \dfrac{s_{fl}}{0.2}}$$

or
$$\frac{(0.2)^2 + s_{fl}^2}{0.2 \times s_{fl}} = \Delta$$

or
$$0.04 + s_{fl}^2 = 0.8\, s_{fl}$$

or $s_{fl}^2 - 0.8\, s_{fl} + 0.04 = 0$

This yields
$$s_{fl} = \frac{0.8 \pm \sqrt{(0.8)^2 - 0.16}}{2} = \frac{0.8 \pm 0.693}{2}$$

Neglecting higher value, $s_{fl} = 0.054$ **Ans.**

Synchronous speed, $\omega_s = \dfrac{120 \times f}{P} = \dfrac{120 \times 50}{4} = 1500$ rpm

$\therefore$ full load speed $= 1500\,(1 - 0.054) = 1419$ rpm **Ans.**

(ii) Since rotational loss is negligible, Mechanical power developed

$$(P_m) = \text{shaft power output} = 1200 \text{ W}$$

$$\text{Power in air gap, } P_G = \frac{P_m}{1-s_{fl}} = \frac{12000}{1-0.054} = 12684.99 \text{ W}$$

$\therefore$ Rotor ohmic loss at full load torque $= s_{fl} \times P_g$

$$= 0.054 \times 12684.99 = 684.99 \text{ W} \quad \textbf{Ans.}$$

(iii) In order to calculate starting torque in terms of full load current Applying the relation,

$$\frac{I_{2st}}{I_{2fl}} = \sqrt{\frac{s_{fl}^2 + s_{max}^2}{s_{fl}^2 + \left(1+s_{max}^2\right)}} = \sqrt{\frac{(0.054)^2 + (0.2)^2}{(0.054)^2 \times 1.04}} = 3.76$$

$$I_{2st} = 3.76 \times I_{2fl} \quad \textbf{Ans.}$$

Problem 6.60 : A three phase, star-connected, 440 volts, 50 Hz, 6 pole induction motor has the following per phase constants in ohms reffered to stator $r_1 = 0.30\,\Omega$, $x_1 = 0.90\,\Omega$, $r_2 = 0.24\,\Omega$, $x_2 = 0.90\,\Omega$, $x_\phi = 52\,\Omega$

Core, friction and windage losses = 500 Watts

Compute the stator current, rotor speed, output torque and efficiency when motor is operated at rated voltage and rated frequency at a slip of 5%.

Solution

Given, $r_1 = 0.30\,\Omega$, $x_1 = 0.90\,\Omega$, $r_2 = 0.24\,\Omega$, $x_2 = 0.90\,\Omega$

$X_\phi = 52\,\Omega$, $P = 6$, $f = 50$ Hz, $P_c + P_{wf} = 500$ W

$$\text{Synchronous speed, } n_s = \frac{120 \times f}{P} = \frac{120 \times 50}{6} = 1000 \text{ rpm}$$

The equivalent circuit per phase can be drawn as

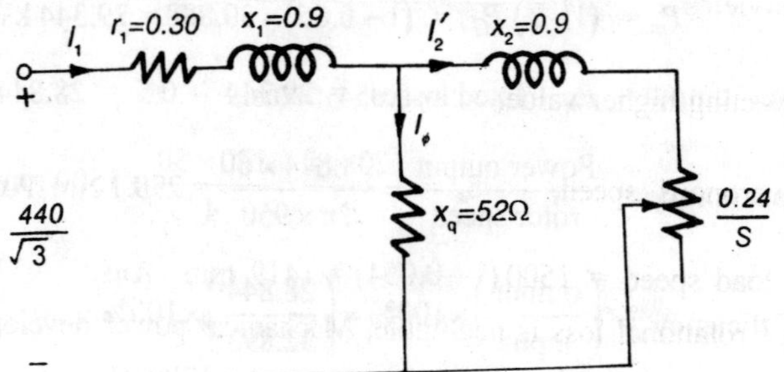

Fig. P. 6.60

Now $\bar{Z}_f = \dfrac{\left(\dfrac{r_2}{s} + j\,x_2\right)(j\,X_\phi)}{\dfrac{r_2}{s} + j\left(x_2 + X_\phi\right)} = \dfrac{\left(\dfrac{0.24}{0.05} + j\,0.90\right)(j\,52)}{\dfrac{0.24}{0.05} + j\,(0.90 + 52)}$

$$= \frac{(4.8 + j\,0.90)(j\,52)}{4.8 + j\,52.90} = \frac{4.88\angle 10.62°\;52\angle 90°}{53.12\angle 84.82°}$$

$$= 4.78\angle 15.8° = (4.6 + j\,1.3)\,\Omega$$

$$\bar{Z}_{total} = \bar{Z}_1 + \bar{Z}_f = 0.30 + j\,0.9 + 4.6 + j\,1.3$$

$$= 4.90 + j\,2.2 = 5.37\angle 24.18°\,\Omega$$

$$\text{Stator current} = \frac{\dfrac{440\angle 0°}{\sqrt{3}}}{5.37\angle 24.18°} = 47.31\angle -24.18°\text{ A}$$

Stator current = 47.31 A **Ans.**

We have, $R_f = 4.6\,\Omega$

Stator p.f. = cos 24.18 = 0.9123 (lag)

$$\text{Power input} = \sqrt{3} \times V_1 \times I_1 \times \cos\theta$$

$$= \sqrt{3} \times 440 \times 47.31 \times 0.9123 = 32.893\text{ kW}$$

Torque air gap power $= 3\,I_1^2\,R_f = 3 \times (47.31)^2 \times 4.6 = 30.888\text{ kW}$

Rotor speed, $n_r = n_s(1 - s) = 1000(1 - 0.05) = 950$ rpm **Ans.**

Mechanical power developed,

$$P_m = (1 - s)\,P_G = (1 - 0.05) \times 30.888 = 29.344\text{ kW}$$

∴ Power output = P_m – fixed losses = 29.344 – 0.5 = 28.844 kW

$$\text{Output torque} = \frac{\text{Power output}}{\text{rotor speed}} = \frac{28.844 \times 60}{2\pi \times 950} = 290.1\text{ Nm}$$ **Ans.**

Now, $\text{Efficiency} = \left(\dfrac{\text{output}}{\text{input}}\right) \times 100\% = \left(\dfrac{28.844}{32.893}\right) \times 100\%$

= 87.69% **Ans.**

Problem 6.61 : A 3-phase star connected squirel cage induction motor has a full load torque 5 Nm at 960 rpm. If this motor is subjected to dynamic breaking with a direct current of 20 A supplied between any two of its stator terminals, then compute the initial dynamic torque. Neglect stator impedance and rotational losses.

The motor has per phase standstill rotor leakage impedance of $0.1 + j\,0.2\,\Omega$ reffered to stator side and magnetising reactance of $16\,\Omega$. For dynamic breaking, one stator phase is connected is series with other two in parallel.

Solution

When one stator phase is connected in series with the other two in parallel, the stator current is given by

$$I_1 = \frac{1}{\sqrt{2}} I_{dc} = \frac{1}{\sqrt{2}} \times 20 = 14.14 \text{ A}$$

For a full load rotor speed of 960 r.p.m, the synchronous speed must be 1000 rpm.

$$\therefore \qquad \text{Slip} = \frac{1000 - 960}{1000} = 0.04$$

Equivalent rotor current,

$$I_2 = I_1 \times \frac{X_\phi}{\sqrt{\left(\frac{r_2}{s}\right)^2 + (X_\phi + x_2)^2}}$$

$$= 14.14 \times \frac{16}{\sqrt{\left(\frac{0.1}{0.04}\right)^2 + (16 + 0.2)^2}} = 13.80 \text{ A}$$

$\therefore$ Dynamic breaking,

$$T_{ed} = \frac{3}{\omega_s} I_2^2 \times \frac{r_2}{s} = \frac{3 \times 60}{1000 \times 2\pi} \times (13.80)^2 \times \frac{0.1}{0.04} = 13.65 \text{ Nm}$$

$\therefore$ Total initial braking torque

$$= T_{efl} + T_{ed} = 13.65 + 5$$

$$= 18.65 \text{ Nm} \quad \textbf{Ans.}$$

Problem 6.62 : A 440 V, 50 Hz, 6 pole, 3-phase star induction motor has following data:

Stator leakage impedance/Phase = $0.2 + j\,1.0\,\Omega$

Standstill rotor leakage impedance per phase reffered to stator $= 0.2 + j\,1.0\,\Omega$

Full load slip = 0.06

No load-current is negligible,

(a) Determine initial braking torque soon after plugging.

(b) Determine the initial dynamic breaking torque when d.c. is fed to the two phases in series to produce same air gap flux.

Solution

Given, $P = 6, \quad f = 50$ Hz

$$r_1 = 0.2\,\Omega,\ x_1 = 1.0\,\Omega,\ r_2 = 0.2\,\Omega,\ x_2 = 1.0\,\Omega$$

$$\text{Synchronous speed, } n_s = \frac{120 \times f}{P} = \frac{120 \times 50}{6} = 1000 \text{ rpm}$$

$$\text{Terminal voltage/Phase} = \frac{440}{\sqrt{3}} = 254 \text{ V}$$

Since no-load current is negligible

$$\text{Rotor current, } I_2 = \frac{V_1}{\left(r_1 + \dfrac{r_2}{s}\right) + j\,(x_1 + x_2)}$$

Electromagnetic torque,

$$T_e = \frac{3}{\omega_s} I_2^2 \times \frac{r_2}{s} = \frac{3}{\omega_s} \frac{V_1^2}{\left(r_1 + \dfrac{r_2}{s}\right)^2 + (x_1 + x_1)^s} \times \frac{r_2}{s}$$

Full load torque at slip 0.06 is given by

$$T_{efl} = \frac{3 \times 60}{2\pi \times 1000} \times \frac{(254)^2}{\left(0.2 + \dfrac{0.2}{0.06}\right)^2 + (1+1)^2} \times \frac{0.2}{0.06}$$

$$= \frac{3 \times 60}{2\pi \times 1000} \times \frac{(254)^2}{(3.53)^2 + 4} \times \frac{0.2}{0.06} = 374.46 \text{ Nm}$$

(a) As soon as any two stator terminals are inter changed for plugging, the slip becomes $2 - s = 2 - 0.06 = 1.94$

Plugging torque,

$$T_P = \frac{3 \times 60}{2\pi \times 1000} \times \frac{(254)^2}{\left(0.2 + \frac{0.2}{1.94}\right)^2 + (2)^2} \times \frac{0.2}{1.94}$$

$$= \frac{180}{2 \times 3.14 \times 1000} \times \frac{(254)^2}{4.092} \times \frac{0.2}{1.94} = 46.59 \text{ Nm}$$

$\therefore$ Total initial braking torque $= T_{efl} + T_P = 46.59 + 374.46$

$= 421.05$ Nm **Ans.**

(b) For negligible no-load current, the air gap flux is proportional to $\bar{E}_2$, which is given by

$$\bar{E}_2 = \frac{\bar{Z}_2}{\bar{Z}_1 + \bar{Z}_2} \times \bar{V}_1$$

Here, $\bar{Z}_1 = 0.2 + j\,1.0, \quad \bar{Z}_2 = 0.2 + j\,1.0$

$$\bar{V}_1 = 254 \angle 0°$$

$$\bar{E}_2 = \frac{0.2 + j\,1.0}{0.4 + j\,2.0} \times 254\angle 0° = 127 \angle 0°$$

Rotor current under dynamic braking is given by

$$I_2 = \frac{E_2}{\sqrt{\left(\frac{r_2}{s}\right)^2 + x_2^2}} = \frac{127}{\sqrt{\left(\frac{0.2}{0.94}\right)^2 + 12}} = 126.33 \text{ A}$$

Since no-load current is negligible, $I_1 = I_2$

D.C. excitation is given by

$$I_{dc} = \sqrt{\frac{3}{2}} \times I_2 = \sqrt{\frac{3}{2}} \times 126.3 = 154.69 \text{ A}$$

Initial dynamic braking torque,

$$= \frac{3}{\omega_s} \times I_2^2 \times \frac{r_2}{s}$$

$$= \frac{3 \times 60}{2 \times \pi \times 1000} \times (154.69)^2 \times \frac{0.2}{1.94} = 70.70 \text{ Nm} \quad \textbf{Ans.}$$

Problem 6.63 : A 12 pole, 3 phase induction frequency converter is driven at 3000 rpm and its stator is connected to 50 Hz system. If the rotor terminal voltage at standstil! is 20 V, then find what voltages and frequencies can be obtained from this frequency converter.

Solution :

Given, $p = 12, \; f = 50 \text{ Hz}, \; V_2 = 20 \text{ V}$

Synchronous speed of this converter $= \frac{120 \times 50}{12} = 500 \text{ rpm}$

The output frequency is given by,

$$= \left(1 \pm \frac{3000}{500}\right) \times 50 = (1 \pm 6) \times 50 = 350 \text{ Hz or } 250 \text{ Hz} \quad \textbf{Ans.}$$

At standstill, rotor voltage is 20 V and frequency is 50 Hz.

$\therefore$ Rotor voltage at 250 Hz $= \frac{20}{50} \times 250 = 100 \text{ V}$ **Ans.**

Rotor voltage at 350 Hz $= \frac{20}{50} \times 350 = 140 \text{ V}$ **Ans.**

Problem 6.64 : A 3-phase, 50 Hz, 12 pole, wound rotor induction motor has its slip ring directly to slip power converter. The induction motor controls the speed of a load requiring proportional to the square of the speed. The speed is adjustable over a range from 20 to 480 rpm and load required as power of 3000 kW at 480 rpm. For negligible losses, find the power handled by the slip-power converter in terms of slip, when slip-power is returned to the supply. Also find the maximum power that the slip-power converter should be capable.

Solution :

Given, $p = 120, \; f = 50 \text{ Hz},$

Synchronous speed, $n_s = \frac{120 \times 50}{12} = 500 \text{ rpm}$

$$T_L \propto n^2 \text{ or } T_L = k\omega^2$$

$\therefore$ Load power, $P_L = \omega T_L = k\omega^3$

Since $\omega = \omega_s (1 - s)$

$$\therefore \quad P_L = k\omega_s^3 (1 - s)^3 \quad (1)$$

$$\text{Now slip} = \frac{500-480}{500} = \frac{20}{500} = 0.04$$

$$\text{For slip} = 0.04,\ P_L = 3000 \text{ kW}$$

or $$k\omega_s^3 \times (1-0.04)^3 = 3000 \text{ or } k\omega_s^3 = 3390.8 \text{ kW}$$

$\therefore$ $$P_L = 3390.8\,(1-s)^3$$

From (1) When the slip power is returned to the supply

Load power = Motor input (1 – *s*)

$\therefore$ $$\text{Motor input } P = \frac{P_L}{1-s} = \frac{3390.8\,(1-s)^3}{(1-s)} = 3390.8\,(1-s)^2$$

The slip power as a function slip is given by,

$$P_{sp} = sP = 3390.8\,\text{s (s. a c e)}\,(1-s)^2$$

$$\text{For 20 rpm, Slip} = \frac{500-20}{500} = \frac{480}{500} = 0.96$$

$\therefore$ Motor input at 20 rpm = $3390.8\,(1-0.96)^2 = 5.43$ kW **Ans.**

Slip power at 20 rpm = 0.96 × 5.43 = 5.21 kW **Ans.**

$$\text{For 480 rpm, Slip} = \frac{500-480}{500} = \frac{20}{500} = 0.04$$

Motor input at 480 rpm = $3390.8\,(1-0.04)^2 = 3124.96$ kW **Ans.**

Slip power at 480 rpm = 0.04 × 3124.96 = 125 kW **Ans.**

The slip at which slip power is maximum can be calculated by differentiating P_{sp} with respect to slip

$$\frac{d\,P_{sp}}{ds} = \frac{d\left(3390.85\,(1-s)^2\right)}{ds}$$

$$= 3390.8\,(1-s)^2 - 3390.8 \times s \times 2\,(1-s) = 0$$

or $$3390.8\,(1-s)\left[(1-s-2s)\right] = 0$$

$$\text{This yields } s = \frac{1}{3}$$

$$\therefore \text{Maximum slip power } (P_{sp})_{max} = 3390.8 \times \frac{1}{3} \times \left(1-\frac{1}{3}\right)^2$$

$$= 3390.8 \times \frac{1}{3} \times \frac{4}{9} = \underline{\underline{502.34 \text{ kW}}}\ \textbf{Ans.}$$

Reader's Notes

Reader's Notes

Reader's Notes